Christian M. Grams

# Quantification of the Downstream Impact of Extratropical Transition for Typhoon Jangmi and Other Case Studies

Wissenschaftliche Berichte des Instituts für Meteorologie und Klimaforschung des Karlsruher Instituts für Technologie (KIT)
Band 56

Herausgeber: Prof. Dr. Ch. Kottmeier

Institut für Meteorologie und Klimaforschung
am Karlsruher Institut für Technologie (KIT)
Kaiserstr. 12, 76128 Karlsruhe

Eine Übersicht aller bisher in dieser Schriftenreihe erschienenen Bände finden Sie am Ende des Buches.

# Quantification of the Downstream Impact of Extratropical Transition for Typhoon Jangmi and Other Case Studies

by
Christian M. Grams

KIT Scientific Publishing

Dissertation, Karlsruher Institut für Technologie (KIT)
Fakultät für Physik, 2011
Referenten: Prof. Dr. Sarah C. Jones, Prof. Dr. Klaus Dieter Beheng

**Impressum**

KIT Scientific Publishing

Karlsruher Institut für Technologie (KIT)
KIT Scientific Publishing
Straße am Forum 2
D-76131 Karlsruhe

KIT Scientific Publishing is a registered trademark of Karlsruhe Institute of Technology. Reprint using the book cover is not allowed.

www.ksp.kit.edu



Print on Demand 2015

ISSN 0179-5619
ISBN 978-3-86644-776-9
DOI: 10.5445/KSP/1000024940

# Quantification of the downstream impact of extratropical transition for Typhoon Jangmi and other case studies

Zur Erlangung des akademischen Grades eines
DOKTORS DER NATURWISSENSCHAFTEN
der Fakultät für Physik des
Karlsruher Instituts für Technologie (KIT)

genehmigte

DISSERTATION

von

Dipl.-Met. Christian Michael Warnfrid Grams
aus Karlsruhe

Tag der mündlichen Prüfung: 21. Oktober 2011
Referent: Prof. Dr. Sarah C. Jones
Korreferent: Prof. Dr. Klaus Dieter Beheng

# Contents

# Kurzfassung

Die Umwandlung eines tropischen Wirbelsturms in ein außertropisches Tief und dessen Wechselwirkungen mit der Strömung der mittleren Breiten bezeichnet man als "außertropische Umwandlung" (extratropical transition - ET). Die außertropische Umwandlung kann Extremwetterereignisse sowohl in der Nähe des sich umwandelnden Tiefs als auch in Regionen stromab wie z.B. Europa oder Nordamerika verursachen. Die komplexen Wechselwirkungen während der außertropischen Umwandlung führen zu großen Unsicherheiten in der numerischen Wettervorhersage. Ein besseres Verständnis der beteiligten physikalischen Prozesse ist daher für Verbesserungen der Vorhersage von ET unabdingbar.

In dieser Arbeit wird der stromabwärtige Einfluss von ET auf die Strömung der mittleren Breiten und auf die Vorhersagbarkeit anhand von fünf Fallstudien untersucht und quantifiziert. Diese Fallstudien sind Taifun Jangmi, welcher während der THORPEX-Messkampagne T-PARC im Jahr 2008 im westlichen Nordpazifik aufgetreten ist, Hurrikan Hanna (2008) im Nordatlantik, sowie die Taifune Choi-wan (2009), Lupit (2009) und Malakas (2010) im westlichen Nordpazifik. Das wichtigste Instrument in dieser Arbeit sind numerische Simulationen mit Hilfe des mesoskaligen COSMO Modells. Die Beobachtungsdaten der T-PARC Messkampagne werden verwendet, um die Modellanalyse zu validieren und um die Wechselwirkung von Jangmi mit der Strömung der mittleren Breiten zu untersuchen. Es wurde weiter eine Methode entwickelt, die es erlaubt einen tropischen Wirbelsturm aus einer Modellanalyse zu entfernen. Diese Methode basiert auf der Inversion der potentiellen Wirbelstärke (potential vorticity - PV) und erlaubt es den Beitrag eines tropischen Wirbelsturms an der Fernwirkung der außertropischen Umwandlung zu quantifizieren sowie die Bedeutung der relativen Lage des Wirbelsturms zur Strömung der mittleren Breiten zu untersuchen. Mittels quasi-geostrophischer Diagnostik kann der Beitrag der Strömung der mittleren Breiten quantifiziert werden.

Die ausführliche Fallstudie zu Jangmi (2008) zeigte den diabatisch verstärkten Netto-Transport von Luft niedriger PV aus unteren Schichten in die obere Troposphäre als den entscheidenden physikalischen Prozess auf, welcher die Rückenbildung direkt stromab des ET-Systems sowie eine Hebung der Tropopause und eine Beschleunigung des Strahlstroms der mittleren Breiten bewirkt. Die Hebung erfolgt dabei über drei verschiedene Wege. Zunächst erfahren Luftpakete konvektiven Auftrieb im inneren Kern des tropischen Wirbelsturms. Zu einem späteren Zeitpunkt der außertropischen Umwandlung erfolgt weiterer nahezu senkrechter Auftrieb innerhalb neuer konvektiver Ereignisse vor der baroklinen Zone der mittleren Breiten. Schließlich wird (sub-)tropische Luft beim Auftreffen auf die barokline Zone zu einem dynamisch forciertem schrägen Aufstieg entlang der geneigten (Feucht-)Isentropen gezwungen. In der Höhe werden die Luftpakete anschließend im Ausströmungsbereich des tropischen Wirbelsturms in Richtung des Strahlstroms der mittleren Breiten advehiert. Somit führt das Zusammenspiel der Zirkulation in den unteren Schichten des tropischen Wirbelsturms mit der baroklinen Zone der mittleren Breiten sowie die Wechselwirkung des Ausflusses des tropischen Wirbelsturms in der Höhe mit dem Strahlstrom der mittleren Breiten zu einem Netto-Transport von Luft niedriger PV in die obere Troposphäre. Dabei spielen diabatische Veränderungen der PV eine entscheidende Rolle. Dieser physikalische Prozess trat in allen hier untersuchten ET-Fällen auf und scheint somit ein gemeinsames Merkmal von außertropischer Umwandlung zu sein.

Weiter wurde die Anfangsposition von Jangmi in einer Reihe von Jangmi-Szenarien verändert. Es zeigte sich, dass südlich eines sich von Westen annähernden Troges ein kritischer Bifurkationspunkt für die Zugbahn von Jangmi existiert. Falls der tropische Wirbelsturm eine kritische Linie überschreitet, welche durch die Summe der trogrelativen nichtdivergenten Windkomponenten der Hintergrundströmung definiert ist, wird der Sturm polwärts ziehen und sich sehr wahrscheinlich wieder verstärken. Andernfalls zieht er zurück in die Tropen und zerfällt. Diese kritische Linie teilt benachbarte Szenarien klar in Szenarien mit polwärts ziehendem Sturm und in Szenarien mit äquatorwärts ziehendem Sturm ein. In den Szenarien, in denen der Sturm polwärts zieht, löst Jangmi einen markanten Rossby-Wellenzug aus und hat somit eine starke Fernwirkung in stromabwärtigen Gebieten.

In allen Fällen, welche diese starke Fernwirkung aufweisen, befindet sich das ET-System vor einem ausgeprägten Trog. Die Rückenbildung direkt stromab des ET-Systems behindert zunächst eine ostwärtige Verlagerung des Troges und des sich bildenden Rückens. Weiter bewirkt die Rückenbildung eine Verstärkung des Rossby-Wellenzuges. Im Fall von Choi-wan (2009) führt dies zu einer westlichen Verschiebung des Rossby-Wellenzugs um rund 10°, was zu einem frühen Kaltlufteinbruch im Westen Nordamerikas und zu einer späten Hitzewelle in der Region um Vancouver Island führt. Im Fall von Malakas (2010) hat die Veränderung der Strömung einen sehr schmalen stationären Trog an der Ostküste der USA zur Folge, welcher dort starke Niederschläge verursacht. In den Fällen von Hanna (2008) und Lupit (2009) tritt eine starke Rückenbildung direkt stromab des ET-Systems auf. Allerdings ist die Fernwirkung jenseits des ersten Rücken-Trog-Paares schwach. Die Wechselwirkung von Hanna (2008) mit dem stromaufwärtigen Trog führt weiter zur Bildung einer Mittelmeerzyklone.

Schließlich wird der Einfluss der außertropischen Umwandlung auf die Vorhersagbarkeit in der numerischen Wettervorhersage quantifiziert. Gemittelt über alle Fälle ergibt sich aufgrund von ET eine Reduktion der Vorhersagbarkeit um 2 Tage. Diese Reduktion ist vor allem der anfänglichen Rückenbildung direkt stromab des ET-Systems zuzuschreiben. In den Fällen, welche eine starke Veränderung der Strömung der mittleren Breiten aufweisen, nimmt die Vorhersagbarkeit weiter ab.

# Abstract

The transformation of a tropical cyclone (TC) into an extratropical system and its interaction with the midlatitude flow, referred to as extratropical transition (ET), has the potential to cause high impact weather in the vicinity of the transforming cyclone as well as in downstream regions such as Europe or North America. The complex interactions during ET result in large uncertainties in numerical weather prediction. A better understanding of the physical processes involved is needed for improvements of ET forecasts.

The downstream impact of ET in terms of the modification of the midlatitude flow and of the reduction in predictability is investigated and quantified in five case studies of real ET events. These are Typhoon Jangmi that occurred in the western North Pacific during the THORPEX Pacific Asian Regional Campaign (T-PARC) in 2008, Hurricane Hanna (2008) in the North Atlantic, and the western North Pacific Typhoons Choi-wan (2009), Lupit (2009), and Malakas (2010). The main tool in this study are numerical simulations using the mesoscale COSMO model. T-PARC observations are used to validate the model analysis and to investigate the interaction of Jangmi with the midlatitude flow. A potential vorticity (PV) surgery technique based on PV inversion is developed that allows the contribution of the TC to the downstream impact of ET to be quantified and the role of the relative position of the TC and the midlatitude flow to be investigated. Quasi-geostrophic diagnostics serve to quantify the contribution of the midlatitude flow.

The detailed case study of Jangmi (2008) revealed the diabatically enhanced net transport of low PV air from the lower troposphere to jet level as the crucial physical process governing ridgebuilding directly downstream of the ET system, a lifting of the tropopause, and an acceleration of the upper-level jet streak.

Thereby ascent occurs along three distinct paths. Initially air parcels experience convective buoyant ascent in the TC inner core. During later stages of ET air parcels rise in upright buoyant ascent within new convective bursts ahead of the midlatitude baroclinic zone. Finally, (sub-)tropical air that impinges on the baroclinic zone experiences dynamically forced slantwise ascent along the tilted (moist-)isentropes. At upper-levels the air parcels are advected by the TC outflow towards the midlatitude jet stream. Thus, the joint interaction of the low-level TC circulation with the midlatitude baroclinic zone as well as the interaction of the upper-level TC outflow with the upper-level midlatitude jet stream results in a net transport of low PV air to jet level. Thereby diabatic PV modification plays a crucial role. This physical process occurred in all ET cases studied here and thus seems to be a common feature of ET.

A set of Jangmi scenarios with relocated initial storm position revealed that a critical bifurcation point for the track of Jangmi exists south of the approaching midlatitude trough. If the TC moves beyond a critical line defined by the sum of trough-relative non-divergent wind components of the background flow, the storm will move poleward and reintensify. Otherwise it will move back to the tropics and decay. It is shown that this critical line sharply separates neighbouring scenarios into reintensifiers and decayers. In the reintensification scenarios Jangmi further triggers a pronounced Rossby wave train and thus has a strong downstream impact.

In all cases that exhibit a strong downstream impact, the ET system becomes located ahead of a pronounced midlatitude trough. The ridgebuilding directly downstream of the ET system initially hinders an eastward propagation of the upstream trough and triggers an amplification of the Rossby wave train. In the case of Choi-wan (2009) this results in a westward shift of the the upper-level Rossby wave train by around 10°, leading to an early cold air outbreak in western North America and a late autumn heat wave in the Vancouver Island region. In the case of Malakas (2010) the modification of the downstream flow results in a stationary PV streamer at the US East coast leading to a heavy precipitation event in that region. In the cases of Hanna (2008) and Lupit (2009) an intense ridgebuilding directly downstream of the ET event occurs. However, the downstream impact beyond the first downstream ridge-trough couplet is weak. Furthermore, the interaction of Hanna (2008) with its upstream trough has implications for the formation of a Mediterranean cyclone.

Finally, the impact of ET on the predictability in numerical weather prediction is quantified. On average over all cases ET leads to a reduction in predictability of 2 days. This reduction is mainly due to the initial ridgebuilding directly downstream of the ET system. The predictability further decreases in the cases that exhibit a strong modification of the downstream flow.

# 1 Introduction

Tropical cyclones are some of the weather systems which attract very high public interest. This is due to the fact that they are accompanied by extreme wind speeds and high rainfall amounts and thus have a large potential for damage to constructions and human activities on land and at sea. A recent tropical cyclone was Hurricane Irene at the end of August 2011. For this particular case the forecast of the track with Irene recurving along the US East Coast was very good. However, the intensity in terms of wind and rainfall was difficult to predict several days in advance. Despite the uncertainty in intensity a media hype about possible devastating impacts started when Irene approached New York City[1]. Less public attention is mostly paid for a tropical cyclone after it has made landfall or weakens. However, a decaying tropical cyclone often continues to move poleward, transforms into an extratropical system and undergoes a complex interaction with the midlatitude flow. This is referred to as extratropical transition (ET - Jones et al. 2003). During ET sustained strong winds and large rainfall amounts in the ET system constitute a high potential for direct impact on human activities and infrastructure. Moreover, the interaction with the midlatitude flow can modify the flow in downstream regions which constitutes a remote impact of ET. In the case of Irene the impacts of the landfall in New York were less than expected, but severe flooding, power outages and destruction of infrastructure occurred in New England and Southeastern Canada during ET. The latter attracted much less media attention. Nevertheless, public awareness that tropical cyclones may have a remote impact on midlatitude weather has increased in recent years (e.g. Figure 1.1).

When moving poleward, a tropical cyclone approaches the midlatitude baro-

[1] see media summary at http://fivethirtyeight.blogs.nytimes.com/2011/08/29/how-irene-lived-up-to-the-hype/?emc=etal

**Ex-„Irene" bringt bis zu 30 Grad**

**Meteorologischer Herbst beginnt warm / Der Sommer war zu trübe und zu nass**

**Offenbach/Karlsruhe** (dpa/BNN). Der meteorologische Herbst beginnt in Deutschland unter dem Einfluss von Ex-Irene. Der ehemalige Hurrikan, der vor Tagen an der US-Ostküste tobte, zieht über den Atlantik und pumpt Warmluft nach Europa. Das bringt laut Vorbersage des Deutschen Wetterdienstes (DWD) ab heute steigende Temperaturen und schönes Wetter, zumindest bis zum Wochenende.

Um etwa zwei Grad steigen die Temperaturen nun täglich, bestätigt auch Dominik Jung, Diplom-Meteorologe beim Wetterdienst q.met, der auch das tägliche BNN-Wetter liefert und das Wetterportal www.wetter.net betreibt. „Am Samstag kommen wir in Karlsruhe bis knapp an die 30-Grad-Marke", kündigt Jung an. „Bestes Grill- und Badewetter." Erst am Sonntagnachmittag sei mit ersten Schauern und Gewittern zu rechnen.

Viele vom Achterbahn-Wetter genervte Menschen dürfte das spätsommerliche Hoch kaum versöhnen. Die Bilanz von Tourismus-Branche und Freibädern für diesen Sommer fällt vielerorts durchwachsen aus. Auch in der Bilanz des DWD fällt der Sommer unbeständig, trüb und bundesweit zu nass aus. An den Küsten litten diesen Sommer besonders Campingplatzbetreiber und Strandkorbvermieter. Bundesweit blieben Freibäder tagelang leer.

Wann nach dem heutigen Beginn des meteorologischen Herbstes auch tatsächlich mit herbstlichem Wetter zu rechnen ist, können Meteorologen noch nicht sagen. Der Herbst, der laut Kalender erst Ende des Monats beginnt, umfasst für die Meteorologen die vollen Monate September, Oktober und November.

Figure 1.1: "Up to 30 degrees (Celsius) in the aftermath of Irene". Headline of an article in a local newspaper from Karlsruhe, Germany (Badische Neueste Nachrichten, 1 September 2011, page 12).

clinic zone that is a distinct temperature gradient separating moist warm (sub-) tropical air from drier cool midlatitude air. The tropical cyclone becomes more asymmetric due to the increasing shear and the mixing-in of drier cooler air, and frontal structures evolve as a result of the interaction with the baroclinic zone. This indicates the gradual transformation of the tropical system into an extratropical cyclone. The transforming cyclone may reintensify, and even reach higher intensity in terms of wind, minimum pressure and rainfall amount than during tropical stage. The ET system may also cross an entire ocean basin. Thus an ET system poses a serious meteorological hazard over land and ocean due to its *direct impact* by heavy winds, ocean swell, and the potential for severe flooding (Jones et al. 2003). Furthermore, the ET system interacts with the upper-level jet stream. Pole- and equatorward deflections of the jet stream - so-called ridges and troughs - play a crucial role in the formation of midlatitude weather systems. A ridge-trough couplet can propagate along the jet stream which is referred to as a upper-level Rossby wave. The interaction of the ET system with the midlatitude jet may trigger upper-level Rossby waves and thus may cause a severe *remote impact*, e.g. extratropical cyclogenesis, in regions downstream from the actual ET system such as Europe or North America.

ET occurs in every ocean basin that exhibits recurving tropical cyclones. The highest total number of ETs is found in the western North Pacific where about 27% of the tropical cyclones undergo ET (6 per year, Klein et al. 2000). In the North Atlantic generally less tropical cyclones occur but about 46% of these transform into an extratropical cyclone (4-5 per year, Hart and Evans 2001). The complex interactions between the tropical cyclone undergoing ET and midlatitude weather systems and the various physical processes involved make ET a severe challenge for numerical weather prediction models. Due to the acceleration of the cyclone during ET the correct track is difficult to predict. However, the ET depends critically on the phasing of the cyclone with the midlatitude flow. As a consequence, in operational deterministic weather prediction models ET is often poorly represented resulting in large forecast errors. In current ensemble prediction systems ET is linked with large forecast uncertainty in the area of the ET system and in downstream regions (Harr et al. 2008; Anwender et al. 2008).

Despite ongoing research many aspects of ET and details of the physical processes involved remain unclear (Jones et al. 2003). One of the major open questions is about the relative contribution of the midlatitude flow and of the tropical cyclone to the ET process and downstream impact. From previous studies it emerged that the relative contribution often highly depends upon the individual case. The reintensification of an ET system is favoured when the upper-level tropical cyclone outflow enhances the equatorward entrance region of a midlatitude jet streak and the interaction of the tropical cyclone low-level circulation with the baroclinic zone alters the region favourable for extratropical development (Klein et al. 2002). This suggests that ET is a dynamic process in which both the midlatitudes and the tropical cyclone have a contribution. However, other studies identified typical midlatitude flow configurations ahead of a trough in which ET occurs and conclude that the primary contribution is from the midlatitudes (Harr et al. 2000; McTaggart-Cowan et al. 2001). Albeit several ET case studies indicated that the tropical cyclone remnants play a crucial role in the reintensification of the ET system (Agustí-Panareda et al. 2004; McTaggart-Cowan et al. 2004). Also in a set of idealised studies it emerged that the strength of the midlatitude trough may be secondary for the final intensity of the ET system compared to the crucial impact of the tropical cyclone remnants (Ritchie and Elsberry 2003). In some cases an extratropical cyclone would not be present without ET (Agustí-Panareda et al. 2005). However, ET may also diminish extratropical development (Agustí-Panareda 2008). The foregoing case studies also mention the interaction of the tropical cyclone outflow with the midlatitude jet stream. During ET a substantial poleward advection of tropical air occurs which results in a lifting of the tropopause and associated upper-level ridgebuilding directly downstream of the ET system (Atallah and Bosart 2003). The variety of possible scenarios for the evolution of the *ET system itself* gives reasons for the uncertainty in the numerical prediction of ET. As a consequence, even larger uncertainty for the evolution of the *downstream flow* exists.

In idealised studies quantifying the relative importance of the midlatitude flow and of the tropical cyclone for the downstream impact of ET, Riemer et al. (2008) and Riemer and Jones (2010) showed that ridgebuilding immediately downstream of an ET system initially hinders the eastward propagation of Rossby waves and that Rossby waves are significantly more amplified in the presence of ET. They also showed that the phasing of the tropical cyclone and the midlatitude flow is crucial for the downstream impact of ET. However, only a few studies (e.g. Cordeira and Bosart 2010; Archambault 2011) give evidence for these results in realistic case studies. Thus, the relative importance of the midlatitude flow and of the tropical cyclone for the downstream impact of ET in realistic case studies remains an open question.

The interaction of the low-level tropical cyclone circulation with the midlatitude baroclinic zone plays a key role in the conceptual pictures of ET of Klein et al. (2000) and Agustí-Panareda et al. (2004). Ascent along the baroclinic zone is supposed to transport tropical air to upper-levels and thus to enhance the upper-level ridgebuilding by lifting the tropopause. Using idealised simulations, Ritchie and Elsberry (2001) demonstrated the validity of the conceptual ET picture of Klein et al. (2000). However, they did not find clear evidence for the ascent along the baroclinic zone. Therefore more work is needed to give evidence and a better understanding of the interaction with the baroclinic zone in real cases of ET.

The remote impact of ET is often described as an amplification of downstream Rossby waves, e.g. Martius et al. (2008) postulate that ET may constitute a far upstream precursor triggering a sequence of Rossby waves that lead to heavy precipitation in the Alps. In predictability studies it is argued that uncertainty in ensemble prediction systems "spreads downstream" of an ET event (Harr et al. 2008; Anwender et al. 2008). However, in this view it remains unclear what would have happened if the ET system was not present, e.g. whether the Rossby wave train or uncertainty are mainly forced from the midlatitudes and would be present anyway.

Finally some idealised and real case studies suggest that the phasing of the tropical cyclone relative to the midlatitude flow is essential for the reintensification and downstream impact during ET (e.g. Klein et al. 2002; Ritchie and Elsberry 2007; Riemer and Jones 2010; Scheck et al. 2011b). Scheck et al. (2011b) argue that the steering during ET is governed by the flow relative to the Rossby wave propagation. In this Rossby wave relative frame a critical bifurcation point exists determining if a storm moves poleward or is repelled back to the tropics. However, if such a bifurcation point exists in realistic case studies still remains unanswered.

This dissertation addresses the open questions about the downstream impact of ET, raised in the foregoing discussion, by case studies of realistic ET events. A potential vorticity (PV) perspective is employed to investigate the physical processes involved in ET. A PV surgery technique is developed for removing and relocating a tropical cyclone. This allows for a quantification of the tropical cyclone contribution to the downstream impact of ET in realistic ET scenarios. Quasi-geostrophic diagnostics help to assess the impact of the midlatitude flow. This quantification is supplemented by a Lagrangian perspective on ET using trajectory calculations.

A first major case study emerges from the THORPEX[2] Pacific Asian Regional Campaign (T-PARC) in the western North Pacific in 2008. Only very view systematic observations of ET exist. Therefore, T-PARC aimed for investigating on what determines the downstream impact of ET in more detail. Data from an enhanced observational network, various research flights, and supplementary numerical weather prediction provide an excellent basis for the investigation of ET. In this study one of the two tropical cyclones that underwent ET during T-PARC, Typhoon Jangmi at the end of September 2008, is investigated in detail. The recurvature and ET of Jangmi was very uncertain in deterministic and

[2] The Observing System Research and Predictability Experiment (THORPEX) is a 10-year international research and development programme to accelerate improvements in the accuracy of one-day to two-week high impact weather forecasts for the benefit of society, the economy and the environment. See http://www.wmo.int/thorpex/ and the European science plan at http://www.wmo.int/pages/prog/arep/wwrp/new/documents/European_Plan_5_July_web.pdf for more information.

in ensemble weather prediction systems, suggesting significant forecast challenges for this case. During ET, Jangmi underwent marked structural changes when approaching the midlatitude baroclinic zone. A distinct outflow-jet interaction occurred during which the jet core wind speed over Japan increased. Model forecasts during T-PARC motivated research flights into the transforming typhoon and enabled to collect observational data of this outflow-jet interaction. Although Jangmi transformed into an extratropical cyclone, the transition was not followed by extratropical reintensification and Jangmi decayed. Instead, a weak extratropical cyclone developed to its northeast. Thus although undergoing a rather unusual ET, several aspects qualify Jangmi as a case for getting a better general understanding of ET.

In the first half of this thesis, a detailed analysis of the ET of Jangmi during T-PARC is given addressing the following research questions.

- Which physical processes govern the structural changes during the ET of Jangmi from the erosion of the tropical cyclone inner core to the interaction with a zonal jet stream?
- Which contribution did Jangmi have to the acceleration of the jet streak over Japan? Which role did the interaction of Jangmi with the baroclinic zone play?
- What are the implications of the modified jet for the evolution of the downstream flow?
- Why did Jangmi decay and an extratropical cyclone develop directly downstream?

The knowledge gained from the detailed analysis of the ET of Jangmi is broadened and more general aspects are investigated with the help of Jangmi scenarios with relocated initial storm positions as well as four additional real ET case studies.

- Which role does the phasing of Jangmi and the midlatitude flow play for the uncertainty in the track forecast prior to recurvature?

- Which role does the position of the tropical cyclone relative to the midlatitude flow play for the reintensification of the ET system and its downstream impact?
- What are the physical processes involved in the interaction with the baroclinic zone and what implications do they have for the modification of the midlatitude flow?
- Is an acceleration of the midlatitude jet streak a common feature of ET? What governs the outflow-jet interaction? What are the implications for the downstream flow?
- Which contribution does the tropical cyclone remnants have to the midlatitude flow evolution beyond the ridgebuilding directly downstream of the ET system?
- Does a Rossby wave train or uncertainty downstream of an ET system originate from ET or is it preexistent in the midlatitude flow?
- To what extent does ET contribute to reduced predictability and which are the physical processes primarily involved?

The outline of this thesis is as follows. In Chapter 2 the meteorological background for the weather systems involved in ET is given. The data and diagnostic tools are described along with some theoretical background in Chapter 3. A detailed analysis of the ET of Jangmi along with the quantification of its impact on the jet acceleration is presented in Chapter 4. A broader view on ET is presented in Chapter 5 by the investigation of scenarios in which the initial position of Jangmi has been relocated and of four other real cases with a different midlatitude flow configuration. Finally the findings of these case studies are brought together and general conclusions about the midlatitude impact of ET derived in Chapter 6. The findings of this thesis are summarised, directions for future work pointed out and answers to the research questions given in Chapter 7.

# 2 The path of tropical cyclones through extratropical transition

As a tropical cyclone moves through ET, transforms into an extratropical cyclone, and causes modifications of the downstream flow, complex interactions occur between different weather systems involving various physical processes. Most apparently, the structure of the ET system itself changes from an organised symmetrical tropical cyclone (Figure 2.1a) to an asymmetric extratropical system (Figure 2.1b). However, ET has also implications for the environmental and downstream flow evolution. In the following an overview of the current knowledge about tropical cyclones and their transformation into extratropical cyclones is given along with a brief introduction of other weather systems discussed in this thesis.

(a) Irene (2011) tropical stage

(b) Irene (2011) undergoing ET

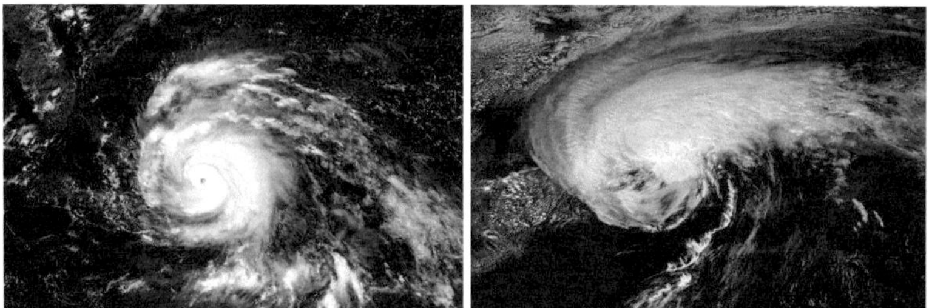

Figure 2.1: Visible GOES East imagery of Irene (a) at 17:45 UTC 24 August 2011 during tropical stage in the Caribbean and (b) at 17:45 UTC 28 August 2011 undergoing ET. Image courtesy of NOAA-NASA GOES Project available at http://goes.gsfc.nasa.gov/goeseast-lzw/fulldisk/fullres/vis/ retrieved, 30 August 2011. The two regions shown have the same spatial extent.

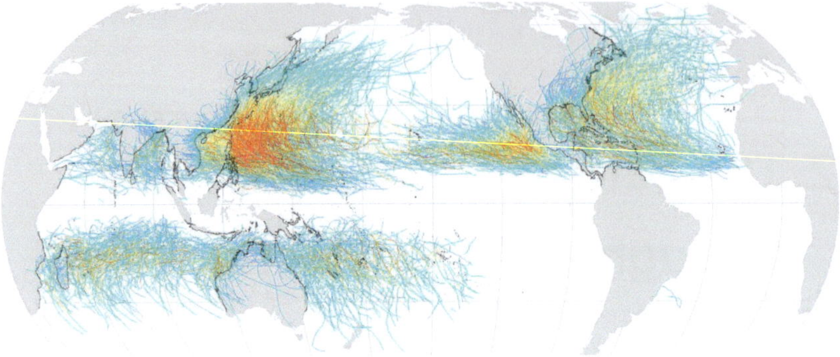

Figure 2.2: Tracks of all tropical cyclones which formed from 1945 to 2006. The colours indicate the intensity on the Saffir-Simpson Scale from tropical depression (dark blue), tropical storm (light blue), category 1 (beige) to category 5 (red). Image retrieved from http://en.wikipedia.org/wiki/File:Tropical_cyclones_1945_2006_wikicolor.png at 30 August 2011

## 2.1 Tropical cyclones

Tropical cyclones (TCs) are some of the most intense weather systems on earth. With wind speed in excess of 250 $\mathrm{kmh}^{-1}$, heavy rain, and storm surge, TCs pose a severe threat to human activities on land and at sea. Some of the most costly and deadly TCs occurred in recent years. In 2008 Cyclone Nargis in the Indian Ocean devastated Burma killing the unbelievable number of more than 130.000 people[1]. Hurricane Katrina in 2005 destroyed New Orleans, USA, killing 1800 people and causing damage at around 81 billion USD (Knabb et al. 2005). Very recently Hurricane Irene (2011 - Figure 2.1) recurved along the US East Coast killing at least 36 people and causing an estimated economic loss of 10 billion USD.

Tropical cyclones occur over all tropical oceans except the southern Atlantic (Figure 2.2). A TC with wind speed of more than 17 $\mathrm{m\,s}^{-1}$ is called tropical storm. Depending on the ocean basin different names for tropical storms with wind speed of more than 33 $\mathrm{m\,s}^{-1}$ are used. TCs are called Hurricanes in the North Atlantic, Eastern and Central Pacific and eastern South Pacific. Typhoons occur in the western North Pacific. In all other ocean basins TCs are simply

[1] http://topics.nytimes.com/topics/news/international/countriesandterritories/myanmar/cyclone_nargis/index.html

called Cyclones. Slightly different classifications are used amongst the institutions monitoring the ocean basins. In this study we employ the classifications of the Japan Meteorological Agency (JMA) for Typhoons in the western North Pacific, and of the National Hurricane Center (NHC) for Atlantic Hurricanes. JMA uses the 10 minute mean of maximum sustained wind and four typhoon categories[2]. A Typhoon has a maximum sustained wind of 18-33 $m\,s^{-1}$, a Strong Typhoon of 33-43 $m\,s^{-1}$, a Very Strong Typhoon of 44-53 $m\,s^{-1}$, and a Violent Typhoon of more than 54 $m\,s^{-1}$. In international standard a Typhoon has a maximum sustained wind of more than 33 $m\,s^{-1}$. NHC uses a 1 minute mean of sustained winds and five categories according to the latest Saffir-Simpson Scale[3]. A Hurricane of category 1 has sustained winds of 33-42 $m\,s^{-1}$, category 2 of 43-48 $m\,s^{-1}$, category 3 of 49-58 $m\,s^{-1}$, category 4 of 59-69 $m\,s^{-1}$, and category 5 of more than 70 $m\,s^{-1}$.

A TC can form over the tropical ocean when the environmental conditions are favourable. The tropical cyclogenesis requires an initial atmospheric disturbance, sea surface temperatures of more than 26°C in a water column of at least 50 m, high values of humidity in the low and mid troposphere, and low vertical wind shear over the entire troposphere (TCs generally weaken when shear is larger than 10 $m\,s^{-1}$). A 5-10° distance from the equator is needed so that Coriolis force allows for an ambient rotation needed for the intensification of the TC. A fully developed TC is nearly axisymmetric and consists of a primary cyclonic circulation extending through almost the entire troposphere (Figure 2.3). The primary circulation is approximately in gradient wind balance and has a radius of 100-1000 km. However, the radius of maximum wind speeds is within 10-100 km. A secondary circulation establishes due to inflow in the planetary boundary that is deflected by surface friction. The surface inflow triggers a process of energy conversion similar to a heat engine. Sensible and latent heat fluxes transfer energy from the warm sea surface to the low-level inflow. Convergence near the TC centre forces the air to ascend in the so called eyewall. Condensation occurs and the energy is released by latent heating. The strong convection in the eyewall enhances the further ascent to the tropopause where

[2] `http://agora.ex.nii.ac.jp/digital-typhoon/help/unit.html.en#id2`

[3] `http://www.nhc.noaa.gov/sshws.shtml`

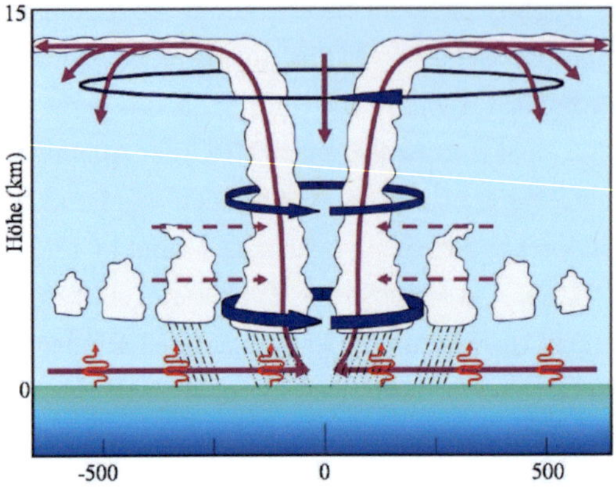


Figure 2.3: Schematic cross section of a tropical cyclone. Figure 2.1 of Riemer (2007) adapted from Jones (2002). The blue arrows indicate the primary TC circulation, the violet arrows the secondary TC circulation. The red meandering arrows indicate latent and sensible heat flux from the ocean surface into the troposphere. The abscissa shows the distance (in km) from the centre of the TC, the ordinate shows the height (in km) above mean sea level (amsl).

divergence results in a shallow layer of upper-level outflow. The outflow is deflected anticyclonically due to the Coriolis force and extends over a much larger region than the primary cyclonic circulation.

The motion of a TC is governed predominantly by the environmental flow. Usually a TC is located equatorward or west of the subtropical anticyclone during tropical stage and exhibits an initially west- and slightly poleward track. However, a TC can alter this steering flow and thereby impact its own track. When a TC moves on the western side of the subtropical anticyclone it recurves and starts moving into the midlatitudes. This is favoured if a midlatitude trough approaches from the west. A detailed review about the occurrence, structure and dynamics of tropical cyclones is given by Emanuel (2003) and in Chapters 1-4 of Chan and Kepert (2010).

## 2.2 Extratropical transition

Extratropical transition (ET) summarises the complex interactions and transformations that occur when a TC moves into the midlatitudes (Jones et al. 2003).

A major motivation for ET research is the reduced predictability in current numerical weather prediction systems for the midlatitude flow evolution in the vicinity of the ET system and in downstream regions (e.g. Anwender et al. 2008; Harr et al. 2008). This uncertainty emerges from the complex interaction of the various physical processes involved in ET and the crucial dependence of the downstream flow evolution on the relative position of the TC and the midlatitude flow (e.g. Ritchie and Elsberry 2007; Riemer and Jones 2010; Scheck et al. 2011b). These studies underline that ET can have a severe impact in remote regions. However, only few studies aimed to quantify the far downstream impact in realistic cases.

Conceptual pictures have been developed to give a more general view of ET. Early work by e.g. Matano and Sekioka (1971) classifies northwest Pacific ET as *complex* when the TC interacts with a preexistent surface baroclinic zone and as *compound* when interaction occurs with the frontal system of a preexistent extratropical cyclone. A third type, in which the TC dissipates after recurvature has been added by Brand and Guard (1979). Today the two stage picture of ET, *transformation stage* and reintensification stage, introduced by Klein (1997) is widely accepted. As noted by Jones et al. (2003) the reintensification stage is better named *extratropical stage* as a cyclone that does not reintensify after transformation can pose a severe hazard while dissipating. Klein et al. (2000) identified a three-step evolution of the transformation stage based on IR satellite imagery and model analyses (Figure 2.4), which is refined in a PV perspective by Agustí-Panareda et al. (2004).

First, when the TC moves polewards the outer cyclonic TC circulation impinges on the midlatitude baroclinic zone. The baroclinic zone is a strong low- to mid-level temperature gradient that separates warm, moist (sub-)tropical air from cool, drier midlatitude air. As a result colder, drier air is advected equatorward on the western side where deep convection decays, often seen as a dry slot penetrating into the TC inner core (left panels in Figure 2.4). On the southern and eastern side the poleward flow of tropical air maintains deep convection. Dynamically forced ascent along the tilted isentropes of the baroclinic zone occurs and a cirrus shield develops to the north when the TC interacts with the polar jet.

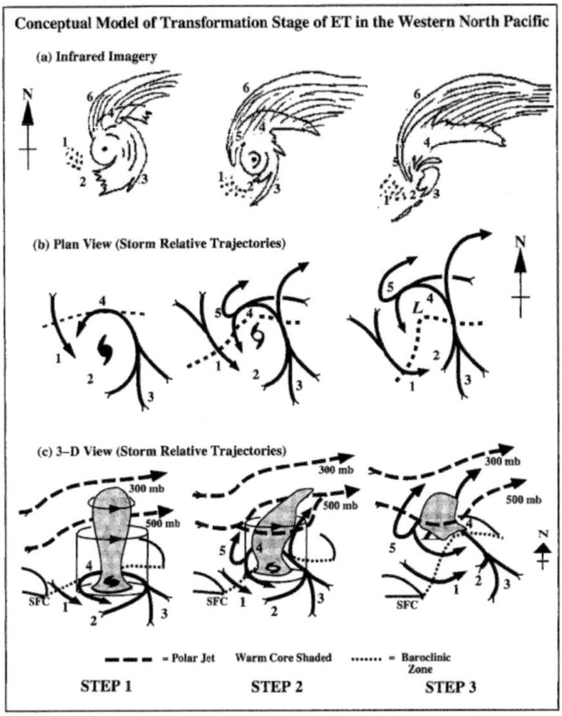


Figure 2.4: Figure 5 from Klein et al. (2000): Conceptual model of transformation stage of ET in the western North Pacific, with labelled areas as follows: 1) environmental equatorward flow of cooler, drier air (with corresponding open cell cumulus); 2) decreased tropical cyclone convection in the western quadrant (with corresponding dry slot) in step 1, which extends throughout the southern quadrant in steps 2 and 3; 3) environmental poleward flow of warm, moist air is ingested into tropical cyclone circulation, which maintains convection in the eastern quadrant and results in an asymmetric distribution of clouds and precipitation in steps 1 and 2; steps 2 and 3 also feature a southerly jet that ascends tilted isentropic surfaces; 4) ascent of warm, moist inflow over tilted isentropic surfaces associated with baroclinic zone (dashed line) in middle and lower panels; 5) ascent (undercut by dry-adiabatic descent) that produces cloudbands wrapping westward and equatorward around the storm centre; dry-adiabatic descent occurs close enough to the circulation centre to produce erosion of eyewall convection in step 3; 6) cirrus shield with a sharp cloud edge if confluent with polar jet.

The slantwise ascent along the baroclinic zone can be regarded as a so-called warm conveyor belt (WCB - Browning 1990; Eckhardt et al. 2004; Grams et al. 2011). Along this rapidly ascending airstream latent heat release due to condensational heating enhances the ageostrophic frontal circulation. This constitutes a positive feedback mechanism for the warm frontogenesis to the northeast of the ET system. The diabatic heating likewise leads to diabatic production of low-level PV. The cyclonic circulation induced by this anomaly can trigger a rapid propagation of low-level PV anomalies along the baroclinic zone. This process is referred to as diabatic Rossby wave propagation (DRW - Parker and Thorpe 1995) and is found to be a precursor anomaly to severe extratropical cyclones (Böttcher and Wernli 2011; Moore and Montgomery 2004, 2005). Above the level of maximum heating PV reduces within a WCB to lower tropospheric values or below. Thus a WCB transports low PV air to tropopause level which modifies the upper-level midlatitude wave guide.

Secondly, when the TC moves further poleward a cyclonic rotation of the baroclinic zone occurs resulting in a frontal wave due to the low-level dipole of cold air advection to the west and warm air advection to the east (middle panels in Figure 2.4). On the eastern side the baroclinic zone advances poleward and ascent along tilted isentropes continues. Some of the air parcels turn anticyclonically to the northeast at upper levels due to the Coriolis force and due to vertical shear associated with the baroclinic zone when moving in the region of the upper-level westerlies. A cirrus shield with a sharp cloud edge appears as the warm upper-level TC outflow becomes confluent with the polar jet (cf. Figure 2.1). Due to the vertical wind shear the upper-level warm core anomaly is also advected downstream. Some parcels may turn cyclonically and descend on the western side. The dry slot subsequently expands in the south (c.f. Figure 2.1b), where weak cold frontogenesis may occur. The subsidence in the west counteracts the slantwise ascent in the east. Deep buoyant convection is reduced on the eastern side, but persists to the north where multi-layer clouds indicate the formation of a warm sector. The warm frontogenesis is typically more vigorous than the cold frontogenesis (Harr and Elsberry 2000).

Lastly, in the conceptual picture of Klein et al. (2000) the TC becomes embedded in the baroclinic zone and is tilted due to strongly increasing vertical shear (right panels in Figure 2.4). As the TC becomes located over cooler sea surface the low-level warm core erodes. A weaker mid-level warm core persists. The dipole of cold and warm advection and the interaction of the outer circulation with the baroclinic zone results in the formation of a frontal system. Descending air west of the storm centre progressively weakens the TC inner core convection, leading to an erosion of the eyewall. Ascent along the baroclinic zone continues and joins the midlatitude westerly jet. The transformed TC exhibits the characteristics, asymmetry, and frontal structures of a baroclinic extratropical cyclone, marking the end of the transformation stage.

The transformation of Hurricane Irene is clearly depicted in the satellite imagery of 28 August 2011 (Figure 2.1b). South of the inner core the eyewall erodes. A small rather meridionally oriented cloud band shows the formation of a weak cold front. Behind this front to the southwest of the TC undergoing ET very dry air intrudes in Irene's centre. To the north the cloud pattern reflects the forming warm sector. In the same region an eastward sheared cirrus shield indicates the interaction with the midlatitude jet.

If the ET system moves into a midlatitude environment favourable for extratropical development, a period of strong reintensification may follow. Harr et al. (2000) identified two characteristic midlatitude circulation patterns associated with western North Pacific ET. In the *Northwest* pattern the primary midlatitude circulation system is a trough ahead of which a strong reintensification occurs. These ET systems move more north and northeastward and the coupling with the baroclinic zone results in a favourable environment for extratropical development. In the *Northeast* pattern the ET system moves into a strong zonal flow between a primary cyclone to its northeast and the subtropical ridge. The cyclone moves rapidly eastward without significant intensification. The strong zonal flow prevents a direct interaction between the TC and the midlatitudes. Harr et al. (2000) conclude that TC remnants have only little influence on the intensification process, and that the midlatitude flow configuration is most important. Klein et al. (2002) describes the reintensification stage of ET by type-B extratropical cyclogenesis (Petterssen and Smebye 1971). In this framework, cyclogenesis occurs when an upper-level trough approaches a baroclinic zone and upper-level advection of positive PV ahead of the trough becomes vertically aligned with low-level warm advection at the baroclinic zone. The reintensification is favoured when the upper-level outflow anomaly enhances the divergence in the equatorward entrance region of the downstream jet streak. The poleward advection of tropical air results in a lifting of the tropopause and a poleward deflection of the midlatitude wave guide (Atallah and Bosart 2003). This diabatically enhanced downstream ridgebuilding leads to an increase of the tropopause temperature gradient. As a consequence, the jet core wind speed must increase to restore thermal wind balance. Strong ridgebuilding can alter and amplify the upstream trough and retard its eastward propagation so that extratropical cyclogenesis and precipitation ahead of the trough are enhanced significantly (Atallah and Bosart 2003). The phasing of the ET system with the approaching upper-level trough plays a crucial role. Also, the interaction of the low-level circulation and the baroclinic zone may alter the region favourable for extratropical development. Thus the interaction of the TC with the midlatitude flow during ET is a dynamic process in which the TC as well as the midlatitude flow contribute to a possible reintensification of the ET system. As outlined in the introduction, various other case studies investigated the role of the midlatitude flow and the TC in the reintensification during ET. The contribution of the TC remnants to the midlatitude flow evolution was highly dependent on the particular case. In some studies the TC remnants were crucial for the downstream development (Agustí-Panareda et al. 2004; McTaggart-Cowan et al. 2004; Agustí-Panareda et al. 2005), in others they only played a minor role (Harr et al. 2000; McTaggart-Cowan et al. 2001) or diminish extratropical development (Agustí-Panareda 2008). All of these studies found the phasing between the TC and the midlatitude flow to be important for reintensification.

An objective classification of ET is given by the cyclone phase space of Hart (2003). This method allows ET systems to be distinguished based on their asymmetry in the thermal and frontal structure and the thermal anomaly in their inner core (warm core, cold core). Evans and Hart (2003) define the onset of ET in this cyclone phase space when a certain threshold of thermal asymmetry is reached and ET is completed when the TC has a cold core. However not all ET systems transform into a cold core system. Hurricane Danielle (1998 - McTaggart-Cowan et al. 2003) and Hurricane Iris (1995 - Thorncroft and Jones 2000) preserved their tropical inner core when undergoing ET on the poleward jet exit region. Therefore McTaggart-Cowan et al. (2003) distinguish between the *baroclinic mode* and the *tropical mode* of ET. In the baroclinic mode of reintensification the TC moves into the equatorward entrance region of a downstream midlatitude jet and becomes a cold core system similar to Harr and Elsberry (2000)'s Northwest type. In the tropical mode of reintensification the TC moves to the poleward exit region of a downstream jet thereby encapsulating the tropical air in the inner core and thus conserving the warm core structure. This is rather consistent with the Northeast type of ET in Harr and Elsberry (2000).

Only a few studies have investigated the downstream impact of ET beyond the direct interaction of the ET system and the midlatitude flow. In their idealised framework Riemer et al. (2008) and Riemer and Jones (2010) showed that ET can trigger a Rossby wave train. The upper-level midlatitude flow is characterised by a strong temperature gradient at the tropopause step (or PV gradient in a $\Theta$ framework) serving as a midlatitude wave guide. Equatorward deflections of cool air (with high PV values) are called troughs, and poleward deflections of warmer air (with low PV values) ridges. A ridge-trough couplet forms a Rossby wave. As shown by Riemer et al. (2008) in strong ET cases the ridgebuilding directly downstream of an ET event results in an amplification of a first downstream ridge-trough couplet. This amplification leads to a faster and stronger downstream cyclogenesis ahead of the first downstream trough (Riemer and Jones 2010) which triggers a second ridge-trough couplet further downstream (Figure 2.5a). Thus the Rossby wave can propagate downstream resulting in a second and third downstream Rossby wave. This sequence of several Rossby

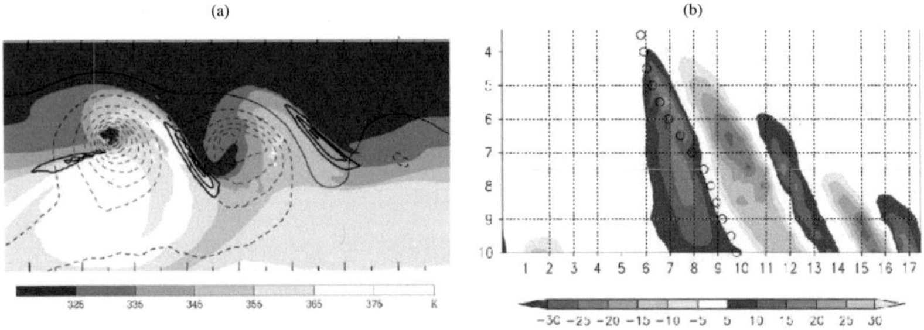


Figure 2.5: (a) Figure 2c of Riemer et al. (2008). Potential temperature (shaded in K) and wind speed >45 m s$^{-1}$ (thick contours every 5 m s$^{-1}$) on the 2 PVU surface along with pressure at mean sea level (black solid 995 hPa, dashed < 995 hPa every 5 hPa). Tick marks every 600 km. (b) Figure 5 of Riemer et al. (2008). Longitude-time diagram (Hovmöller) of 200 hPa meridional wind (shaded, in m s$^{-1}$). Location of the initial TC vortex marked by circles. Dotted grid lines in x-direction every 1000 km.

waves is called Rossby wave train. A longitude-time (Hovmöller) diagram of the upper-level meridional wind component highlights the Rossby wave train emerging downstream of the ET event (Figure 2.5b). The evolution of the third downstream ridge is indicated by the southerly flow at 16000 km after day 8. Idealised studies of ET showed also that in the presence of an ET system the amplification of a Rossby wave train extends further downstream and is more pronounced (Riemer and Jones 2010).

If a trough narrows it is often referred to as "PV streamer". The latter may cut off and form an upper-level cut-off low, which usually conserves upper-level cold air and is therefore a long-lasting system. If a PV streamer extends far equatorward it may trigger tropical cyclogenesis. This is called tropical transition (TT). Such a PV streamer can likewise cut off in (sub-)tropical latitudes. Such a cut-off low is referred to as tropical upper tropospheric trough (TUTT) cell. TUTT cells are often precursors of tropical cyclogenesis in the western Pacific.

A more detailed review on the current understanding of ET, its implications for the downstream flow evolution and numerical weather prediction is given by Jones et al. (2003) and Harr (2010).

# 3 Data and Methods

## 3.1 Model and observational data

In this study ET is investigated based on model and observational data. The global analyses of the European Centre for Medium-Range Weather Forecasts (ECMWF) provide a model data basis. This analysis data is used as the initial and boundary data for simulations with the regional numerical weather prediction model COSMO. Various atmospheric measurement data, mainly collected during the T-PARC campaign, give observational evidence for the model data and deeper insight in some physical processes during ET.

### 3.1.1 ECMWF analyses

The ECMWF is an independent intergovernmental organisation supported by 34 European states[1]. The objectives of the ECMWF are to develop and provide global model and data assimilation systems for the medium to extended range weather forecast, facilitate super computing facilities for scientific research and to foster scientific and technical collaboration within the organisations and scientific community involved in numerical weather prediction.

The six-hourly analysis of ECMWF's Integrated Forecast System (IFS, ECMWF 2010) severe as the basic model data set in this study. The global analysis has a T799 spectral resolution and 91 vertical levels, and is interpolated to a regular horizontal grid with 0.25° horizontal resolution. A research version of the ECMWF analysis is used for the case study of Typhoon Jangmi during T-PARC (experiment DROP from Weissmann et al. 2010). This analysis has assimil-

[1] `http://www.ecmwf.int/about/overview/` retrieved at 25 August 2011

ated all quality controlled dropsonde data collected during T-PARC. Weissmann et al. (2010) showed that typhoon track forecasts based on these analyses were more accurate than using the original analyses. The ECMWF analyses are interpolated on a COSMO grid with 0.25° horizontal resolution and 79 vertical levels (see Section 3.1.2) which enables us to use the same diagnostic tools for ECMWF and COSMO data. In the case of Jangmi short-range COSMO simulations (forecast hour 1-5 h) are combined with the six-hourly ECMWF analysis to provide a one-hourly ECMWF "pseudo-analysis", needed for the verification against airborne measurements (Section 4.2). For all other ET case studies the six-hourly operational ECMWF analyses interpolated on the COSMO grid are used.

### 3.1.2 COSMO model simulations

Several case studies of ET are investigated in this study based on three-dimensional numerical simulations with the COSMO model (Consortium for Small-scale Modeling).

COSMO is a non-hydrostatic limited-area weather prediction model. It was originally developed at the Deutscher Wetterdienst (DWD) as "Lokal-Modell (LM)" and has been used operationally at DWD since December 1999. Over time the model became a community model and is today operationally run and further developed by the national weather services and research institutions of the consortium in more than seven countries. The design of the COSMO model aims to resolving weather systems at the meso-β (5 km- 50 km) and meso-γ (500 m- 5 km) scales in both operational forecasts and scientific applications. In this study a horizontal resolution of 0.25° (approximately 28 km) is used due to the need for covering large areas. The model is based on the non-hydrostatic hydro-thermodynamical equations for fully compressible flow. In the following the model implementation is summarised briefly. A detailed description of COSMO is given by (Doms and Schättler 2002) and information on the Consortium for Small-scale Modeling is available at the COSMO web-site `http://www.cosmo-model.org`.

#### 3.1.2.1 Basic equations

The governing equations in the COSMO model start from the full Eulerian equations for a moist atmosphere. Reynolds and mass-weighted averaging (for wind components, temperature and water constituents) are used to separate the resolvable mean flow on the grid scale from the irresolvable perturbations on the sub-grid scale. The application to small-scale meteorological systems requires a prognostic equation for vertical motion as vertical acceleration can not be neglected compared to the pressure gradient and buoyant forces. This results in a set of non-hydrostatic fully compressible equations for the moist atmosphere, including sound waves. The equation set has the advantage to be valid for weather systems on different scales and of the adaptability to large model domains. Numerical efficiency is assured by a time-splitting scheme following Klemp and Wilhelmson (1978). Although no scale-dependent approximation is used, approximations of the thermodynamics are made for simplicity: molecular fluxes, except for diffusion of liquid and solid forms of water, are neglected. The molecular fluxes of liquid water and ice are represented as sedimentation fluxes (precipitation). The specific heat of moist air is approximated by that of dry air. The effect of diffusion fluxes of water and phase changes on the pressure tendency equation are neglected. In addition the buoyancy term in the heat equation and the mean dissipation rate due to viscous stress are neglected.

Finally a set of prognostic equations for the evolution of the non-hydrostatic compressible mean flow results. The prognostic variables are the three-dimensional wind components, pressure, temperature, and the water constituents.

- The momentum equation

$$\rho\frac{d\vec{v}}{dt} = -\nabla p + \rho\vec{g} - 2\vec{\Omega} \times (\rho\vec{v}) - \nabla \cdot \mathbb{T} \tag{3.1}$$

  describes the three-dimensional wind. The density of air is given by $\rho = \sum_{x=d,v,l,f} \rho^x$, with mixture constituents $\rho^x$, x=d,v,l,f index for mixture constituents dry air ($d$), water vapour ($v$), liquid ($l$), and frozen water ($f$). The total (Lagrangian) time derivative operator is $d../dt = \partial../\partial t + \vec{v} \cdot \nabla..$ and $\partial../\partial t$ is the local (Eulerian) time derivative operator, $\vec{v} = (u, v, w)$

is the barycentric velocity relative to the rotating earth, $\nabla$ is the gradient operator, $p$ is pressure, $\vec{g}$ the apparent acceleration due to gravity, $\vec{\Omega}$ the constant angular velocity of earth rotation, and the Reynolds stress tensor $\mathbb{T} = \overline{\rho \vec{v}'' \vec{v}''}$ describes the turbulent flux of momentum as the Reynolds average (denoted by the overline $\overline{x}$) of the product of density and the deviations of the wind components from their mass-weighted average (marked by $x''$).

- The pressure tendency equation is given by

$$\frac{dp}{dt} = -\frac{c_{pd}}{c_{vd}} p \nabla \cdot \vec{v} + \left( \frac{c_{pd}}{c_{vd}} - 1 \right) Q_h \tag{3.2}$$

  with $c_{pd}$ and $c_{vd}$ being the specific heat of dry air at constant pressure and volume, and $Q_h$ denotes diabatic heating or cooling.

- The thermodynamic equation

$$\rho c_{pd} \frac{dT}{dt} = \frac{dp}{dt} + Q_h \tag{3.3}$$

  describes the temporal change of temperature $T$.

- The budget equations for water constituents are

$$\rho \frac{dq^v}{dt} = -\nabla \cdot \vec{F}^v - (I^l + I^f), \tag{3.4}$$

  for specific humidity $q_v = \rho_v / \rho$ with the turbulent flux of the water constituents $x = v, l, f$ given by $\vec{F}^x = \overline{\rho \vec{v}'' q^x}$ and the phase transition rate for liquid water and ice $I^{l,f}$. Similarly,

$$\rho \frac{dq^{l,f}}{dt} = -\nabla \cdot (\vec{P}^{l,f} + \vec{F}^{l,f}) + I^{l,f} \tag{3.5}$$

  describes the budget equations for liquid $q_l = \rho_l / \rho$ and frozen $q_f = \rho_f / \rho$ cloud water. The precipitation flux for liquid water and ice is given by $\vec{P}^{l,f} = \rho q^{l,f} \vec{v}_T^{l,f}$ with the terminal velocity $\vec{v}_T^{l,f}$ of constituents $l, f$.

- The set of prognostic equations is completed by the diagnostic equation of state

$$\rho = p\left[R_d(1+q^v\left(\frac{R_v}{R_d}-1\right)-q^l-q^f)T\right]^{-1} \tag{3.6}$$

with the gas constants for dry air $R_d$ and water vapour $R_v$, and by the diabatic heating term

$$Q_h = L_V I^l + L_S I^f - \nabla\cdot\left(\vec{H}+\vec{R}\right) \tag{3.7}$$

with $L_V, L_S$ being the latent heat of vaporisation and sublimation, $\vec{H} = c_{pd}\overline{\rho\vec{v}''T}$ the turbulent flux of sensible heat and the $\vec{R}$ flux of solar and thermal radiation.

#### 3.1.2.2 Model implementation

For the numerical solution the continuous equations are transformed into a discretised form using finite differences of second order and solved on an Arakawa-C/Lorenz staggered grid (Arakawa and Lamb 1981). A generalised terrain following vertical coordinate $\zeta$ for $n$ full and $n+1$ half model levels is used in COSMO. This vertical coordinate is time-independent and defined from the top (half level index $k_{1/2}$) to the bottom (half level index $k_{n+1/2}$). Scalars are defined at the centre of a grid box at full model levels and the velocity components normal to the corresponding box faces (Figure 3.1).

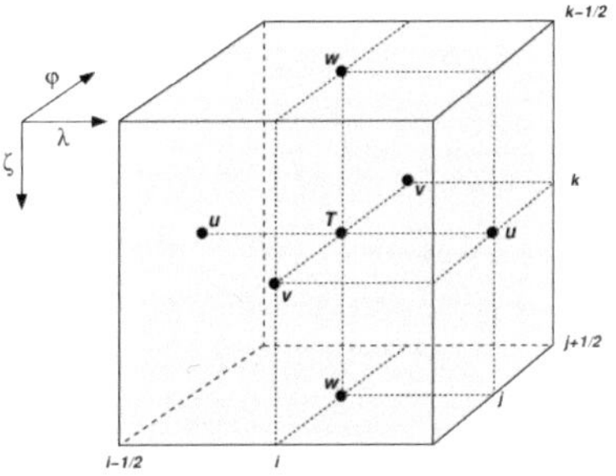


Figure 3.1: COSMO model variables in a grid box of the Arakawa-C/Lorenz staggered grid. Figure 4.1 from Doms and Schättler (2002).

A model reference state is introduced assuming horizontal homogeneity, stationarity, and prescribed to be in hydrostatic balance. Thus a variable $\psi$ at a specific location is described as the sum of its reference state and a deviation:

$$\psi(\lambda, \varphi, z, t) = \psi_0(z) + \psi^*(\lambda, \varphi, z, t)$$

where $\lambda$ denotes geographical longitude, $\varphi$ latitude, $z$ height, and $t$ time. Furthermore, the reference state is prescribed to be dry and at rest. Thus

$$u, v, w, q^x(\lambda, \varphi, z, t) = u^*, v^*, w^*, q^{*x}(\lambda, \varphi, z, t)$$

and

$$T, p, \rho(\lambda, \varphi, z, t) = T_0(z), p_0(z), \rho_0(z) + T^*, p^*, \rho^*(\lambda, \varphi, z, t).$$

The profiles of reference temperature $T_0(z)$, pressure $p_0(z)$, and density $\rho_0(z)$ fulfil the gas equation: $p_0(z) = \rho_0(z) R_d T_0(z)$.

The temporal integration in the COSMO setup used in this study is effected with a second order leapfrog method. A time-splitting scheme following Klemp and Wilhelmson (1978) is employed to fulfil the Courant-Friedrich-Levi (CFL) criterion $u\Delta t/\Delta x < 1$ ($u$ propagation speed, $\Delta t$ time step, $\Delta x$ horizontal grid scale) while assuring numeric efficiency. The prognostic equations are separated into terms which are directly linked to fast sound waves and into comparatively slow modes of motion. The time step of the slow modes is divided into a number of small time steps, and fast terms are computed every small time step while slow terms are computed only for the large time step. This reduces the computational cost while ensuring numerical stability.

#### 3.1.2.3 Model configuration and parametrisations used

Not all physical processes on the entire range of scales can be computed with current computer resources. For an as complete as possible representation of the atmosphere while limiting computational cost some physical processes in particular sub-grid scale processes are parametrised. Doms et al. (2007) give details on the parametrisations available in COSMO.

In this study a single setup based on COSMO Version 4.13 is used. Rotated coordinates are not used to ensure the applicability of the model output to various post-processing utilities. Consequently, the horizontal grid with a resolution of 0.25° varies from around 5 km in higher latitudes (80°) to 28 km at the equator, limiting the northward extent of the model domain. The 79 constant vertical COSMO levels are defined to be as close as possible to the lowest 79 vertical levels of the deterministic IFS ECMWF model[2], so that the planetary boundary layer and the layer around the tropopause have a higher vertical resolution than the usual COSMO setup. The data is interpolated on pressure levels from 1000 hPa to 25 hPa every 25 hPa.

The parametrisation of radiative processes follows Ritter and Geleyn (1992). Eight spectral intervals are introduced, in each of which absorption is assumed to be constant. Sub-grid scale moist convection is parametrised with the Tiedtke mass flux scheme (Tiedtke 1989) and the formation of precipitation on the grid scale is represented by a Kessler-type bulk formulation (Kessler 1969) including snow, cloud water, graupel, and cloud ice. The turbulent fluxes represent the exchange of momentum, heat, and humidity between the surface and the free atmosphere and are thus crucial for a correct numerical simulation. Sub-grid scale turbulence links the resolvable to the irresolvable fluctuations of motion. The turbulence parametrisation is based on a 2.5-order 1-D TKE diagnostic closure (in the notation of Mellor and Yamada 1982), combined with a TKE based surface transfer scheme for parametrisation of the the surface fluxes (Raschendorfer 2001, and section 3.3 in Doms et al. (2007)). Horizontal homogeneity is assumed so that the horizontal components are neglected. The scheme consistently couples the atmosphere and the underlying surface by a stability and roughness-length dependent surface flux formulation (Raschendorfer 2001). Therefore the lowest atmospheric layer is subdivided into a laminar surface layer just above the rigid surface and a turbulent Prandtl layer above. The laminar surface layer allows for discriminating the model variables at the rigid surface (soil) and at the lower boundary for the turbulent atmosphere. In the Prandtl layer, the TKE-scheme is applied, with the help of vertical gradients to

[2]The IFS uses time dependent pressure levels. The COSMO levels are defined using the IFS reference pressure of the lowest 79 of these pressure levels.

the logarithmic Prandtl layer profiles (Raschendorfer et al. 2003). In contrast to other schemes, using empirically derived parameters for estimating stability and roughness-length, these parameters are derived from the coefficients of the turbulence scheme. Temperature and humidity at the ground needed for the calculation of the surface fluxes are provided by the multi-layer soil and vegetation model TERRA-ML. TERRA-ML solves the heat conduction equation directly. In total 8 soil layers are used. The sea surface temperature is provided by the initial data and remains constant throughout a simulation.

#### 3.1.2.4 External, initial and boundary data

In COSMO, surface properties e.g. surface height, vegetation, or land use are prescribed with an external constant data set. Limited area weather prediction models further require initial and boundary data to account for larger scale weather systems outside the model domain. This data is provided by a (lower-resolved) global model or another COSMO run (nesting). In this study, initial and boundary data are provided by the six-hourly analysis from the deterministic global IFS model of the ECMWF. The application of the various post-processing tools on analysis data requires an interpolation of the ECMWF analysis onto the COSMO grid. This is computed by the preprocessor programme INT2LM (Version 1.15).

### 3.1.3 Observational data

Beside standard synoptic measurements collected at airport stations and official weather stations communicated via the Global Telecommunication System (GTS) we use special observations collected during T-PARC.

Research flights performed with the research aircraft DLR-Falcon 20 D-CMET of the German Aerospace Center (DLR) during T-PARC provide an excellent data set for studying the interactions of the TC and the midlatitude flow that occur during ET. The DLR-Falcon aircraft was equipped with in-situ instruments and drop sondes for measuring the standard meteorological quantities (temperature, humidity, three-dimensional wind, pressure) along the flight track and in vertical profiles below the aircraft. Additionally measurements along and below the flight track are provided by two remote sensing instruments[3]. A Doppler wind lidar system (WIND) provides profiles of the wind vector below the aircraft using the velocity-azimuth display technique (for details of the wind lidar system and an error assessment see Weissmann et al. 2005). "The scanning wind lidar yields profiles between 0.5 and 12 km altitude with an accuracy of better than 1 m/s at a horizontal resolution of 5 to 10 km. The vertical resolution of the measurements is 100 m". A water vapour differential absorption lidar (DIAL) provides high spatial resolution measurements of water vapour below the air craft (for details of the DIAL system see Wirth et al. 2009). "For Water vapour profiles from 0.5 to 12 km altitude with 10 % accuracy, the horizontal resolution will be between 1 and 5 km, and the vertical resolution between 300 and 500 m. Both lidar ranges are limited in the near and the far field: there are no data below 100 m above surface and 200 m (wind: 500 m) below flight altitude." In the post-processing the raw data is interpolated on a regular grid determined in horizontal direction by the profile number (defined by the time of measurement) and in vertical direction by the altitude. For the WIND lidar one profile every 32 s is computed corresponding to a horizontal resolution from 6 km to 9 km given an aircraft speed ranging from $190\,\mathrm{m\,s^{-1}}$ to $280\,\mathrm{m\,s^{-1}}$ and for the DIAL lidar one profile every 65 s corresponding to a horizontal resolution from 12 km to 18 km. The vertical resolution of both processed lidar data sets is 100 m. The one-hourly ECMWF pseudo-analysis data is then interpolated on this lidar grid using the method of Schäfler et al. (2010) and Schäfler et al. (2011) (see Section 4.2.1).

MTSAT satellite data are used. The brightness temperature for IR channel 1 ($10.3 - 11.3\,\mu m$) gives an estimate of the spatial extent of high clouds and temporal sequences of this data help to identify deep convection during ET. The twelve-hourly data with a 0.05° horizontal resolution was provided by JMA via their T-PARC support page [4]. Satellite wind data are derived from atmospheric motion vectors (AMVs - e.g. Velden 1996; Velden et al. 1997, 2005).

[3] `http://www.pa.op.dlr.de/tparc/system.html` retrieved at 25 August 2011, cited text passages from that webpage are indicated by quotation marks

[4] `http://tparc.mri-jma.go.jp/special/special.html` retrieved at 13 August 2010

The AMVs are computed by tracking edges in the satellite imagery (at different wavelengths) for subsequent timesteps. The height assignment varies around $\pm$ 25-35 hPa (Chris Velden, personal communication). Hourly AMV data at irregular locations were provided by the Cooperative Institute for Meteorological Satellite Studies (CIMSS) via the T-PARC data archive [5].

Observations of the ocean surface wind speed are given by QuikScat twelve-hourly data with 0.25° horizontal resolution. The data is also provided by JMA via their T-PARC support page.

## 3.2 The conceptual model of potential vorticity

### 3.2.1 Potential vorticity

Potential vorticity (PV) is a meteorological variable, that combines thermodynamic and dynamic characteristics of atmospheric motion. In this study we use the definition of Ertel (1942) for PV in a fully baroclinic, compressible flow:

$$PV = \frac{1}{\rho}\vec{\eta}\cdot\nabla\Theta. \tag{3.8}$$

where $\rho$ is the density of the fluid, $\vec{\eta}$ the absolute vorticity, and $\nabla\Theta$ the three-dimensional (Cartesian) gradient of potential temperature $\Theta$. The unit of PV is $\mathrm{m^2\,s^{-1}\,K\,kg^{-1}}$ and usually given in PV units, $1\,\mathrm{PVU}=10^{-6}\mathrm{m^2\,s^{-1}\,K\,kg^{-1}}$.

The PV equation in the presence of a frictional force $\vec{F}$ and a diabatic potential temperature source $d\Theta/dt = \dot{\Theta}$ is given in height coordinates (Eq.(70a) in Hoskins et al. 1985) by:

$$\frac{dPV}{dt} = \frac{1}{\rho}\vec{\eta}\cdot\nabla\dot{\Theta} + \frac{1}{\rho}(\nabla\times\vec{F})\cdot\nabla\Theta. \tag{3.9}$$

From Equation (3.9) it emerges directly that PV is conserved in a Lagrangian sense following three-dimensional, non-hydrostatic, adiabatic, inviscid flow.

[5] http://www.eol.ucar.edu/projects/t-parc/ and http://data.eol.ucar.edu/codiac/dss/id=110.028 retrieved at 13 January 2009

This is the well-known "conservation principle" of PV. It shows that PV constitutes an air mass tracer when advective processes dominate frictional and diabatic ones. Furthermore, when frictional processes are negligible, such as in the free atmosphere, non-advective changes in the PV field give an indication of the diabatic modification of PV (e.g. Stoelinga 1996; Pomroy and Thorpe 2000).

The second important characteristic of PV is expressed in the "invertibility principle". In analogy to the absolute vorticity, that contains all relevant information for a barotropic flow, PV contains all information about the wind field, the pressure and temperature (mass) field associated with the baroclinic flow. Provided that a balance condition defining the relation of wind and mass field, a reference state, and suitable boundary conditions are given and that the inversion problem is solved globally, one can deduce the wind and mass field for a given PV distribution by inverting a Laplacian operator. In particular the impact of a PV anomaly on the wind and mass field extends over a much larger region than the actual anomaly.

### 3.2.2 PV thinking

Applying the conservation principle and the invertibility principle of PV, permits a powerful view of the dynamics of weather systems. As weather systems usually have a characteristic PV signature, the conservation principle gives a conceptual picture of their evolution in time. The invertibility principle allows for quantifying the consequences of this evolution for the global flow and mass field. The investigation of weather systems entirely in terms of their distinct PV signature goes historically back to Kleinschmidt (1950a,b, 1951, 1955) and was reintroduced by Hoskins et al. (1985) as the so-called "PV thinking".

The flow and temperature field associated with an isolated upper-level positive PV anomaly is illustrated in Figure 3.2. The cyclonic circulation anomaly extends over a much larger area than the positive PV anomaly itself. The impact of the PV anomaly in the vertical is notably determined by the stratification. The circulation and temperature anomaly extends less above the PV anomaly in

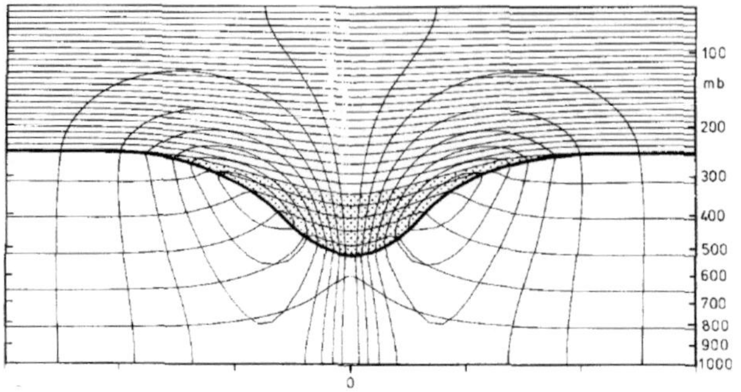


Figure 3.2: Figure 15 from Hoskins et al. (1985) courtesy A. J. Thorpe: Circular symmetric cyclonic flow (wind speed across the section in black thin contours every 3 m $s^{-1}$, zero contour omitted) associated with a positive PV anomaly (stippled). The thin black lines represent isentropes every 5 K. The reference static stability (and PV) was defined uniform in the troposphere (below black thick line) and six times larger in the stratosphere above. The PV anomaly is constructed by deflecting the troposphere downward by 24 K. The anomaly calculation is detailed in Thorpe (1985).

the more stably stratified stratosphere but below the PV anomaly the circulation and temperature anomaly reaches down to the surface. The temperature and wind field reflect an anomalous high static stability and absolute vorticity in the positive PV anomaly, below and above the positive PV anomaly static stability is reduced relative to the reference state. The same conclusions - with opposite sign - are valid for a negative upper-level anomaly.

For near surface temperature anomalies Bretherton (1966) showed, in the context of quasi-geostrophic theory, that a positive or negative potential temperature anomaly can be regarded as a cyclonic or anticyclonic PV anomaly that is con-

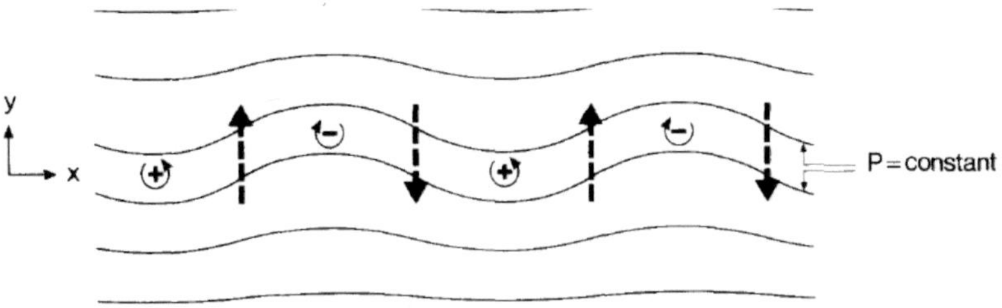


Figure 3.3: Figure 17 from Hoskins et al. (1985): Schematic IPV map of a x-periodic Rossby wave train in an initial northward IPV gradient in a westerly base flow. The contours of IPV are deflected, leading to positive (+) and negative (-) IPV anomalies. The dashed arrows show the induced wind field, leading to a westward propagation of the Rossby wave train with respect to the base flow.

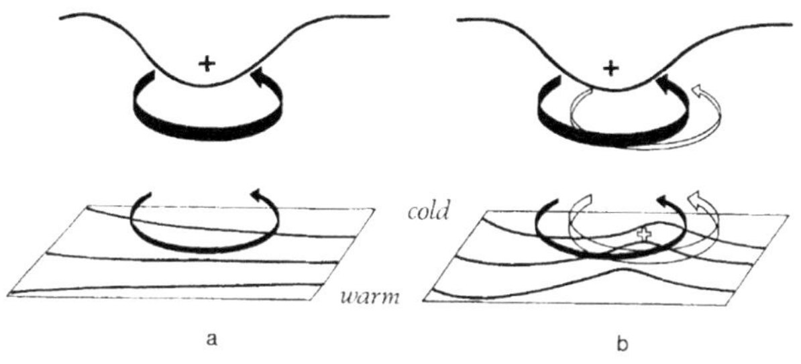


Figure 3.4: Figure 21 from Hoskins et al. (1985): Schematic picture of the interaction of an upper-level positive IPV anomaly (black contour and + at upper levels) over a low-level baroclinic zone (black contours at lower levels). The flow induced by the upper IPV anomaly (black arrows in (a) and (b)) and by the developing lower IPV anomaly (white arrows in (b)) is indicated.

centrated at the surface. In terms of PV thinking Hoskins et al. (1985) following Bretherton (1966) pointed out that a near surface positive or negative potential temperature anomaly is associated with a cyclonic or anticyclonic circulation.

On an isentropic upper-level surface, the transition from the relatively high tropopause in the tropics to a lowered tropopause at high latitudes is reflected in a pronounced PV gradient in the midlatitudes. This PV gradient serves as the midlatitude wave guide and separates tropospheric low PV air equatorward from stratospheric high PV air polewards. Deflections of the PV gradient result in the triggering of a Rossby wave train (Figure 3.3). In a purely barotropic (horizontal) view, equatorward deflections constitute a positive PV anomaly associated with the midlatitude troughs, poleward deflections a negative PV anomaly associated with the midlatitude ridges. The PV anomalies induce poleward flow east of a trough and west of a ridge and equatorward flow east of a ridge and west of a trough. As a consequence, poleward advection of low PV air occurs between a trough and a ridge and equatorward advection of high PV air between a ridge and a trough. This results in a propagation of the Rossby wave train to the west relative to the base flow.

The standard cyclogenetic situation of an upper-level trough approaching a surface baroclinic zone can be regarded in the view of "PV thinking" (Figure 3.4). The flow induced by the upper-level positive PV anomaly leads to a deflection of the baroclinic zone at the surface (Figure 3.4a). As a consequence, a warm thermal surface anomaly develops at the southern side of the baroclinic zone.

This thermal anomaly, regarded as a positive PV anomaly, induces its own cyclonic circulation slightly ahead of the upper-level trough (Figure 3.4b). This allows for a subsequent propagation of the low-level anomaly along the baroclinic zone. Also there may be a positive feedback for the downstream propagation of the upper-level positive anomaly. When both anomalies couple, a backward tilted column of positive PV extends through the troposphere. Due to vertical shear the upper-level anomaly becomes collocated with the low-level anomaly during the subsequent time, marking the mature stage of the cyclonic development. Lacking the interaction of the upper-level anomaly with the baroclinic zone, the cyclone starts to decay.

The PV equation 3.9 describes how PV anomalies can be produced by diabatic processes (1st term RHS) when the gradient of the temporal change of potential temperature $\Theta$ has a parallel component to the absolute vorticity $\eta$. The temporal change $\Theta$ is primarily due to the latent heating by condensational processes. On the synoptic scale and in mature TCs the vertical component of $\eta$ plays the major role and consequently diabatic PV modification is mainly linked to the vertical gradient of the temporal change of potential temperature $\partial\dot{\Theta}/\partial z$. Wernli and Davies (1997) noted that two situations have to be distinguished (Figure 3.5). In the case of an impulsive diabatic heating, a positive PV anomaly will develop below the level of maximum diabatic heating, as long as $\dot{\Theta}$ increases with height (Figure 3.5a). Above the level of maximum heating, $\partial\dot{\Theta}/\partial z$ is negative and a negative PV anomaly will occur. However, if the

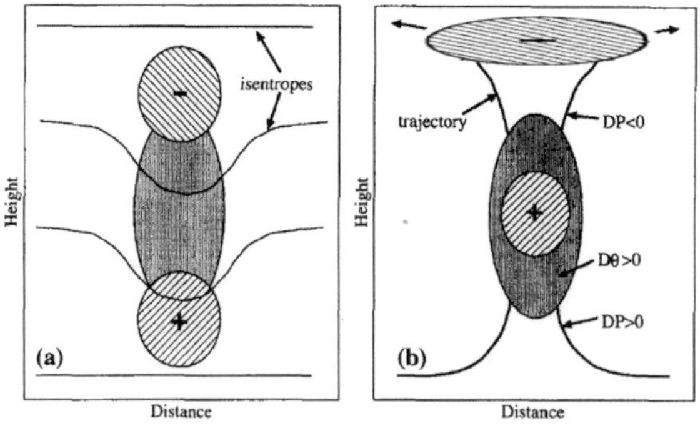


Figure 3.5: Adapted from Figure 4 in Wernli and Davies (1997): Illustrative vertical cross sections of diabatically produced PV anomalies (+ or - signs). The region of diabatic heating is grey shaded. (a) "impulsive diabatic heating", (b) "steady diabatic heating".

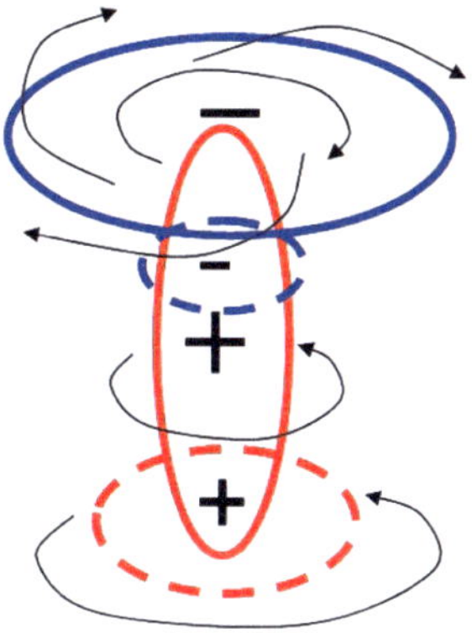


Figure 3.6: Schematic view of the PV anomalies in a tropical cyclone. The tall red contour represents the positive, cyclonic PV tower in the TC inner core, where steady diabatic heating occurs. The shallow blue coloured contour indicates the upper-level negative, anticyclonic outflow anomaly. The dashed contours encircle the regions of in-situ diabatic PV production (red,+) and reduction (blue,-). The induced circulations are indicated by black arrows.

timescale of diabatic heating is comparable to the timescale for vertical advection of saturated air parcels, the positive PV anomaly will be vertically advected to higher levels and reside at the level of maximum heating (Figure 3.5b). The negative anomaly is likewise advected vertically and becomes a shallow upper-level negative PV anomaly below the tropopause (cf. Figure 1 in Pomroy and Thorpe 2000).

This conceptual model of diabatic PV anomalies in a steady state situation of diabatic heating is comparable with the role of diabatic processes in a tropical cyclone. In the TC inner core, the upright buoyant ascent causes steady condensational heating leading to the evolution of a strong cyclonic PV anomaly,

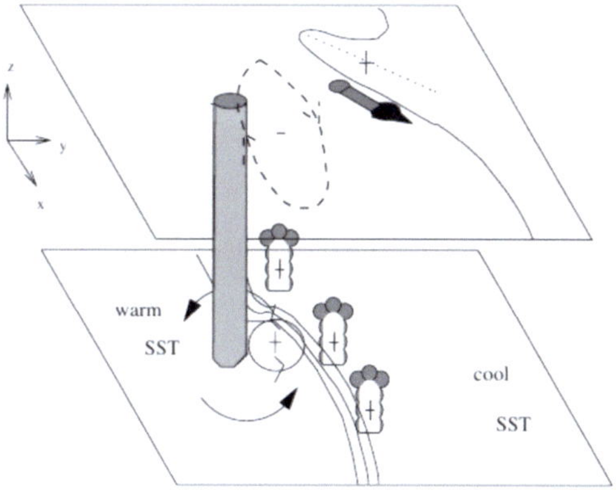


Figure 3.7: Figure 18 (b) from Agustí-Panareda et al. (2004): Illustration of the ET process. The upper plane represents the upper troposphere and the lower plane the surface. See text for details.

the positive PV tower (Figure 3.6). At upper-levels, below the tropopause, divergent anticyclonic outflow is present that constitutes a negative PV anomaly above the level of maximum diabatic heating.

A schematic picture of the PV anomalies involved in the ET processes has been proposed by Agustí-Panareda et al. (2004). During ET the tropical cyclone approaches a low-level baroclinic zone (black contours, surface in Figure 3.7). The flow (black arrows, surface) associated with the positive PV tower of the TC inner core (grey upright cylinder) initiates frontogenesis at the baroclinic zone reflected in a deflection of the low-level isotherms and the associated surface thermal anomaly (encircled +, surface). Dynamically forced ascent of tropical warm moist air along the tilted isentropes of the baroclinic zone and upright ascent in convection triggered ahead of the baroclinic zone leads to steady condensational heating resulting in positive PV anomalies aligned along the baroclinic zone (clouds with plus sign). This is overlayed at upper levels with a region of diabatic PV reduction that merges with the anticyclonic negative PV anomaly of the outflow. A midlatitude jet streak (horizontal grey arrow) is located at the maximum PV gradient between the outflow anomaly and the positive PV anomaly associated with an approaching upper-level trough (+ sign in the black PV contour and dotted trough axis).

### 3.2.3 PV inversion

The invertibility principle of PV allows for calculating the wind and temperature field associated with a PV anomaly, provided that a balance condition, a reference state for the mass distribution (temperature field), and suitable boundary conditions are given and that the differential equation is solved globally.

In this study, non-linear balance (Charney 1955) is used to specify the relation of the flow field (stream function $\Psi$) and mass distribution (geopotential $\Phi$). Non-linear balance was designed for investigating "meteorologicaly significant large-scale motions" in the primitive hydrodynamical equations without the "noise" due to "spurious large-amplitude inertio-gravitational oscillations" (Charney 1955). The latter are due to high-frequency modes for which horizontal divergence is in the order of relative vorticity. The non-linear balance equation is derived by taking the horizontal divergence of the horizontal momentum equation, effecting a Helmholtz-partitioning of the horizontal wind field $\vec{v}_h = \vec{v}_\Psi + \vec{v}_\chi$ in a non-divergent part $\vec{v}_\Psi = \vec{k} \times \nabla_h \Psi$ associated with the stream function $\Psi$ and an irrotational part $\vec{v}_\chi = \nabla_h \chi$ associated with the velocity potential $\chi$. Scaling assumptions for large-scale flows allow the irrotational (divergent) part to be neglected: ${\nabla_h}^2 \chi = \nabla \cdot \vec{v}_h \ll \zeta = \vec{k} \cdot \nabla \times \vec{v}_h = {\nabla_h}^2 \Psi$. Thus the basic assumption of non-linear balance is that horizontal divergence is much smaller than relative vorticity in the meteorologicaly significant large scale motions. This results in the non-linear balance equation in spherical coordinates:

$$\nabla_h^2 \Phi = \nabla_h \cdot (f \nabla_h \Psi) + \frac{2}{a^4 cos^2\varphi} \frac{\partial(\partial\Psi/\partial\lambda, \partial\Psi/\partial\varphi)}{\partial(\lambda, \varphi)}, \tag{3.10}$$

where $\nabla_h$ is the horizontal gradient operator, $\Phi$ is the geopotential, $a$ the Earth radius, and $\lambda, \varphi$ the longitude, latitude in geographical coordinates. Hence, non-linear balance initially excludes inertio-gravitational motions and thus constitutes a filter for meteorological "noise". Albeit neglecting the divergent flow relative to the vorticity is not properly valid in tropical cyclones, especially in the inner core, non-linear balance has been successfully applied to TCs (e.g. Wu and Kurihara 1996). Divergent flow components can be expected to be in the same order of magnitude as rotational components in the entrance and exit region of jet streaks also. Nevertheless, non-linear balance has been successfully applied in various studies of ET (e.g. McTaggart-Cowan et al. 2001; Agustí-Panareda et al. 2004, 2005; Agustí-Panareda 2008; Riemer et al. 2008; Riemer and Jones 2010), and of extratropical cyclones (e.g. Davis and Emanuel 1991; Davis et al. 1996; Ahmadi-Givi et al. 2004).

A second diagnostic relation of $\Psi$ and $\Phi$ with link to the PV field is deduced from the definition of PV (3.8). Using standard meteorological scaling assumptions the absolute vorticity is approximated by $\vec{\eta} = f\vec{k} + \nabla \times \vec{v}_h$, where $f = 2\Omega sin\varphi$ is the planetary vorticity, $\Omega = 2\Pi/86400s$ the angular velocity of the earth rotation, and $\vec{v}_h = (u, v, 0)$ the horizontal wind vector. In Equation (3.8) reformulated in spherical coordinates, the horizontal wind is replaced by the non-divergent wind, the vertical wind is neglected, and the Exner function $\pi = c_p\,(p/p_o)^{\frac{R_d}{c_p}}$ used as vertical coordinate ($R_d$: gas constant for dry air, $c_p$: specific heat capacity of dry air with constant pressure, $p_0 = 1000\,\text{hPa}$: reference pressure) along with the associated hydrostatic approximation $\partial\Phi/\partial\pi = -\Theta$. This yields in the definition of PV:

$$PV = \frac{g\kappa\pi}{p}\left[\left(f+\nabla_h{}^2\Psi\right)\frac{\partial^2\Phi}{\partial\pi^2} - \frac{1}{a^2cos^2\varphi}\frac{\partial^2\Psi}{\partial\lambda\partial\pi}\frac{\partial^2\Phi}{\partial\lambda\partial\pi} + \frac{1}{a^2}\frac{\partial^2\Psi}{\partial\varphi\partial\pi}\frac{\partial^2\Phi}{\partial\varphi\partial\pi}\right]. \tag{3.11}$$

For a given PV distribution Equation (3.11) along with the non-linear balance equation (3.10) and suitable boundary conditions form a closed system for the stream function $\Psi$ and geopotential $\Phi$.

The system is solved with the code of Davis and Emanuel (1991) using successive over-relaxation (SOR). The first guess is determined by the model geopotential and the barotropic stream function calculated from the model relative vorticity. At the lateral boundaries Dirichlet conditions are specified from the model data. The lateral boundary condition for $\Psi$ has to be modified so that any net divergence around the boundaries vanishes. At the horizontal boundaries von Neumann boundary conditions are used by prescribing the vertical derivatives of $\Psi$ and $\Phi$. The horizontal boundary condition is thus prescribed by potential temperature via the hydrostatic approximation $\partial\Phi/\partial\pi = -\Theta$. Optionally potential temperature from the model data can be prescribed instead of the vertical derivative, which yields in a better convergence of the solver in some cases. The SOR solver requires Equations 3.10 and 3.11 to be elliptic which is given if static stability $\partial^2\Phi/(\partial\pi)^2 = -\partial\Theta/\partial\pi$ and absolute vorticity are positive. Therefore negative PV values are set to small positive values during the inversion and unstable layers set to very small stable stratification. Details on the numerical solution are given in the appendix of Davis and Emanuel (1991).

The PV inversion is solved on pressure level data. We typically use the data every 25 hPa from $p_{bot} = 900\,\text{hPa}$ to $p_{top} = 100\,\text{hPa}$. The top and bottom boundary conditions are then given at 887.5 hPa and 112.5 hPa. The lowest level $p_{bot}$ should be located above the top of the planetary boundary layer, where non-linear balance equations are expected to be fulfilled. In the case studies presented here, the SOR solver will only converge if the highest level $p_{top}$ is below the tropopause. Therefore $p_{top}$ was varied between 200 and 25 hPa. In some cases convergence was only achieved when potential temperature from model data was prescribed at the horizontal boundaries, rather than the vertical derivatives.

The PV inversion code can be applied on gridded data with a horizontal resolution of 20 km or coarser. In this study model data on a geographical grid with 0.25° horizontal resolution is available without rotated coordinates. To achieve convergence of the SOR solver despite the convergence of meridians in midlatitudes (0.25°longitude $\approx$ 14 km at 60°N), prior to the PV inversion the model data is interpolated onto a coarsened geographical grid with 0.75° horizontal resolution. We found also, that the SOR solver only converges if the southern boundary is at least at 3°N. The PV inversion domain must enclose a region at least as large as the Rossby radius of deformation around the considered PV feature to reduce the impact of the lateral boundaries on the results.

The output of the PV inversion provides the stream function $\Psi_b$ and the geopotential $\Phi_b$, that are associated with the given initial PV distribution and that fulfil non-linear balance. The associated horizontal wind components $u_b, v_b$ and temperature $T_b$ on the coarsened horizontal grid can be derived from the following relations:

$$u_b = -\frac{1}{a}\frac{\partial \Psi_b}{\partial \varphi}, \quad v_b = \frac{1}{a cos\varphi}\frac{\partial \Psi_b}{\partial \lambda} \tag{3.12}$$

$$T_b = -\frac{p}{R_d}\frac{\partial \Phi_b}{\partial p} \tag{3.13}$$

### 3.2.4 PV surgery

Weather systems such as TCs, are often accompanied by a distinct signature in the PV field. This PV field can be modified prior to PV inversion. For example, the PV signature of a distinct weather system can be removed and the balanced stream function $\Psi_b$ and geopotential $\Phi_b$ can be computed without the signature of the weather system. In a second step, this stream function $\Psi_b$ and geopotential $\Phi_b$ can be used to remove the anomaly in meteorological fields that is associated with the weather system from the analysis while leaving the larger scale environment unchanged. Finally, the modified analysis can be applied as initial condition for a COSMO simulation. This allows us to study the temporal evolution of the meteorological fields without the weather system of interest. Comparing this COSMO simulation to a control COSMO simulation initialised from the unmodified analysis, enables us to quantify the impact of the weather system on the flow evolution.

In the following, a PV surgery technique is outlined that is designed to remove the positive PV anomaly associated with the PV tower of a mature TC as thus the primary cyclonic circulation and the TC warm core as cleanly as possible.

Non-linear balance (Charney 1955) is based on the assumption that horizontal divergence is of a much smaller magnitude than relative vorticity $\nabla \cdot v_h \ll \zeta$. While this is a good estimation for the large-scale flow, divergence and vorticity are often of the same order of magnitude in tropical weather systems. Nevertheless in previous studies on Atlantic ET systems (e.g.- McTaggart-Cowan et al. 2001; Agustí-Panareda et al. 2004, 2005; Agustí-Panareda 2008) sensitivity experiments were initialised successfully directly from the balanced fields. In TCs divergence is large but relative vorticity is even larger and also the model domain in these studies was at most as far south as 20°N, where Coriolis force already plays some role. Thus rotating systems may occur that fulfil $\nabla \cdot v_h \ll \zeta$. However, for simulating ET in the Pacific in this study the COSMO model domain needs to cover the tropics south of 20°N and the midlatitudes. Balanced fields in the entire domain would neglect an important contribution to the total flow by divergent components in the tropics. As gravity waves (e.g. equatorial trapped waves) triggered by divergent processes belong to the solution of the model equations errors occur through neglecting them. These errors were manifested, in gravity waves emerging from the equatorial regions with wind speed amplitudes in the order of the actual variable and propagating through the entire model domain for at least the first 24 hours of integration. We found also that the balanced flow and mass field of a TC (computed from the unmodified PV distribution) was broader and deeper, and consequently the TC was more intense. Agustí-Panareda et al. (2004) reports similar problems when using balanced fields as initial fields leading to significant differences of the TC *and* upper-level midlatitude flow in her control experiment from the analysis. As we could neither compute a realistic TC (from an unmodified PV field) nor a realistic background flow (from a modified PV field) when using solely non-linear balance in the entire model domain we aimed for a better suited TC removal technique.

The basic idea of our PV surgery method is to replace the meteorological fields in the vicinity of a PV anomaly by the background flow. This background flow can be derived by subtracting from the analysis an anomaly in model variables that is linked to the PV anomaly. In a first attempt, the TC anomaly was defined as the balanced anomaly, that is the difference of a PV inversion on the unmodified PV field and of a second PV inversion on a PV field in which the PV anomaly of the TC inner core had been removed. However, it turned out that this balanced anomaly does not represent the TC entirely. In a next step the TC anomaly was defined as the balanced anomaly plus the divergent wind components and an estimate of the flow in the TC planetary boundary layer. Although this method achieved a more advanced and more complete removal of the TC from the analysis, some spurious remnants persisted which required about 24 hours of COSMO simulation to decay. Also an at most too intense balanced TC yielded in an overcompensation of the TC flow.

Finally, the cleanest TC removal technique arises when the background flow is prescribed to fulfil non-linear balance and is limited to a region in the vicinity of the storm. The vicinity of the PV anomaly is thereby defined as the region where a remote action of the PV anomaly occurs. For a localised PV anomaly[6] like a TC this region is approximately defined by a circle with the Rossby radius of deformation

$$L_R = \frac{NH}{f}, \quad N^2 = \frac{g}{\Theta}\frac{\partial \Theta}{\partial z}, \tag{3.14}$$

centred around the storm, with the Brunt-Väisälä frequency $N$, a characteristic height $H$, and the planetary vorticity $f = 2\Omega sin\varphi$. For a TC $L_R$ is about 2000 km ($H = 10$ km, $\varphi = 25°$, $\Theta = 300$ K, $g = 9.81$ m s$^{-2}$, $\partial\Theta/\partial z = 5 \cdot 10^{-3}$ K m$^{-1}$). Outside this region the analysis is left unchanged. This avoids unrealistic gravity wave behaviour due to unbalanced weather systems in the tropics and due to

[6] on a horizontal surface

boundary conditions. The edge of the modification region transitions smoothly to the analysis field (details later). The use of meteorological fields that fulfil non-linear balance in the region of the storm is justified as we aim to represent the large scale background flow in the vicinity of the TC that is primarily determined by the subtropical ridge. The subtropical ridge is mostly not dominated by divergent components and the flow fulfils the non-linear balance equations (Charney 1955).

In the following, details about the various steps for the computation of the modified analysis are given.

In our configuration COSMO requires the three-dimensional fields on model levels of temperature $T$, pressure perturbation $p^*$, zonal, meridional, and vertical wind components $u, v, w$, and of the humidity variables specific humidity $q_v$, cloud water $q_c$, cloud ice $q_i$ together with some one-dimensional surface variables as initial fields. The major task for the PV surgery is to define an anomaly in $T, p^*, u, v, w, q_v, q_c, q_i$ linked to the distinct PV anomaly. Once the anomaly is defined it can be subtracted from the original analysis to give an analysis without the system of interest and subsequently added at another location to displace the system.

1. **PV inversion:** The first step is to apply PV inversion to a modified PV field. Therefore, at grid points at which PV exceeds a threshold $PV_{thres}$, PV is set to $PV_{thres}$ prior to the PV inversion. This is effected only within a selected sub domain (in terms of longitudinal, latitudinal, and vertical extent). In the case of the removal of a negative anomaly within a selected sub domain at grid points at which PV undercuts a threshold $PV_{thres}$, PV is set to $PV_{thres}$ prior to the PV inversion. At each level the removed PV is distributed equally to all grid points of that level to keep the total PV constant. This is required for numerical consistency. Two definitions of the threshold are implemented. It can either be a constant value for all levels or the level average PV of the total domain. The latter method would account for higher stratospheric PV values at upper levels when the PV anomaly is close to the midlatitudes. For an ET event the positive PV tower in the inner core of a TC is removed prior or at the beginning of ET.

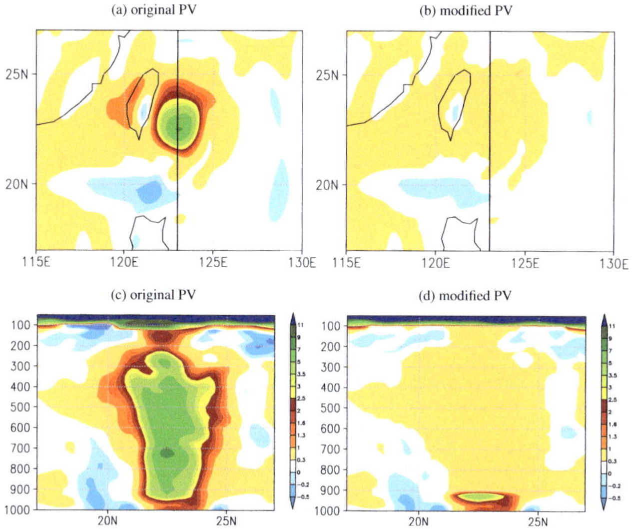


Figure 3.8: Original (a,c) and modified (b,d) PV (shaded) before computing the PV inversion in the case of Jangmi. A horizontal cross section at 700 hPa is shown in (a,b). The black line at 123°E indicates the location of the vertical cross section shown in (c,d). The time is 00 UTC 28 September 2008.

An example for the PV modification in the case of Jangmi is presented in Figure 3.8. The horizontal and vertical cross section show the positive PV anomaly of Jangmi's inner core with PV exceeding 0.3 PVU over a horizontal region of $\approx 3°$radius and between 1000 hPa and 200 hPa with maximum values of 9 PVU (Figure 3.8a,c). In this case the selected sub-domain has the horizontal extent of the domain shown and extends from $p_{bot}$ =900 to $p_{top}$ =100 hPa. PV has been set to the constant threshold $PV_{thres}$ where PV is larger than $PV_{thres} = 0.3$ PVU (Figure 3.8b,d).

After the modification of the PV field a PV inversion is calculated on the specified sub domain and will give - on pressure levels (index "$p$") and on the coarsened horizontal grid required for PV inversion - the balanced geopotential ${\Phi_b}^p$ and stream function ${\Psi_b}^p$ that are associated with this PV field.

2. **Preparation for the calculation of the anomaly:** Before defining the anomaly, the balanced fields ${\Phi_b}^p$ and ${\Psi_b}^p$ are interpolated back onto the ori-

ginal horizontal grid and the balanced temperature ${T_b}^p$ (Eq. 3.13) and horizontal wind components ${u_b}^p, {v_b}^p$ (Eq. 3.12) are calculated from ${\Phi_b}^p$ and ${\Psi_b}^p$. Then ${T_b}^p$,${u_b}^p$,${v_b}^p$, and the (constant) pressure ${p_b}^p$ are interpolated linearly onto the COSMO model levels (which have a fixed geopotential height at each grid point) using ${\Phi_b}^p$ as a vertical coordinate for the values at each pressure grid point. Thus the balanced fields ${T_b}^m$,${u_b}^m$,${v_b}^m$,${p_b}^m$ are available on model levels (index "$m$"). The pressure is further split into the reference state and the pressure perturbation ${p_b}^m = {p_0}^m + p_b^{*m}$.

The PV inversion is calculated from $p_{bot}$ to $p_{top}$ which are typically above the planetary boundary layer (see Section 3.2.3) and below the tropopause. The anomaly is defined only for the corresponding model levels below $p_{top}$ and thus the unmodified analysis is used at upper levels above $p_{top}$. This is justified, as the positive PV anomaly linked to the inner core of the TCs studied here turned out to be very small in the upper levels of the PV inversion domain if present at all (cf. Figure 3.8c,d). Furthermore, the remote action of a PV anomaly is significantly reduced at upper-levels and in the stratosphere due to the strongly stable stratification (cf. Figure 3.2). Nevertheless the interpolation from the pressure levels onto the model levels below $p_{top}$ still requires an extrapolation to levels below $p_{bot}$.

For the extrapolation of the wind components in the lower troposphere the directional and magnitudinal wind change in the planetary boundary layer must be taken into account. A profile linear in the height dependent model levels is prescribed for levels below a selected model level index $k > k_{PBL}$. In our TC case studies $k_{PBL}$ is the index of the model level closest to $p_{bot}$. The linear directional wind change and the linear wind speed reduction from the full model level $k_{PBL}+1$ to full model level $k_n$, ($n$ number of lowest full model level) are computed based on the wind field at $k_{PBL}$. We use a linear directional wind change of 30° counterclockwise and a linear wind speed reduction of 30% as an approximation of the planetary boundary layer profile (personal communication with Chris Davis).

After these prepatory steps the fields ${T_b}^m$,${u_b}^m$,${v_b}^m$,$p_b^{*m}$ derived from the PV inversion output are available on all necessary model levels.

3. **Definition of the $T, p^*, u, v$ anomaly:**

The definition of the $T^m, p^{*m}, u^m, v^m$ anomaly is implemented using a spatial mask. The spatial mask limits the balanced fields to grid points in a sub domain, at which a distinct criterion is fulfilled. Similar to the PV inversion domain this sub domain should cover the region around the PV anomaly where a remote action occurs which is approximately defined by the Rossby radius of deformation $L_r$ (Equation (3.14)). Vertically the sub domain reaches from the ground to the model level, that corresponds to the upper pressure level $p_{top}$ of the PV inversion domain (see step 2).

Technically the spatial mask is implemented by defining an anomaly $X'$, $X = T^m, p^{*m}, u^m, v^m$ at each model level as the difference between ECMWF analysis interpolated onto the COSMO grid $X^m$ and the balanced fields ${X_b}^m$. Inside the box and at grid points where a distinct criterion is fulfilled $X' = X^m - {X_b}^m$, otherwise $X' = 0$. For $T'$ and $(p^*)'$ the criterion is $(p^*)' < p^*_{crit}$. Using the same criterion for $T'$ and $(p^*)'$ emerges from the fact that $p_b^{*m}$ corresponds to ${\Phi_b}^p$ and ${T_b}^{m,p}$ is the vertical derivative of ${\Phi_b}^p$ (Equation (3.13)). Analogously, for $u'$ and $v'$ the criterion is $\Psi' = \Psi^m - {\Psi_b}^m < \Psi_{crit}$ as both wind components are derived from ${\Psi_b}^p$ (Equation (3.12)). The "natural" values for $p^*_{crit}$ and $\Psi_{crit}$ would be zero as this describes the outer limit of an impact of the PV anomaly. Empirically we found that the anomalies will be more cleanly constrained to the PV anomaly, if $p^*_{crit}$ and $\Psi_{crit}$ are slightly less than zero. Typically, values of -0.5 hPa$\geq p^*_{crit} \geq$-2.0 hPa and -1.0$\cdot10^6$m$^2$s$^{-1} \geq \Psi_{crit} \geq$ -3.0$\cdot10^6$m$^2$s$^{-1}$ are used for the positive PV anomaly linked with a TC. The masks defined by $(p^*)' < p^*_{crit}$ and $\Psi' < \Psi_{crit}$ are linearly smoothed at the edges over a selected number of grid points to avoid sharp transitions from analysis to balanced fields. In summary the anomaly is defined according to

$$X' = \begin{cases} X^m - {X_b}^m & \text{, where } (p^*)' < p^*_{crit} \text{ for } (p^*)', T' \text{and where } \Psi' < \Psi_{crit} \text{ for } u', v' \\ 0 & \text{, where } (p^*)' \geq p^*_{crit} \text{ for } (p^*)', T' \text{and where } \Psi' \geq \Psi_{crit} \text{ for } u', v' \end{cases} \quad (3.15)$$

An example for the pressure perturbation anomaly $(p^*)'$ and the meridional wind component anomaly $v'$ associated with Jangmi on the lower mid-troposphere is presented in Figure 3.9. The horizontal domain in Figure 3.9a-c, e-g is the modification sub domain that covers the region defined by the Rossby radius of deformation. A critical value of $p^*_{crit} = -0.5$ hPa is used. This contour clearly outlines the anomaly of pressure perturbation linked to Jangmi (Figure 3.9a,b). In the modified analysis (Figure 3.9c) the pressure perturbation linked to the subtropical anticyclone to the east of Jangmi extends westward over the region of the removed TC. The pressure perturbation linked to Jangmi in the analysis (black) has the shape of a steep cone which is reflected in the zonal profile of $p^*$ (Figure 3.9d). The balanced fields (red) constitute a "lid" on this cone (the spike at about 122°E reflects the impact of Taiwan on $p^*$ in the terrain following vertical coordinates). The smoothing at the edges of the anomaly (here over 10 grid points) lets the modified analysis (blue) merge from the analysis to the balanced fields. Similar structures are evident for $v$ (Figure 3.9e-h). A threshold of $\Psi_{crit}$ = -2.0·$10^6$m$^2$s$^{-1}$ is used.

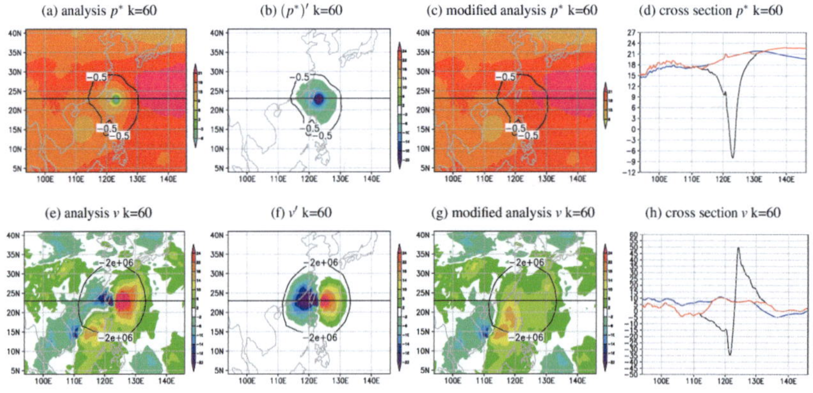


Figure 3.9: Illustration of the TC removal for $p^*$ (in hPa, top, a-d)) and $v$ (in m s$^{-1}$, bottom, e-h). Data is shown at model level $k = 60$ (approximately 700 hPa). Shown are the unmodified analysis (a,e), the anomaly after applying the spatial mask (b,f), the modified anomaly (c,g), and a cross section at 23°N through Jangmi (d,h) of the analysis (black), balanced fields (blue) and modified analysis (red, dashed). In (a-c) the black contour of $p^* = -0.5$ hPa and in (e-g) the black contours of $\Psi' = -2.0 \cdot 10^{-6}\, m^2 s^{-1}$ indicate the boundary of the region where balanced fields are used; the black horizontal line marks the location of the cross section in (d,h). Time of the PV surgery is 00 UTC 28 September 2008.

The anomaly $v'$ constitutes a dipole (Figure 3.9f) with northerly flow west and southerly flow east of Jangmi's centre. It is remarkable that the PV surgery leaves the circulation associated with a second typhoon at about 112°E, 15°N (Figure 3.9g) unchanged in the modified analysis. The zonal profiles of $v$ (Figure 3.9h) highlight how the PV surgery method cancels out the dipole of northerly/southerly winds in the region of Jangmi in the modified analysis (here smoothing of the edges is over 15 grid points).

The definition of $T^m, p^{*m}, u^m, v^m$ anomalies as the difference between analysis and balanced fields without the PV anomaly implicitly describes the entire meteorological fields associated with a PV anomaly including the divergent components. Non-linear balanced flow is thereby only prescribed in the region where the anomaly affects the flow.

Having defined these anomalies, the modified analysis fields $X_{MODPV}{}^m$ without the contribution of the PV anomaly are computed as the difference of ECMWF analysis $X^m$ and the anomaly $X'$: $X_{MODPV}{}^m = X^m - X'$. The considered PV anomaly and associated fields can be displaced by adding $X'$ to $X_{MODPV}{}^m$ at a different location.

4. **Definition of the $w$ and $q_x$ anomaly:** The $T^m, p^{*m}, u^m, v^m$ anomalies computed from the balanced fields of PV inversion and the ECWMF analysis have to be complemented by the anomalies in vertical velocity $w$ and humidity fields (specific humidity $q_v$, cloud water $q_c$, and cloud ice $q_i$) associated with the PV anomaly. A TC is typically accompanied by a region of strong ascent and high values of humidity.

   The vertical velocity $w^m$ in a specified sub domain (in terms of longitudinal, latitudinal extent, and model half levels) is either set to zero, the level average in that domain, or the level average in the total domain (the latter is usually close to zero). After removing the $w^m$ field that is linked to the PV anomaly at its original location the $w^m$ field can be displaced. Therefore the vertical velocity field at the new location can either be overwritten by the original $w^m$ field in the sub domain or the latter can be added at the new location.

For the humidity variables at grid points within a specified sub domain at which $q_x^m$ exceeds a certain threshold $q_x^{thres}$, $q_x^m$ is set to $q_x^{thres}$. As for vertical velocity the threshold either is the level average in the specified sub domain, or the level average in the total domain. For further manipulation this level average can be multiplied by a factor $qv_{fact}$. This method implements a spatial mask to define and remove the $q_x^m$ features that are linked to a distinct PV anomaly. When displacing the PV anomaly these masked humidity features can either be moved to the new location after removing them in the specified sub domain. At the new location they will overwrite the existing humidity variables. Alternately the humidity variables from the entire sub domain can be displaced to the new location where the existing values are overwritten. The latter method would move the $q_x^m$ field linked to the PV anomaly and to its larger environment (in the limits of the specified sub domain) but this is not used in this study.

For the one dimensional fields required to initialise a COSMO run no modifications are computed. This is justified for the application of this PV surgery method on TCs over the ocean, as COSMO has a constant sea surface temperature throughout the simulation.

In summary, all necessary fields for the initialisation of a new COSMO run without the meteorological fields linked to a distinct PV anomaly are given by this PV surgery method as a modified analysis data set. The definition of an anomaly in $T, p^*, u, v, w, q_v, q_c, q_i$ linked to the PV anomaly enables to displace the anomaly after having extracting it from the original analysis. Both removing and displacing a PV anomaly and associated fields in the analysis and initial fields of a COSMO run offers multiple possibilities for sensitivity experiments. In this study we employ the PV surgery to remove and to relocate the cyclonic circulation, vertical velocity field, warm anomaly, pressure perturbation and humidity fields associated with the positive PV tower of a mature tropical cyclone.

### 3.2.5 The TC outflow anomaly

The PV surgery method outlined here removes the cyclonic circulation and warm core anomaly of a TC that is linked to the positive PV anomaly in the TC inner core by prescribing a non-divergent background flow that fulfils non-linear balance. Thus implicitly the balanced and un-balanced anomaly, in particular the divergent wind components, are removed. If the PV inversion domain and the modification sub domain extend vertically close to the tropopause (upper model level index of sub domain $k_{top} < 39 \approx 220\,\text{hPa}$), as it does in our case studies[7] the PV surgery method will remove implicitly the divergent outflow component.

However, similar to the technique of McTaggart-Cowan et al. (2001) our method does not remove the balanced anticyclonic outflow anomaly. It could be removed by removing the associated negative PV anomaly using our PV surgery method. The definition of the negative PV anomaly is difficult as it merges with environmental low PV air. A deviation from a background flow would be needed. Davis and Emanuel (1991) and Davis (1992) inverted the PV of individual anomalies and also formally not valid successfully quantified the contributions of the anomalies to the total flow by adding the anomalies linearly. Riemer et al. (2008) employed the same method and further estimated the divergent outflow anomaly by a Helmholtz partitioning. Implementing such a sophisticated outflow removal technique was beyond the scope of this study.

When developing our method for the removal of the positive PV tower we found, that the remnant balanced TC outflow rapidly decayed during the first hours of simulation when the positive PV anomaly had been removed at initialisation time. Thus if we neglect removing the balanced TC outflow we only have to ensure that the low PV air that is remnant from the TC outflow prior to initialisation time does not interact with the midlatitude flow. This can be assured by choosing the initialisation time distinctly prior to any interaction of the TC with the midlatitude flow. An interaction can subjectively be determined when low PV air from the tropics moves poleward and deflects the upper-level

[7] Jangmi, Choi-Wan $k_{top} = 25, 27 \approx 100\,\text{hPa}$; Hanna, Lupit, Malakas $k_{top} = 38, 35, 38 \approx 200\,\text{hPa}$ see Appendix A

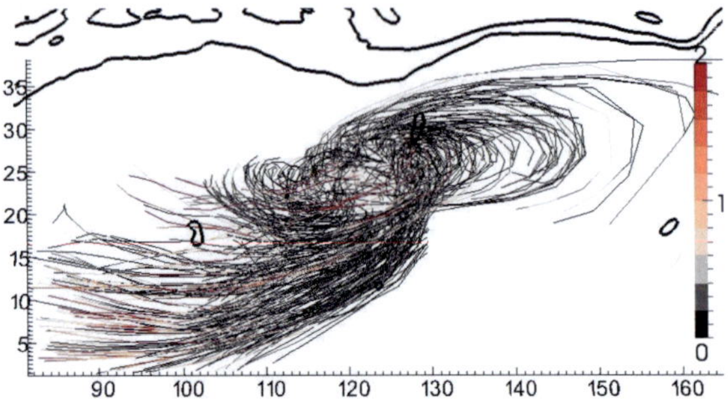


Figure 3.10: Every 100th 96 h forward trajectory emerging from the box (115-130°E, 15-30°N, 8000-15000 m amsl) and fulfilling the criterion PV<0.2 PVU at start time (00 UTC 28 September 2008). Black-red shading indicates the PV of an air parcel along the trajectory (in PVU). Additionally PV (2 and 5 PVU contours, black thick) on the 340 K isentropic level at 12 UTC 30 September 2008 is shown. View is from the top.

midlatitude wave guide. We do this for each case study with the help of PV on an upper-level (ranging between 335-345 K) isentropic level. Tracking visually low PV air on an upper-level isentropic level allows for subjectively determining the outflow anomaly and choosing a time prior to an interaction with the midlatitude flow.

Exemplarily this is detailed here more objectively with the help of trajectory calculations in the case of Jangmi. Trajectories of air parcels emerging from a box above the upper-part of Jangmi and its environment and having PV $< 0.2$ PVU at initialisation time should mainly represent the TC outflow (Figure 3.10). Most of these trajectories move anticyclonically around Jangmi and turn to the southwest into the tropics. The trajectories that initially move northeastward become embedded in the anticyclonic circulation of the subtropical ridge to the east of Jangmi. Only very few trajectories actually move further north in the vicinity of the midlatitude wave guide, represented by the 340 K 2 PVU contour. This justifies the removal of solely the positive PV anomaly as long as the initialisation time has been selected carefully before interaction with the midlatitudes occurs.

We conclude that almost no impact of an outflow anomaly that existed prior to the PV surgery time is conserved in the modified analysis. Thus removing the TC inner core PV anomaly yields in a collapse of the outflow anomaly after PV surgery time and consequently to an implicit removal of the balanced outflow

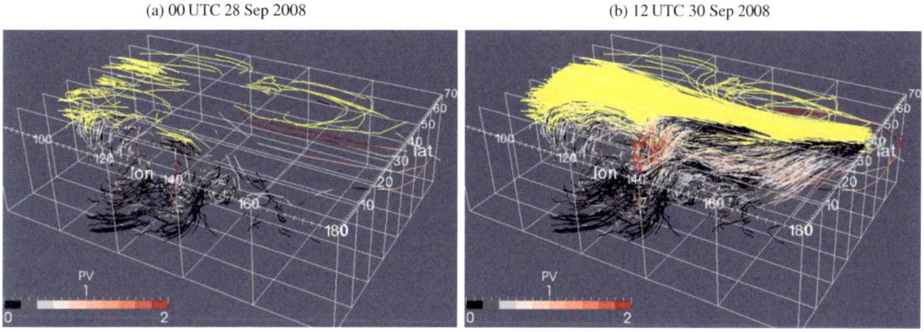


Figure 3.11: Every 100th trajectories emerging from the region 100-179°E, 10-60°N, 8000-15000 m amsl at 12 UTC 30 September 2008 calculated backward until 12 UTC 26 September 2008. The trajectories fulfil the criterion $|\vec{v}_h| > 40\,\mathrm{m\,s^{-1}}$ at 12 UTC 30 September 2008. Black-red shading shows PV along the trajectories. Yellow colours mark trajectories residing above 8000 m through the entire calculation period. Jangmi is located at about the location of high low- and mid-level PV values (red colours in the centre of the domain). The vertical grid extends from mean sea level (bottom) to 15000 m amsl (top). The times are indicated above the panels. Data taken from six hourly ECMWF analysis.

anomaly. If an outflow anomaly persisted in the modified analysis, the impact of ET would rather be underestimated by our TC removal technique.

Trajectory calculations reveal that an important outflow jet-interaction starts *after* the time of PV surgery (Figure 3.11, see also Chapter 4).

When comparing the trajectories at the PV surgery time 00 UTC 28 September 2008 (Figure 3.11a) and at the end of the calculation period (Figure 3.11b) it is obvious that none of them emerge from an upper level low PV region in the tropics. The trajectories that emerge from southern latitudes and reach the jet are

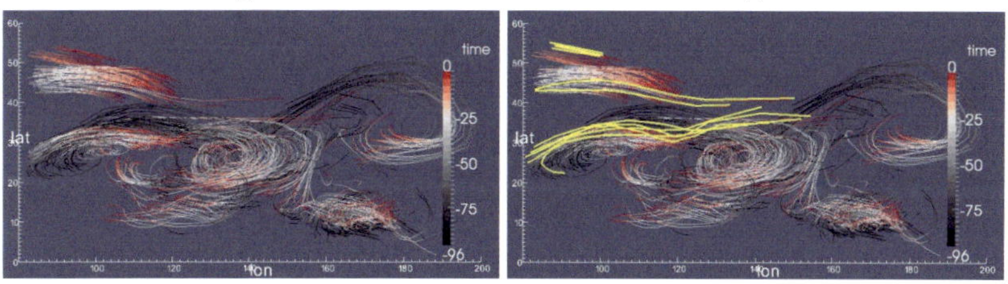


Figure 3.12: Every 100th trajectories similar to Figure 3.11 but fulfilling the criterion that PV<0.5 PVU and to remain above 8000 m amsl through the entire calculation period. Black-red shading indicates the time (in hours before 12 UTC 30 September 2008). Yellow colours in (b) mark every 50th trajectory that additionally fulfils $|\vec{v}_h| > 40\,\mathrm{m\,s^{-1}}$ at 12 UTC 30 September 2008. View is from the top and the jet streak is located in the centre of the domain at 12 UTC 30 September 2008.

lifted in the region of Jangmi from low and mid levels after the PV surgery time (00 UTC 28 September 2008). Trajectories that reside above 8000 m throughout the calculation period (yellow coloured) exclusively emerge from the midlatitudes. Although the criterion $|\vec{v}_h| > 40\,\mathrm{m\,s^{-1}}$ defines a rather weak jet, one may argue that these trajectories would not show low PV air being advected to the southern side of the jet. Therefore a second set of trajectories has been computed (Figure 3.12). These trajectories fulfil the criterion PV$<$ 0.5 PVU and to reside above 8000 m throughout the entire calculation period. The shading indicates the time, with red colours being closer to the jet streak at the end of the calculation period (12 UTC 30 September 2008). It becomes clear that trajectories with low PV values, that originate from tropical latitudes and thus may be part of an outflow anomaly prior to PV surgery, remain in the tropics. Trajectories that reach the jet (yellow trajectories, additionally fulfilling $|\vec{v}_h| > 40\,\mathrm{m\,s^{-1}}$), exclusively emerge from the midlatitudes.

## 3.3 Additional diagnostic tools and methods

### 3.3.1 Trajectories using the LAGRANTO model

In the Lagrangian view the flow is described by the temporal change of the characteristics of air parcels along their path. The Lagrangian frame of reference moves with the flow which is described by the position and physical properties of air parcels at a given time, and the trace in position and physical properties of the same air parcels prior and following a given time (Wernli and Davies 1997). In contrast to the Eulerian frame of reference in which the physical properties in a selected subset of the entire atmosphere are investigated at a fixed time, the Lagrangian frame enables us to investigate coherent structures in non-stationary weather systems.

In this study weather systems are investigated in the Lagrangian frame of reference with the help of trajectories calculated with the model LAGRANTO (LAGRangian ANalysis TOol, Wernli and Davies 1997) provided by Maxi Böttcher and Heini Wernli (Uni. Mainz / ETHZ). The position $\vec{r}^n$ of an air parcel (in the

Eulerian frame of reference) at a given time $t^n$ is used to predict iteratively the new position $\vec{r}^{n+1} = \vec{r}^n + \vec{v} \cdot \Delta t$ at time $t^{n+1}$. Initially $\vec{v} = \vec{v}(\vec{r}^n, t^n)$ is the three-dimensional wind vector at the position $\vec{r}^n$ of the air parcel at time $t^n$. It follows a successive correction (three iterations) of the new position using a mean wind $\vec{v} = \frac{1}{2}(\vec{v}^n(\vec{r}^n, t^n) + \vec{v}^{n+1}(\vec{r}^{n+1}, t^{n+1}))$. One time-step ($\Delta t$) equals a twelfth of the data time step, thus 30 min for the ECMWF analysis and 5 min for the COSMO simulations.

An air parcel is defined by the extent of a model grid box. LAGRANTO allows for calculating trajectories forward and backward in time. In a first step trajectories emerging from each grid point in a given box are calculated for a given time period. Various physical properties are traced along a trajectory. In a second step a subset of this ensemble is selected that fulfils one or a combination of various selection criteria. There are three types of selection criteria. A priori criteria are based upon the position or the physical properties of the air parcels at a given time, e.g. at start time PV is larger than 2 PVU in a box centred around a TC to select only trajectories emerging from the TC or at the end time horizontal wind speed is larger than 60 m s$^{-1}$ to detect all air parcels that end in a jet core. A posteriori criteria are based upon the time-trace of the position or the physical properties along the path, e.g. ascent is larger than 8500 m during the calculation period. Finally, a combination of a priori and a posteriori criteria is possible, e.g. at start time PV is larger than 2 PVU in a box centred around a TC and the ascent is larger than 8500 m during the calculation period to select only those trajectories emerging from the TC and exhibiting strong ascent to the tropopause level. Wernli (1997) uses LAGRANTO to detect coherent ensembles of trajectories, e.g. representing warm conveyor belts. For trajectories emerging from the low-levels (below 2000 m) the criterion of ascent larger than 8500 m in a 48 hour calculation period allows for selecting warm conveyor belts (cf. Figure 4.4 from Böttcher 2010).

### 3.3.2 Quasi-geostrophic vertical velocity diagnostics

In this study the contribution from synoptic-scale weather systems at different levels to vertical motion on a specific level, is assessed by a quasi-geostrophic (QG) approach. A height attributable solution of the QG omega equation is introduced by Clough et al. (1996). Their QG diagnostic allows for distinguishing at a specific level vertical motion induced by low-level and upper-level weather systems. The method has been provided by Sue Gray (U. Reading) and was supported by Maxi Böttcher (U. Mainz / ETHZ) who used the diagnostics extensively in her study of diabatic Rossby waves (Böttcher 2010; Böttcher and Wernli 2011).

The quasi-geostrophic approximation (cf. Holton 2004) assumes the midlatitude synoptic flow $\vec{v}_h = \vec{v}_g + \vec{v}_{ag}$ to be geostrophic $\vec{v}_g \gg \vec{v}_{ag}$ and the hydrostatic balance to be fulfilled. In the equation of motion the acceleration depends on the difference between Coriolis force and pressure gradient force. In QG approximation this depends on the difference of the actual wind from the geostrophic wind which is the ageostrophic wind $\vec{v}_{ag}$, so that the latter is not negligible in the Coriolis term. The equation of continuity under QG approximation indicates that vertical motion is caused by the divergence of the ageostrophic wind component, driving a secondary circulation. The equation of motion and of continuity under QG approximation yield in the barotropic vorticity equation. Combining this with the QG heat equation yields in the omega equation under QG approximation.

The QG omega equation describes the vertical motion resulting from the temperature and vorticity field under quasi-geostrophic approximation. For their height attributable solution Clough et al. (1996) use the Q-vector form of the QG omega equation introduced by Hoskins et al. (1978):

$$N^2 {\nabla_h}^2 w + {f_0}^2 \frac{\partial}{\partial z}\left(\frac{1}{\rho_s}\frac{\partial}{\partial z}(\rho_s w)\right) = 2\nabla_h \cdot \vec{Q}, \qquad (3.16)$$

where $N$ is the Brunt-Väisälä-frequency, $\rho_s$ a reference density profile, $w = \frac{dz}{dt}$ the vertical velocity, and $\vec{Q}$ the Q-vector. The Q-vector

$$\vec{Q} = (Q_1, Q_2) = -\frac{g}{\Theta_0}(\partial\vec{v}_g/\partial x \cdot \nabla_h\Theta, \partial\vec{v}_g/\partial y \cdot \nabla_h\Theta) \tag{3.17}$$

describes the vector rate of change of the potential temperature gradient following geostrophic flow. Convergence of the Q-vector is linked to ascent, divergence to descent.

The differential equation is solved using equally spaced pressure level data from 1000-100 hPa (every 25 hPa). The static stability $N^2$ is thereby kept constant on each pressure level. The solution of Clough et al. (1996) allows for calculating at each grid point the contribution to vertical motion that is induced by (QG PV-) anomalies remote in horizontal and vertical direction. Following Böttcher (2010) and Dixon et al. (2003) the contribution to vertical velocity of synoptic-scale waves in the upper troposphere from 650 to 100 hPa is distinguished from the contribution of weather systems in the lower troposphere from 1000-750 hPa. The contribution from these two layers to vertical motion at 700 hPa is investigated (cf. Figure 2.3 in Böttcher 2010). It is assumed that at this level vertical motion is determined by contributions from both layers and that in extratropical cyclones vertical motion is prominent at this level (Böttcher and Wernli 2011; Deveson et al. 2002).

Albeit valid in synoptic scale weather systems, the QG diagnostics do not entirely capture the contribution to vertical motion by mesoscale systems like tropical cyclones (cf. Böttcher and Wernli 2011). Nevertheless, it is justified to use the QG diagnostics in our study, as we aim to assess the upper-level forcing for ascent in the vicinity of the transforming TC or in downstream regions. This forcing is mainly induced by upper-level flow features like synoptic-scale waves or the jet stream. Vertical motion in regions with strong condensational heating (e.g. within the TC or at the baroclinic zone) is strongly underestimated by the QG diagnostics and therefore the low-level QG contribution to ascent is too small. However, parts of the larger-scale upper-level TC outflow and parts of the dynamically forced slantwise ascent along the baroclinic zone at low and mid levels during ET are described by the QG balance. Therefore the QG diagnostics enables us qualitatively to correctly indicate the regions where the circulations associated with mesoscale weather systems induce ascent or descent.

### 3.3.3 Forecast verification: the anomaly correlation coefficient

The anomaly correlation coefficient (ACC) is a measure for forecast quality in numerical weather prediction. As it is sensitive "to similarities in forecasts and analysed patterns, rather than their absolute values"[8] it is particularly well suited to investigate the representation of the midlatitude flow. The ACC (c.f. Eq. (7.39) in Wilks 1995) is the correlation between the forecast anomaly $(f_i - c_i)$, defined as the difference of a forecast variable $f_i$ and a climatology $c_i$ at each grid point $i = [1, M]$ ($M$ : number of grid points), and the analysis anomaly $(a_i - c_i)$, $a_i$ : analysis at grid point $i$:

$$ACC = \frac{\sum_{i=1}^{M} [(f_i - c_i)(a_i - c_i)]}{\sqrt{\sum_{i=1}^{M} (f_i - c_i)^2 \sum_{i=1}^{M} (a_i - c_i)^2}} \quad . \tag{3.18}$$

Persson and Grazzini (2007, their section 7.1.6, page 70) summarise : "The ACC has a tendency to score large and "good" values during meridional flow situations, small and "bad" values during periods of predominantly zonal flow. This is particular the case in zonal situations when the forecast and observed positions of shallow waves are out of phase." An ACC of 1.0 shows that the forecast and the analysis are identical. A negative ACC indicates anti-correlation between forecast and analysis. Persson and Grazzini (2007) further note that "it has been found empirically that the level ACC=60 % corresponds to the limit where the forecast does not exhibit any significant synoptic skill".

In this study the ACC is calculated 12 hourly for geopotential at different heights to validate COSMO simulations. The forecast is from the COSMO simulations, the analysis is from the deterministic ECMWF analysis, and for the climatology

[8] ECMWF, `http://www.ecmwf.int/products/forecasts/guide/Interpretation_of_the_ACC.html`, retrieved on 4 July 2011.

the daily climatology from the ERA-INTERIM (Dee et al. 2011) project extrapolated on a 0.25°x0.25° horizontal grid is used. Additionally the mean ACC of the operational ECMWF IFS forecasts over the relevant ocean basin and during the August/September/October season were kindly provided by Simon Lang. The COSMO ACCs are interpreted with respect to the mean IFS-ACC which is regarded as a typical operational ACC. Also ACCs of the COSMO ET scenario simulations are calculated using the relevant COSMO reference simulation as analysis in Equation (3.18) (see Section 6.3). This allows for estimating the impact of ET in forecast skill reduction.

### 3.3.4 Additional post-processing tools

The intensity and position of cyclones are estimated by tracking the minimum pressure at mean sea level (pmsl) in the ECMWF analyses or COSMO simulations.

Other meteorological quantities are calculated from the ECMWF analyses and COSMO model fields. If not otherwise stated the data on 25 hPa equally spaced pressure levels is used. These quantities are PV, (equivalent) potential temperature $\Theta_{(e)}$, condensational heating, and the cloud top temperature.

PV is computed according to Equation (3.8) reformulated in pressure coordinates:

$$PV = -g(f\vec{k} + \nabla_p \times \vec{v}_h) \cdot \nabla_p \Theta, \tag{3.19}$$

where g is the acceleration due to gravity, f the Coriolis parameter, $\vec{k}$ the unit vertical vector, and $\nabla_p$ the three-dimensional gradient operator in pressure coordinates (cf. Equation (13) in Hoskins et al. 1985).

Equivalent potential temperature $\Theta_e$ is calculated with respect to temperature and specific humidity according to Equation (43) of Bolton (1980):

$$\Theta_e = T\left(\frac{1000}{p}\right)^{0.2854(1-0.28r)} exp\left[\left(\frac{3.376}{T_L} - 0.00254\right) r(1+0.81r)\right], \tag{3.20}$$

$$T_L = \frac{1}{\frac{1}{T-55} - \frac{ln(RH/100)}{2840}} + 55, \tag{3.21}$$

where $T$, $p$, $r$ are the absolute temperature (in K), pressure (in hPa) and mixing ratio (in $\mathrm{kg\,kg^{-1}}$) at the considered level, $T_L$ the temperature at the lifting condensation level (in K), and RH is the relative humidity (in %).

The calculation of the condensational heating rate is provided by the LAGRANTO preprocessor programme "p2s" of H. Wernli and follows Berrisford (1988):

$$\dot{\Theta} = \frac{-\frac{L}{c_p}\frac{\kappa\Theta\omega}{p}\frac{dq_s}{dt}}{1+\frac{L}{c_p}\frac{dq_s}{dt}}\left\{1-\exp\left(\frac{h_0-h}{5}\right)\right\}, \tag{3.22}$$

if $dp/dt = \omega < 0$ (ascending motion) and the relative humidity $h > h_0 = 80\%$. Here $L$ denotes the latent heat of condensation of water, $c_p$ the heat capacity of water vapour at constant pressure, and $q_s$ the saturation mixing ratio. The basic assumption is that condensation occurs where ascending air is (nearly) saturated. The heuristic threshold function $(1-\exp((h_0-h)/5)$ has been introduced by Berrisford (1988) to account for potential saturation on the sub-grid scale.

Cloud top temperature is calculated from model level data by interpolating the temperature to the level of unit optical depth into the cloud (starting at the model top). Expressions for absorption cross-section for cloud ice and cloud water are obtained from Dudhia (1989). The calculation has been provided by A. Penny (NPS) who adopted the programme cttcalc.f from Mark T. Stoelinga's Read/Interpolate/Plot package [9].

[9] http://www.mmm.ucar.edu/wrf/users/docs/ripug.htm

# 4 The extratropical transition of Typhoon Jangmi (2008)

Typhoon Jangmi was one of the major tropical cyclones in the West Pacific in 2008 and one of the two typhoons that underwent extratropical transition during the THORPEX Pacific Asian Regional Campaign (T-PARC). The enhanced observational network in the West Pacific region during T-PARC as well as various research flights into Jangmi undergoing ET provide an excellent observational data source. The additional data has been assimilated at ECMWF to produce an improved analysis. In this chapter a detailed study of the ET of Jangmi is presented based on this enhanced analysis (experiment DROP from Weissmann et al. 2010, see Section 3.1.1), and based on a subset of the T-PARC observations.

Several aspects qualify Jangmi as a case for getting a better general understanding of ET: When Jangmi underwent ET its outflow was directed towards a jet streak over Japan and the jet core wind speed increased. This leads to the question as to what extent Jangmi had an impact on this acceleration and if this has implications for the evolution of the downstream flow. Although Jangmi transformed into an extratropical cyclone, the transition was not followed by extratropical reintensification. Instead, Jangmi weakened steadily and decayed while to its northeast a weak extratropical cyclone developed. This raises the question as to which physical mechanisms determine the intensification or decay during ET. According to the JMA best track data Jangmi formed on 24 September 2008 in the Philippine Sea southwest of Guam. It headed west-northwestward and reached maximum intensity on 27 September 2008 with a minimum central pmsl below 905 hPa. After landfall on 28 September, Jangmi started to recurve north of Taiwan at 15 UTC 28 September 2008. At 00 UTC 01

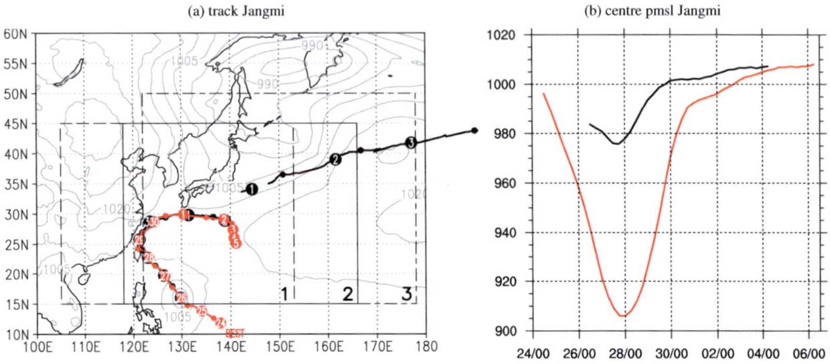


Figure 4.1: (a) Track of Jangmi from six-hourly best track data (courtesy of JMA) starting at 00 UTC 24 September 2008 (red, labelled BEST) and track of Jangmi and of the downstream cyclone (black) from six-hourly ECMWF analyses starting at 00 UTC 26 September 2008. Large dots mark 00 UTC times, labelled by the day of September/October 2008. Small dots mark 12 UTC times. Pmsl from ECMWF analysis at 00 UTC 26 September 2008 (grey with a 5 hPa contour interval). The boxes labelled "1, 2, 3" mark regions shown in Figure 4.5. (b) Minimum central pmsl (in hPa) of Jangmi from best track (red) and ECMWF analyses (black).

October 2008 Jangmi was officially declared extratropical. In deterministic and in ensemble weather prediction systems the recurvature and ET of Jangmi was very uncertain. On 25 September the major part of the ECMWF ensemble members predicted that Jangmi would make landfall on the Chinese mainland and subsequently decay. One day later some members indicated recurvature. However, there was a large spread of possible scenarios ranging from Jangmi moving along the southwestern Chinese coast and making landfall in Hong Kong to recurvature south of Japan. Starting at 27 September all members indicated the recurvature south of Japan. This raises the question as to which physical processes govern this uncertainty in the track forecast.

The complete life cycle of Jangmi and its implications for the downstream flow evolution is described in Section 4.1. Then a verification of important features during Jangmi's ET with the help of observational data is given (Section 4.2). A more detailed investigation of the outflow-jet interaction and a physical mechanism that explains the strong acceleration of the jet streak are presented in Section 4.3. Finally, the outflow-jet interaction is quantified in Section 4.4 using PV surgery experiments. The reasons for the uncertainty of the forecasts of Jangmi's ET and for the lack of reintensification are addressed in more general view in Chapter 5, Section 5.1.

# 4.1 Synoptic overview

## 4.1.1 The tropical stage of Jangmi and impacts on human activities

Typhoon Jangmi formed in the Philippine Sea on 24 September 2008 and intensified whilst following a northwestward track (Figure 4.1). Shortly before landfall in Taiwan it reached peak intensity (minimum central pressure <905 hPa, maximum sustained wind (10-min mean) $v_{sfc}$ >60 m s$^{-1}$; category 4 Saffir-Simpson scale). Between 06 and 09 UTC 28 September 2008 Jangmi made landfall at the northeastern tip of Taiwan where it caused strong winds and heavy rain. The maximum gust measured at Taipei Airport was 46 m s$^{-1}$ and the 12 hour rainfall until 12 UTC 28 September 2008 was 135 mm. Larger rainfall

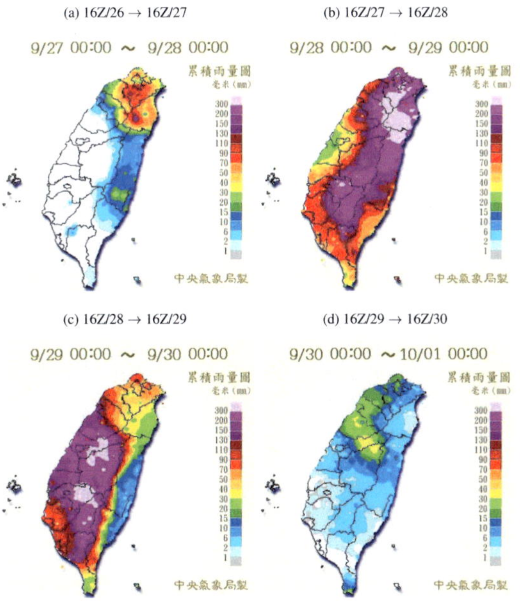


Figure 4.2: Daily precipitation in Taiwan during Jangmi (in mm, 27-30 September 2008). The time indicated is taiwanese time (UTC+8h) thus the period corresponds to the 24 hour period starting at 16 UTC on day-1. The legend is as follows: blue < 15 mm, green 15-30 mm, yellow 30-70 mm, red 70-130 mm every 20 mm, violet 130-200 mm, pink 200-300 mm, rose > 300 mm. Data courtesy Central Weather Bureau (CWB) in Taiwan.

amounts occurred in the centre of the island and widely exceeded 300 mm in 24 hours from 27-30 September 2008 (Figure 4.2, note the change of maximum precipitation in the eastern slopes of Taiwan on 28 September (b) to the western side on 29 September (c)). In Taiwan, Jangmi had a major impact due to flooding, landslides and power outages. At least 2 fatalities were reported.

After landfall, the typhoon weakened rapidly. At 00 UTC 29 September 2008 Jangmi was downgraded to a severe tropical storm and 18 hours later to a tropical storm. Jangmi recurved in the East China Sea on 29 September 2008, transformed into an extratropical system on 30 September 2008 and passed south of the Japanese mainland before it decayed. During ET and recurvature Jangmi caused heavy precipitation on the southwestern Japanese main islands, in particular on the island Kyushu during 30 September 2008 (Figure 4.3). At the station of Satsumakashiwabaru total precipitation in the period from 28 September 2008 to 01 October 2008 reached 380 mm and a daily maximum of 255 mm occurred on 30 September 2008. The maximum daily precipitation was observed in Higashiichiki at 30 September 2008 reaching 268 mm. The accumulated rainfall amount is illustrated in Figure 4.3a and the heavy precipitation at 30 September 2008 in Figure 4.3b.

(a) total precipitation

(b) radar 14 UTC 30 Sep 2008

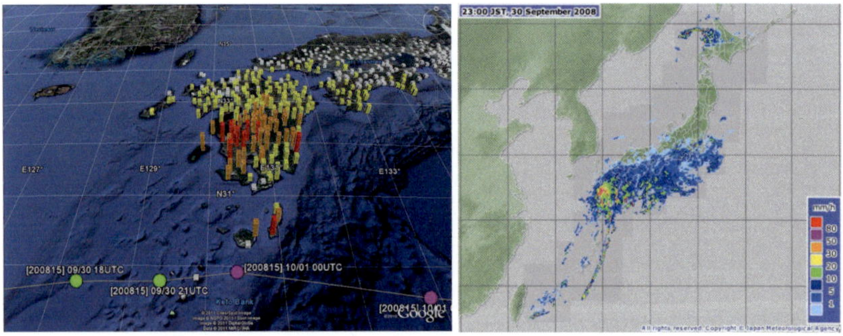


Figure 4.3: (a) Total precipitation in southwestern Japan in the period 04 UTC 27 September 2008 to 04 UTC 2 October 2008. The colour shading is as follows: 0-75 mm: white, 75-200 mm: yellow, 200-330 mm: orange, >330 mm: red. Image courtesy: Digital Typhoon KITAMOTO Asanobu at National Institute of Informatics, Japan. Retrieved from `http://agora.ex.nii.ac.jp/digital-typhoon/kml/disaster/typhoon/prec/total/200815.en.kmz` and Google Earth. (b) Radar composite for Japan at 14 UTC 30 September 2008. Data courtesy Japanese Meteorological Agency.

### 4.1.2 The ET stage of Jangmi

Jangmi was officially declared extratropical 00 UTC 01 October 2008 by JMA. Nevertheless, significant structural changes, beginning two days earlier, indicate that the tropical cyclone is undergoing ET, starting with the "transformation" in the conceptual model of Klein et al. (2000). In this section an overview of the ET stage of Jangmi is given starting at 00 UTC 28 September 2008, a few hours prior to landfall in Taiwan.

We use a combination of model analysis and satellite data. The best track data of JMA is used to estimate intensity in terms of pmsl and low-level winds for Jangmi. The intensity of the extratropical cyclone which develops downstream of Jangmi is specified using pmsl in the ECMWF analysis. The track and central pressure in the ECMWF analysis are summarised in Figure 4.1. Compared to the best track data, the track of Jangmi is represented very well in the model analysis data. However, the intensity is much weaker with the minimum central pressure only reaching 972 hPa compared to 905 hPa in the best track data. This is due to the coarse resolution of the analysis (0.25° lat/lon) as well as insufficient data to define the inner core structure of the TC and known as a common problem in the numerical analysis of tropical cyclones.

The analysis data is shown on the $\Theta = 340$ K isentropic surface which is well-suited to illustrate the mid- and upper-level flow. This isentropic level is located

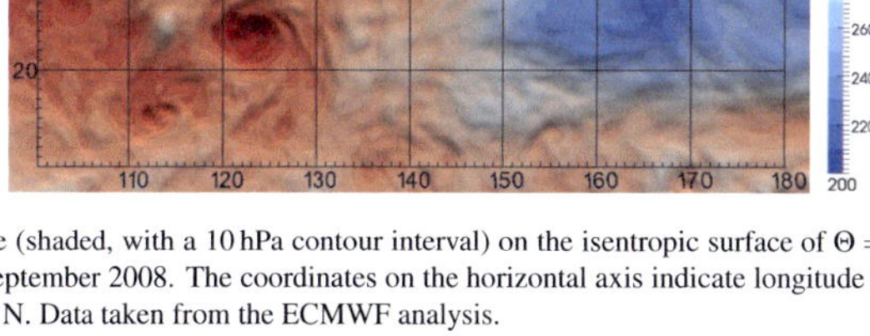


Figure 4.4: Pressure (shaded, with a 10 hPa contour interval) on the isentropic surface of $\Theta = 340$ K. Date shown is 00 UTC 28 September 2008. The coordinates on the horizontal axis indicate longitude in °E, on the vertical axis latitude in °N. Data taken from the ECMWF analysis.

at about 350 hPa well south of the midlatitude jet and at about 200 hPa well north of the jet (Figure 4.4). The 340 k surface is lower ($\approx$ 400 hPa) in tropical weather systems (e.g. Jangmi at 123°E, 23°N in Figure 4.4). The midlatitude wave guide is characterised by a strong isentropic pressure gradient at 39°N centred around the 250 hPa isobar. The wind and PV at the 340 K isentropic surface are calculated for the ECMWF analysis and discussed along with intersection points of 48 hour forward trajectories with that surface (panels on the left of Figure 4.5). The trajectories originate in the lower troposphere (0-2000 m above mean sea level) in a box extending from 100-179°E, 10-60°N and fulfil the criterion of ascent larger than 8500 m$(48h)^{-1}$. The trajectory calculations are restarted every 6 hours and intersection points of all calculations are shown which are valid at the respective times. The criteria used for the trajectory calculations correspond to those of Wernli (1997) to detect so-called warm conveyor belts (WCBs). The MTSAT channel IR1 (infrared wavelength at $10.3 - 11.3\,\mu m$) brightness temperature $T_b$ (panels on the right of Figure 4.5) illustrates the top of clouds along with QuikScat winds showing the observed low-level winds over the ocean.

At 00 UTC 28 September 2008 Jangmi is located at about 123°E, 23°N off the coast of Taiwan (Figure 4.5a, b). The minimum sea level pressure is 910 hPa and maximum sustained winds (10-min mean) exceed 55 m s$^{-1}$. In the previous 24 hours Jangmi has reached maximum intensity with a minimum central pressure of 905 hPa and maximum sustained winds of more than 60 m s$^{-1}$. The QuikScat winds indicate strong surface inflow into the typhoon centre (Figure 4.5b). The brightness temperature of the cloud top shows Jangmi as a TC with symmetric eyewall convection and a matured eye. Slightly to the east of the surface centre, $T_b$ has a minimum of less than 200 K (Figure 4.5b). At the same location the isentropic PV reaches values of 7-10 PVU (inset in Figure 4.5a). Lifting is concentrated in Jangmi's centre as indicated by the trajectory intersection points (Figure 4.5a). At this time the wind at the isentropic surface indicates only weak outflow northeast of Jangmi. In the South China Sea a second tropical storm, Mekkhala, is forming at about 113°E, 15°N. Important subtropical features are the upper tropospheric anticyclonic circulation east of Jangmi between 127-

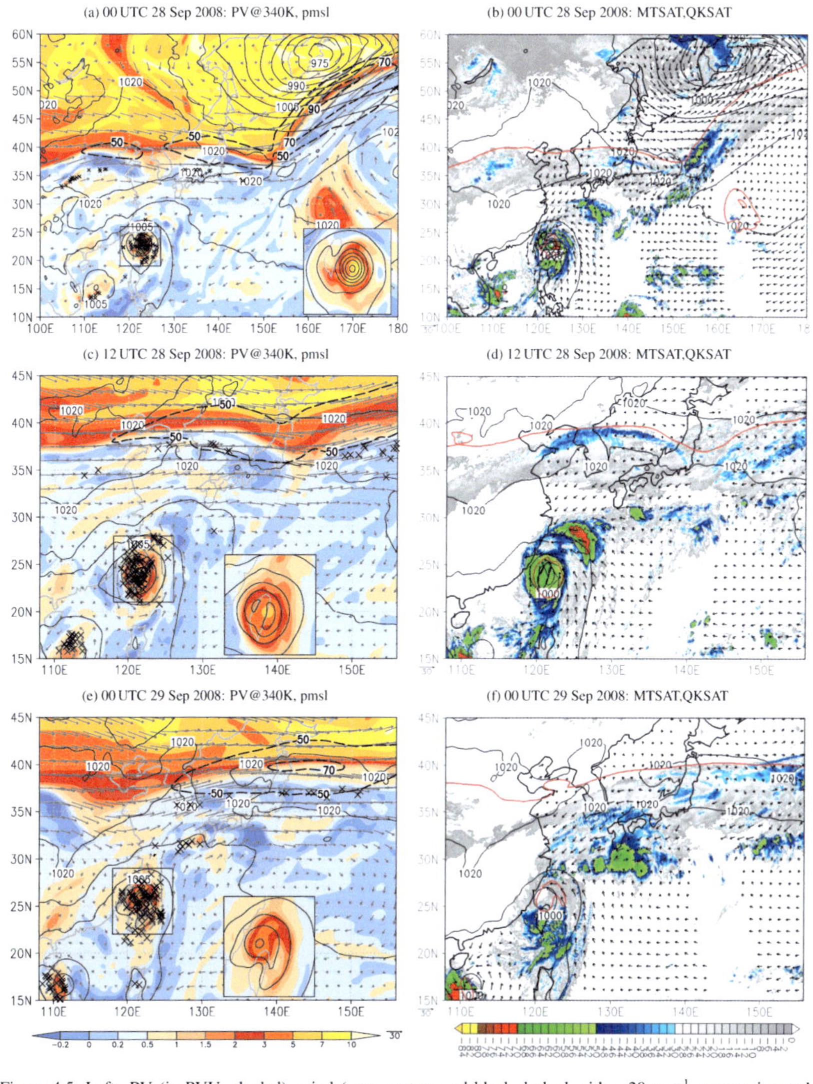


Figure 4.5: Left: PV (in PVU, shaded), wind (grey vectors and black dashed with a $20\,\mathrm{m\,s^{-1}}$ contour interval for $|\vec{v}| > 50\,\mathrm{m\,s^{-1}}$) at the 340 K isentropic level, and pmsl (black contours every 5 hPa). The insets in (a,c,e,i) show PV and pmsl in a box (black) around Jangmi. Right: MTSAT IR channel 1 ($10.3 - 11.3\,\mu m$) brightness temperature (only $<0°$C shown) with QuikScat wind vectors; contours of pmsl (black contours every 10 hPa) and 2 PVU contour of 340 K PV (red) from the ECMWF analysis. ... *(continued on next panel)*

155°E, 15-30°N and a TUTT cell further to the east centred around 162°E, 27°N (Figure 4.5a). The strong PV gradient north of Jangmi at about 39°N marks the transition from tropospheric to stratospheric air masses at the polar front and serves as the midlatitude wave guide (approximately marked by the 2 PVU contour). It remains almost zonal west of 152°E with weak ridges at about 115°E and 140°E. A strong extratropical cyclone is evident centred over Kamchatka with a distinct midlatitude jet streak and ridgebuilding to its southeast (Figures 4.5a, b). Between 110-120°E, 39°N a weak jet streak approaches from the west. The midlatitude flow configuration with a zonal flow to the north and a primary extratropical cyclone to the Northeast resembles Harr et al. (2000)'s Northeast type of ET and thus gives hint that Jangmi will not reintensify.

Twelve hours later, this jet streak is located northeast of Jangmi over the Sea of Japan (Figure 4.5c). The jet core wind speed has increased to more than $60\,\mathrm{m\,s^{-1}}$. On the isentropic surface, southerly and southwesterly outflow northeast of Jangmi starts to become directed towards this jet streak. Jangmi made landfall in Taiwan and the isentropic PV values decrease to 5-7 PVU (inset in Figure 4.5c). Most of the lifting occurs in Jangmi's centre. However, a concentration of air parcel intersection points becomes apparent northeast of the low-level centre. In this region satellite imagery reveals a separation of a rain band from Jangmi (Figure 4.5d), probably due to the landfall in Taiwan. The QuikScat wind vectors show a strong cyclonic circulation around Jangmi's surface centre and convergence in the region of the separated rain band. In the official JMA data the central pressure of Jangmi increases to 950 hPa and maximum sustained winds reach $44\,\mathrm{m\,s^{-1}}$.

At 00 UTC 29 September 2008 Jangmi has weakened dramatically (central pressure 985 hPa, maximum sustained winds $26\,\mathrm{m\,s^{-1}}$). The high brightness temperatures in satellite imagery reveal that convection collapsed, in particular in the centre of Jangmi ($T_b > 250\,\mathrm{K}$, Figure 4.5f). To the northeast of Jangmi's surface centre remnants of the separated rain band start forming a cloud shield. Here, the QuikScat low-level winds indicate a frontal zone with westerly flow to the south and easterly flow to the north. At the location of the separated rain band (at about 132°E, 30°N) the QuikScat winds indicate strong convergence and the coldest cloud top temperature occurs. Along with the persistent convective clouds to the south this underlines the more asymmetric shape of Jangmi at this stage. The satellite winds also indicate the closed surface circulation of Typhoon Mekkhala which heads westward whilst intensifying. The isentropic analysis fields highlight a first outflow-jet interaction (Figure 4.5e). Northeast of Jangmi the isentropic wind vectors show strong southwesterly flow towards the midlatitude jet streak, which reaches a wind speed of more than $70\,\mathrm{m\,s^{-1}}$. At about 130°E, 32°N in the region of the low-level convergence in QuikScat winds, the trajectory intersection points along with a local PV maximum indicate new lifting, presumably at a baroclinic zone. In the inner core of Jangmi the isentropic PV further decreases (3-5 PVU). East of Jangmi the anticyclonic circulation has increased slightly during the last 24 hours and the upper-level low (TUTT-cell) moves southwestward slowly (TUTT cell not seen in Figure 4.5e).

Twelve hours later, the intersection points northeast of Jangmi align to a narrow band over southern Japan (Figure 4.5g). Here the isentropic wind vectors are directed straight from the tropical cyclone undergoing ET to southern Japan. Along their path low PV air is advected to the midlatitudes, maintaining the PV gradient over the Sea of Japan and resulting in an extension of the jet. Details of this outflow-jet interaction will be discussed in Section 4.3. The jet streak has moved east of Japan and weakened. A new wind speed maximum is located over the Yellow Sea at around 120°E, 38°N. Major lifting, as indicated by the trajectory intersection points, still occurs in the centre of Jangmi. Two concentrations of intersection points are noteworthy (both enclosed by the magenta coloured box in Figures 4.5g, h). In the region north of Jangmi at 122-124°E, 28-30°N PV reaches more than 1.5 PVU. A second concentration is located to the northeast at 128-130°E, 30-32°N. When compared to the satellite brightness temperature, both regions coincide remarkably with the beginning of new convective bursts ($T_b < 220\,\mathrm{K}$) in the northeast quadrant of Jangmi (Figure 4.5h). In the first region where high PV values occur, the minimum brightness temperature drops below 200 K. Further to the east, Jangmi's clouds merge with a cloud band extending along the midlatitude wave guide. Convection in the southern part of Jangmi decays along with the overall weakening of Jangmi. The central pressure is 992 hPa and maximum sustained winds reach 23 $m\,s^{-1}$.

The start of new convection northeast of Jangmi marks the beginning of a 48 hour period of strong interaction between the tropical cyclone undergoing ET and the midlatitude flow. This interaction results in an important acceleration of the midlatitude jet and a moderate cyclogenesis downstream.

At 00 UTC 30 September 2008 the first jet streak is located east of 150°E (Figure 4.5i). The second jet streak seen over the Yellow Sea 12 hours earlier has expanded and intensified strongly ($|\vec{v}| > 60 m\,s^{-1}$). It extends from Korea over the Sea of Japan to northern Japan. A region of low PV air which has been partly advected from the west and partly by Jangmi's outflow is evident south of the jet core. At the southern edge of this region the lowest PV values occur where isentropic wind vectors and trajectory intersection points indicate ongoing advection and ascent of tropical air from Jangmi to the midlatitudes (Figure 4.5i). In the northeastern quadrant of Jangmi the convective burst has intensified reflected in the extension of the 1010 hPa isobar of pmsl and a local isentropic PV maximum of up to 2 PVU (inset in Figure 4.5i). Low values of brightness temperature ($T_b < 200$ K) at 127°E, 30°N mark the convective burst in satellite imagery (Figure 4.5j). The TUTT cell has moved further southwestward and is located at around 158°E, 21°N (Figure 4.5i). To its south tropical convection is evident (Figure 4.5j). After steadily heading westward Typhoon Mekkhala makes landfall in Vietnam (not shown).

At 12 UTC 30 September 2008 the outflow-jet interaction of Jangmi and the midlatitude jet becomes obvious. The 340 K isentropic wind vectors show Jangmi's outflow as southwesterly flow from Jangmi at 130°E, 30°N directly into the jet core at 145°E, 41°N (Figure 4.5k). Very low PV air of the TC outflow (PV<0.0 PVU) is advected northeastward, which results in a weak ridge-building and enhances the PV gradient at the midlatitude wave guide. As a consequence the jet core wind speed increases to more than 70 $m\,s^{-1}$. The narrow band of negative PV from Jangmi to the jet streak indicates the presence of symmetric instability and that air ascending slantwise to the jet will be accelerated (Bennetts and Hoskins 1979; Jones and Thorpe 1992). The elongated

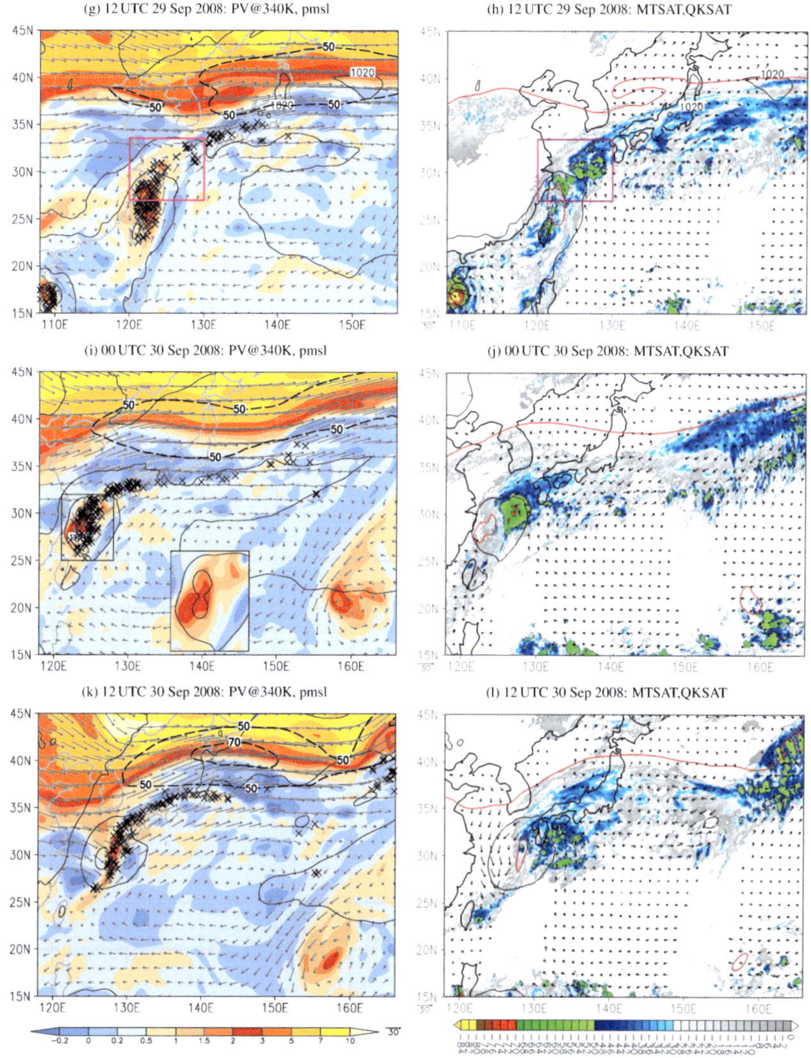


Figure 4.5: ... *(continued)*. Satellite data was provided by JMA. The crosses on the left mark every 10th location (20th in a) where a trajectory intersects the isentropic layer between 339.5 K and 340.5 K. For selected times, regions discussed together with Figure 4.6 are marked by a magenta box for Jangmi and a dark violet box for the downstream cyclone. ... *(continued on next panel)*

region with a high concentration of trajectory intersection points over southern Japan marks ascent at a baroclinic zone. The PV maximum of Jangmi is sheared to the east and becomes separated from the low-level centre of Jangmi (996 hPa pmsl, maximum sustained winds 18 m s$^{-1}$). Instead, new PV is produced diabatically at the baroclinic zone. Details of the interaction with the baroclinic zone will be discussed in Section 4.3. Satellite imagery reveals that Jangmi transforms into an extratropical cyclone (Figure 4.5l). The convective burst has decayed. However, a cloud shield marks the formation of a warm sector northeast of the low-level centre (125-150°E, 30-40°N). Further to the east this cloud shield remains merged with a cloud band which extends along the midlatitude wave guide. A (cold-) frontal structure evolves south of Jangmi from 120°E, 23°N to 130°E, 30°N. This is also reflected in a band of higher isentropic PV (Figure 4.5k) and in the advection of very dry air west of Jangmi (high values of $T_B$, Figure 4.5l). Other features are important for the subsequent development. Centred around an axis from about 122°E, 35°N to 130°E, 45°N the curvature of the PV contours in Figure 4.5k indicates a weak midlatitude trough approaching almost in phase with Jangmi. Between the TUTT cell and the subtropical anticyclone east of Jangmi (147-157°E, 20-30°N, Figure 4.5k) the northeasterly flow accelerated.

At 00 UTC 01 October 2008 a large number of trajectories end in the ridge downstream of Jangmi (Figure 4.5m). Most of the air parcels align along the southern edge of the strong PV gradient and appear in the region of lowest PV. This suggests that these air parcels are responsible for a significant transport of low PV air towards the midlatitudes and a lifting of the tropopause. The very strong PV gradient results in a further acceleration of the midlatitude jet to more than 80 m s$^{-1}$. A second very dense linear concentration of trajectory intersection points occurs from 140°E, 35°N to 157°E, 39°N (Figure 4.5m). Here air ascends along a baroclinic zone (see discussion in Section 4.3, Figure 4.23). The low-level centre of Jangmi (1002 hPa, maximum sustained winds no longer documented) moves eastward slowly and now has become in phase with the midlatitude trough. Lacking the midlatitude forcing for cyclogenesis ahead of the trough this hinders an extratropical reintensification of Jangmi. In-

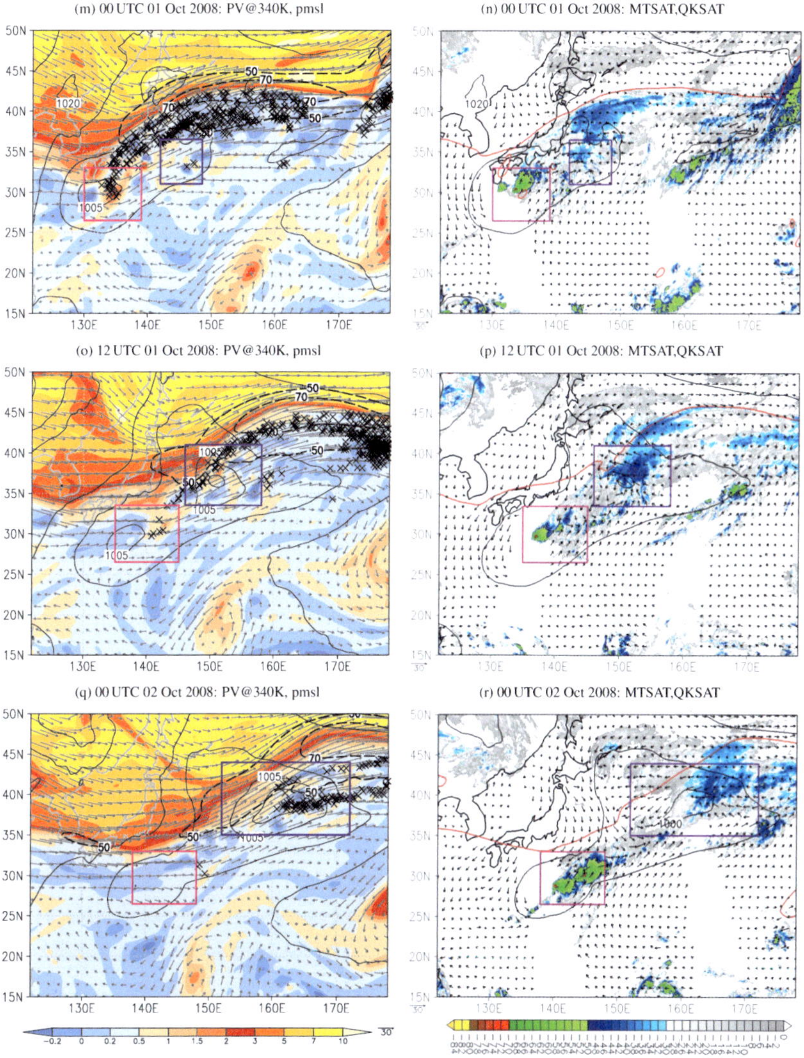


Figure 4.5: ... *(continued)*. The model data is valid at the times indicated. The MTSAT data is valid 15 minutes and the QuikScat data 3 hours prior to the time indicated. The horizontal resolution of the model data and the QuikScat data is 0.25°, of the MTSAT data 0.05°. In (c-r) the focus is on subsets of the region shown in (a,b). The subsets are marked in Figure 4.1, labelled "1,2,3", and the same for 1 (c-h), 2 (i-l), and 3 (m-r).

stead, ahead of the trough the 1010 hPa pmsl contour indicates a pressure drop downstream of Jangmi in the region of the warm sector (135-150°E, 27-35°N, Figure 4.5m). The high isentropic PV in the centre of Jangmi has vanished. However, new PV of more than 3 PVU appears at 135°E, 30°N east of Jangmi's centre (enclosed by the magenta box in Figure 4.5m, n). This is where another new convective burst appears (Figure 4.5n). The cloud shield of the warm sector becomes separated from Jangmi as it moves faster to the east than Jangmi. Convergence of the QuikScat wind vectors at about 145°E, 34°N (encircled by the violet box in Figure 4.5n) slightly south of the cloud shield indicates the start of an extratropical cyclogenesis. Only remnants of the (cold-) frontal structure south of Jangmi persist close to Taiwan from about 120°E, 22°N to 135°E, 30°N (Figure 4.5n,m). South of the TUTT cell tropical convection reappears at 158°E, 15°N (Figure 4.5n,m).

During the next 12 hours downstream of Jangmi a new extratropical cyclone develops ahead of the midlatitude trough and in the southwestern jet entrance region (violet box in Figure 4.5o). Its centre is located at 151°E, 36°N with a central pressure of 997 hPa while Jangmi's central pressure remains at 1002 hPa. Trajectories initiated partly by Jangmi and partly by the new cyclone continue to advect low PV air in the downstream ridge where the jet remains very strong ($|\vec{v}| > 80 \mathrm{m\,s^{-1}}$). Satellite imagery shows that the convection close to Jangmi's centre persists (magenta box in Figure 4.5p). Jangmi's warm sector has become the warm sector of the new cyclone. Surface satellite winds indicate the closed circulation of the new cyclone.

At 00 UTC 02 October 2008 the new cyclone moved rapidly eastward (by 11° longitude in 12 hours) and is now centred at 162°E, 39°N (997 hPa, violet box in Figure 4.5m). In the downstream ridge fewer trajectory intersection points occur. However, a new linear concentration of intersection points appears from the centre of the downstream cyclone to the east marking ascent along the eastward extension of the baroclinic zone. The remnants of Jangmi remain stationary and to their east convective activity continues (magenta box in Figure 4.5r).

In the following days the convective activity associated with the remnants of Jangmi finally decays. The new cyclone remains a rather weak frontal wave with a peak intensity of 992 hPa centre pmsl at 06 UTC 02 October 2008. Thereafter it lacks further development, remains located in the southwestern jet entrance region and crosses the Pacific along with the jet streak. It decays at the North American coast on 06 October 2008.

### 4.1.3 The evolution of the downstream flow after ET

The evolution of the downstream flow is described with the help of the quasi-geostrophic forcing at 700 hPa calculated for ECMWF analyses (see Section 3.3.2). The vertical motion at this level is well suited for the investigation of extratropical cyclogenesis (Deveson et al. 2002; Böttcher and Wernli 2011) and thus helps us to understand why Jangmi did not reintensify and how the downstream cyclone formed.

At 12 UTC 29 September 2008, a dipole at 122°E, 27°N of low-level QG forcing for ascent to the northeast in the region of warm air advection and for descent to the (south)west in the region of cold air advection of up to $\pm 1.5 \cdot 10^{-2} \mathrm{m\,s^{-1}}$ marks Jangmi (Figure 4.6a). The isohypses at 200 hPa to the north of Jangmi indicate a rather zonal flow. However, further to the north from 120°E, 40°N to 126°E, 50°N the curvature in the isohypses indicates a weak trough. Along this trough axis a dipole in upper- and low-level QG forcing with negative values to the west and positive values to the east evolves. When comparing Figure 4.6a to Figure 4.5g, h the region of positive low-level QG forcing northeast of Jangmi coincides with the region of the new convective bursts. This behaviour persists until the final decay of Jangmi on 04 October 2008 (Figures 4.6, 4.7a, b).

One and a half days later, at 00 UTC 01 October 2008 the upper-level QG forcing increases to more than $\pm 1.5 \cdot 10^{-2} \mathrm{m\,s^{-1}}$ with descent to the west and ascent to the east of an line reaching from Jangmi at 130°E, 30°N to 140°E, 52°N (Figure 4.6b). The trough axis in the 200 hPa height appears slightly westward of the line separating descent and ascent. This is typical for developing troughs and helps the trough to deepen and to propagate eastward. Thus the entire region ahead of the trough and downstream of Jangmi is influenced by ascent forced from the upper levels. As a consequence, low-level cyclogenesis is initiated centred at about 144.5°E, 33.75°N. Here a dipole of low-level QG forcing of up to $\pm 1.5 \cdot 10^{-2}\mathrm{m\,s^{-1}}$ appears (marked by the violet box in Figure 4.6b). The dipole centre is exactly at the location where low-level convergence was observed in the surface winds (Figure 4.5n) and a northeastward extension of the 1010 hPa isobar occurs (cf. Figures 4.5m, n, 4.6b). The low-level dipole of QG forcing in the vicinity of Jangmi persists while the QG forcing for ascent to

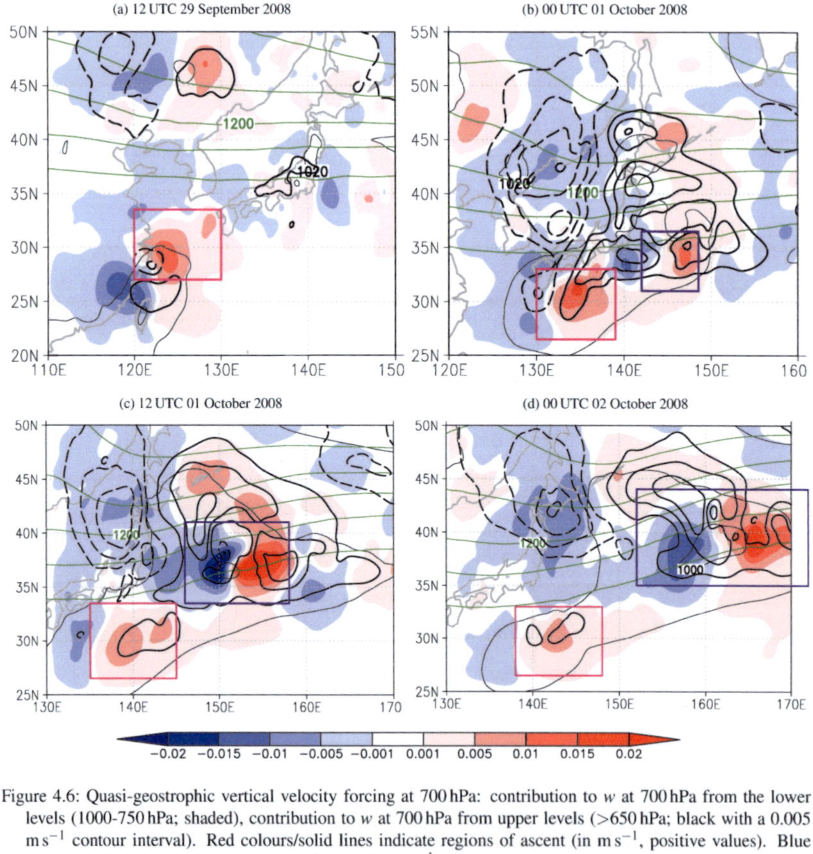


Figure 4.6: Quasi-geostrophic vertical velocity forcing at 700 hPa: contribution to $w$ at 700 hPa from the lower levels (1000-750 hPa; shaded), contribution to $w$ at 700 hPa from upper levels (>650 hPa; black with a 0.005 $\mathrm{m\,s^{-1}}$ contour interval). Red colours/solid lines indicate regions of ascent (in $\mathrm{m\,s^{-1}}$, positive values). Blue colours/dashed lines indicate region of descent (in $\mathrm{m\,s^{-1}}$, negative values). pmsl (black thin contours with a 10 hPa interval) and geopotential $\Phi < 1232$ gpdm at 200 hPa (green contours with a 16 gpdm contour interval). Magenta and dark violet boxes (identical to those in Figure 4.5) highlight the location of Jangmi and of the downstream cyclone. Data based on ECMWF analysis. Time is indicated above each panel. The domain shown in (a-d) focuses on the systems discussed.

the northeast is supported by positive upper-level QG forcing (magenta box in Figure 4.6b). Convective activity continues in this region (Figure 4.5m, n).

In the next 24 hours QG forcing for ascent continues to favour convection northeast of Jangmi, while the forcing for descent to the west weakens (compare magenta boxes in Figure 4.6c and Figure 4.5o, p, and in Figure 4.6d and Figure 4.5q, r).

At 12 UTC 01 October 2008, a new cyclone formed centred at 150.75°E, 36.5°N in the region of positive upper-level QG forcing (marked by the violet box in Figures 4.6c, 4.5o, 4.5p). The low-level dipole of QG forcing intensified in a larger region (Figure 4.6c). It is located in the centre of the new closed surface circulation (Figure 4.5p). To the north $T_b < 230$ K indicates a concentration of higher clouds (Figure 4.5p) in approximately the region of positive (upper- and) low-level QG forcing. Positive and negative values of upper-level QG forcing continue to be separated by a distinct line, which is reflected slightly westward as a trough axis in the 200 hPa isohypses.

At 00 UTC 02 October 2008, the new cyclone, located in the southwestern jet entrance region (Figure 4.5q), starts propagating eastward with the jet, while Jangmi rather remains stationary. The low-level QG forcing dipole propagates with the cyclone (Figure 4.6d). However, the upper-level QG forcing lags behind and is stronger in the western part of the new cyclone. This is consistent with the slower eastward propagation of the trough at 200 hPa. As a consequence, upper-level QG forcing east of the new cyclone weakens and the new cyclone centred at 161.5°E, 39°N does not intensify further (997 hPa). Nevertheless, the high cloud pattern (Figure 4.5r) correlates remarkably with regions of positive low-level QG forcing.

The new cyclone propagates faster than the midlatitude trough and is separated from the associated QG forcing at 00 UTC 03 October 2008 (Figure 4.7a). A smaller zone of positive upper-level QG forcing in the equatorward jet entrance region persists over the cyclone centre (177°W, 42°N), but central pressure remains almost constant. The 200 hPa trough axis is located at around 152°E and approximately separates a region of upper-level QG forcing for descent to the west and for ascent to the east.

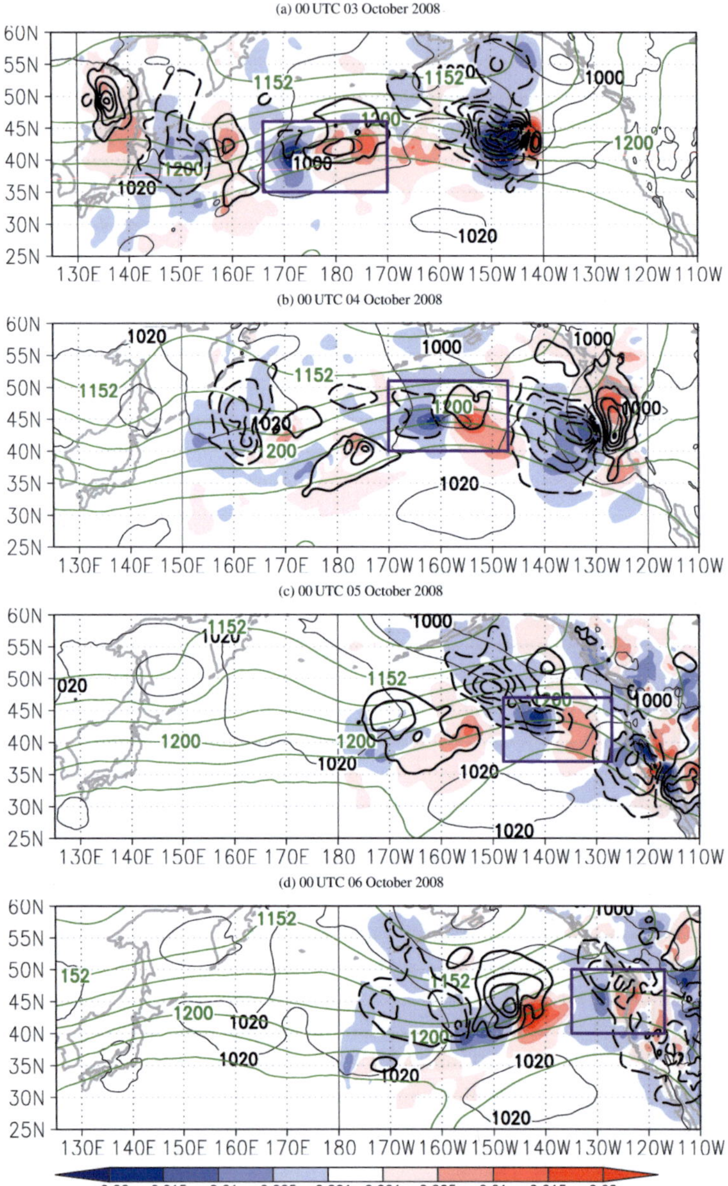


Figure 4.7: As Figure 4.6 but showing the propagation across the Pacific during 03 October 2008 and 06 October 2008. QG forcing is only calculated in the subdomain limited by the black vertical lines.

At 00 UTC 04 October 2008 the dipole of low-level QG forcing persists and is centred on the low-level centre of the cyclone at about 157°W, 45°N (Figure 4.7b). In the absence of strong upper-level QG forcing the cyclone has not altered its intensity. It became a secondary low to a steering cyclone over Alaska. The midlatitude trough axis has moved to about 165°E. Its positive upper-level QG forcing region vanishes.

At 00 UTC 05 October 2008 the upper-level trough reestablishes a zone of positive upper-level QG forcing east of its axis at about 175°W (Figure 4.7c). However, the cyclone propagated to a region of strong negative upper-level QG forcing associated with a midlatitude ridge. The cyclone is located on the ridge axis at this time (139°W, 44.5°N). The low-level QG forcing dipole persists, but forcing for ascent to the east weakens. As a result the cyclone weakens and the central pressure increases to 1001 hPa.

One day later the cyclone makes landfall at the coast of Washington and Oregon, USA (Figure 4.7d). A weak low-level QG forcing dipole is still present. As the cyclone is located continuously underneath a region of strong upper-level QG forcing for descent, its central pressure has increased to 1010 hPa and the cyclone became absorbed into the steering low over Alaska. In the next 12 hours the signature of the cyclone remnants vanishes.

In summary, an extratropical cyclone formed east of Japan immediately downstream of Jangmi undergoing ET. The cyclogenesis occurred in a region favourable for large-scale ascent ahead of a midlatitude trough and in the southwestern entrance region of the jet, which was accelerated during the outflow-jet interaction of Jangmi. This cyclone propagated quickly over the Pacific as a frontal wave. During its propagation the cyclone was continuously located in the southwestern jet entrance region and its propagation speed was the same as the propagation speed of the jet streak. However, due to the slower propagation speed of the midlatitude trough, the cyclone lacked upper-level QG forcing and did not further intensify. It became secondary to a steering low over Alaska and dissipated at the North American coast when the upper-level flow and the jet streak were perturbed at the continent.

## 4.2 Observations

During the T-PARC campaign numerous land-, sea-, air-, and spaceborne meteorological measurement platforms were based in the western North Pacific region. This enabled a unique observational data set of the entire lifecycle of tropical cyclones to be collected. In this study we use a small subset of this data to validate the ECMWF analysis during the extratropical transition of Jangmi and its interaction with the midlatitude flow around 30 September 2008. Data from two DLR Falcon research flights and the MTSAT satellite are investigated. It has to be noted that the Falcon dropsonde data has been processed by the ECMWF data assimilation system. Thus the ECMWF analysis contains some of this observational information.

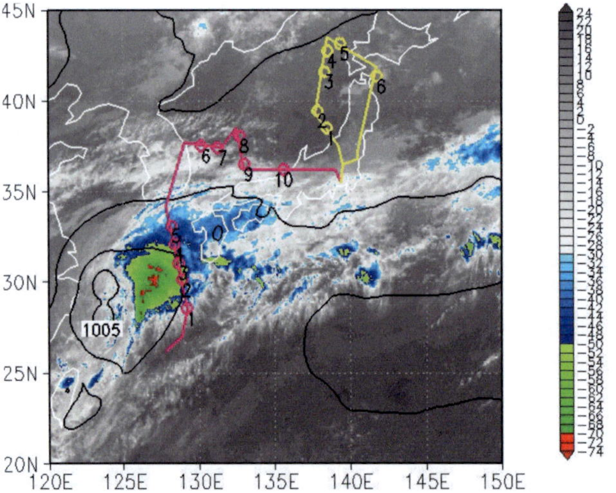


Figure 4.8: Flight tracks and dropsonde locations of DLR-Falcon flights #22 centred 00Z/30 (magenta) and #23 centred 05Z/30 (yellow). MTSAT IR channel 1 (10.3 – 11.3$\mu m$) brightness temperature (in °C) is plotted at 23:45 UTC 29 September 2008. The mean sea level pressure from the ECMWF analysis (black with a 5 hPa contour interval) is shown at 00 UTC 30 September 2008.

### 4.2.1 Aircraft observations

Two DLR Falcon research flights were performed on 30 September 2008 as Jangmi approached the midlatitudes. The DLR Falcon was equipped with in-situ instruments, a dropsonde launch system, a wind lidar, and a DIAL water vapour lidar (see Section 3.1.3 for details). Dropsondes could not be launched over land. Unfortunately meteorological conditions were not always favourable for lidar measurements. This resulted in a poor lidar data availability. To use the observations optimally the quality controlled data from all different aircraft measurements are combined and interpolated on the section beneath the aircraft. Similarly, the model analysis data is interpolated onto the same section and compared with the observations (Figures 4.9, 4.12, 4.14, 4.15).

For wind speed and wind direction the dropsonde, in-situ, and wind lidar data are interpolated on a grid defined in the horizontal direction by the profile number (which is uniquely defined by its time) and the 100 m equally spaced vertical wind lidar levels in the vertical direction. The dropsonde data is linearly interpolated onto these vertical levels, and the profile number which is closest to the dropsonde launch time is assigned. The in-situ data is linearly interpolated onto the times of the profile numbers. In a final step this preprocessed dropsonde and in-situ data is combined with the lidar data and interpolated on the lidar grid using a Delaunay-triangulation-based cubic interpolation. The same procedure is used to interpolate the specific humidity $q_v$. In that case the grid is defined by the same 100 m equally spaced vertical water vapour lidar levels and the different profile numbers of this lidar. The calculation of equivalent potential temperature $\Theta_e$ requires values of temperature and humidity. Because of the lack of temperature measurements the lidar data has to be omitted for this interpolation. Due to very low humidity values at flight level the interpolation of in-situ *and* dropsonde data results in strong gradients where both data sets merge. Therefore the cross section of $\Theta_e$ is interpolated on the (wind) lidar grid using dropsonde data only. However, for the cross section of $q_v$ and $\Theta_e$ temperature and humidity measurements of so-called "fast-fall" sondes, whose parachute failed, were included, if the quality control for these variables was good. The interpolation of model analysis data on the lidar grid follows the method used by Schäfler et al. (2010) and Schäfler et al. (2011) in their intercomparison of lidar measurements and ECMWF analysis. As ECMWF analyses are only available every 6 hours, short-range COSMO simulations were performed on a grid similar to the ECMWF analysis. Hourly output for the 1-5 hour forecasts of these model runs are combined with the ECMWF analysis at the synoptic times, to form a set of hourly "ECWMF pseudo-analyses". In a first step a bilinear interpolation of this pseudo-analysis (on model levels) onto the horizontal (lat/lon) locations of the lidar profiles is performed. Then a linear interpolation of the model level data on the vertical lidar levels is computed. Finally, a linear interpolation of the hourly model data onto the profile times is calculated.

In the following the vertical cross sections are discussed along with a statistical analysis of the deviations and correlations of model analysis and observational data. For the statistical analysis the absolute difference AD is defined as $AD = y_{model} - y_{obs}$, where $y$ stands for a meteorological variable of either model analysis or observational data. The relative difference RD is defined as $RD = 2 \cdot AD/(y_{model} + y_{obs})$.

Falcon flight #22 started at Kadena on the island of Okinawa at 22:20 UTC 29 September 2008 and ended on the Japanese mainland in Atsugi at 01:40 UTC 30 September 2008 (magenta coloured track in Figure 4.8). In the first half of the flight the aircraft headed north towards Korea and passed the eastern edge of the convective burst in the vicinity of Jangmi. On this leg 7 dropsondes were launched in total, 2 of them had a parachute failure but passed quality control. The second half of the flight captured the jet region with the aircraft heading from Korea eastward to Japan. On this leg another 5 dropsondes were successfully launched. Due to cloud cover the water vapour lidar measurements are only available during the very first part of this flight. However, the wind lidar captured a layer between 11000 and 8000 m above mean sea level (amsl) at jet level during the entire flight. By combining the different aircraft measurements this flight enables us to investigate the convective burst and the midlatitude jet.

The sections of wind speed and wind direction along the flight track reveal two important features. During the first half of the flight (up to drop #5 at around

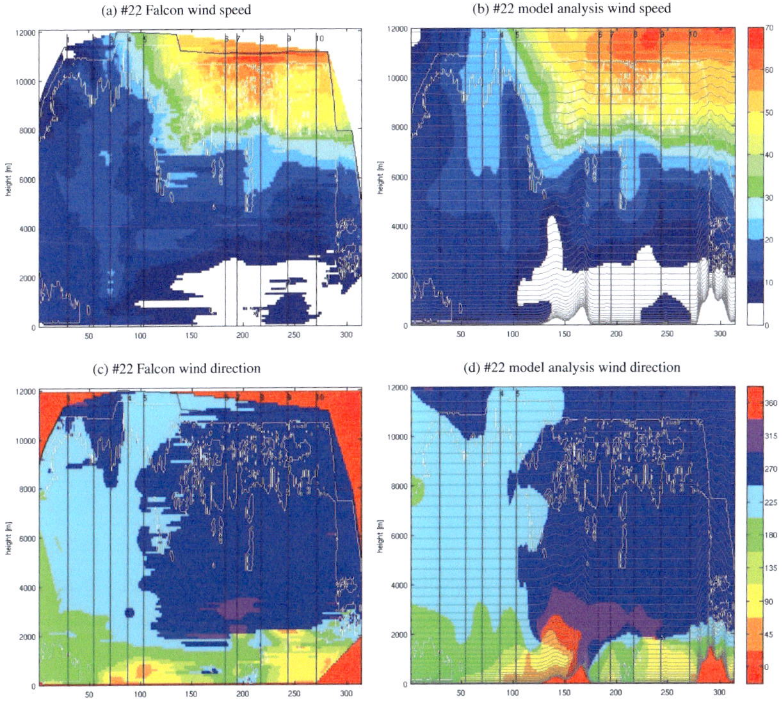


Figure 4.9: Wind speed (top, in $\mathrm{m\,s^{-1}}$) and wind direction (bottom, in °) in the section of Falcon flight #22 from 22:20 UTC 29 September 2008 to 01:40 UTC 30 September 2008. The vertical axis is height in m amsl, the horizontal axis the "profile number" which is similar to time (in average 1 profile every 32 s). The black vertical lines indicate the location of dropsondes, labelled at the top with the dropsonde number. Lidar measurements are available in the white outlined regions. The position of the aircraft is indicated by the black line at the top and no measurements are available above this line. The left panels show observational data interpolated in the plane of section. The right panels show ECMWF pseudo-analyses model level data with the grey lines indicating the model surfaces. See text for details of the interpolation method.

profile 100) moderate wind speeds of 10-30 $\mathrm{m\,s^{-1}}$ occur through the entire troposphere (Figure 4.9a,b). Higher wind speed (20-30 $\mathrm{m\,s^{-1}}$) is detected around drop #3 (profile 70) in particular above 5000 m. This is the region of the new convective burst and at upper levels of the outflow of Jangmi (Figure 4.5i,j). The wind direction at upper levels during this first flight leg gives an indication of the anticyclonic outflow of Jangmi (Figure 4.9c,d). A predominantly westerly flow is evident above 9000 m from drop #1 to #3. This is the northeastern edge of Jangmi (cf. Figure 4.8) where westerly outflow is expected. Further north,

(a) #22 scatter plot wind speed (b) #22 scatter plot wind direction (c) #22 scatter plot specific humidity

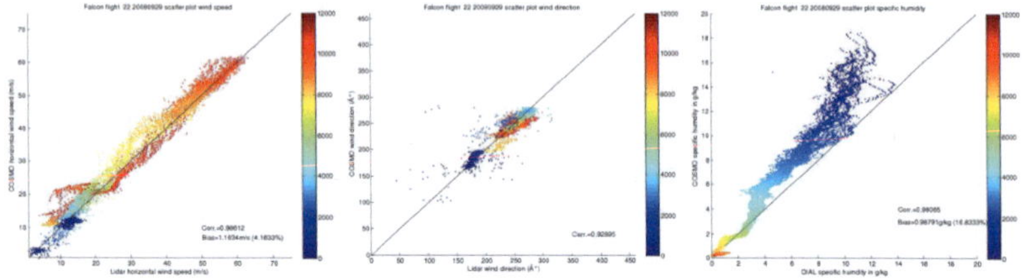

Figure 4.10: Scatter plots of wind speed, wind direction and specific humidity during Falcon flight #22 colour-coded by height in m amsl. Lidar data is on the abscissa and the model analysis data interpolated to the corresponding location in the lidar grid on the ordinate. The numbers indicate the correlation coefficient and the total mean absolute (relative) difference.

the wind direction changes to southwesterly flow (from drop #4 (profile 85) to about profile 120). This region is located north of Jangmi and thus the observed wind direction is consistent with the expected anticyclonic outflow. In terms of wind speed an increase is detectable in this region, indicating that the aircraft approaches the jet core region and that the jet core extends towards Jangmi's outflow region. The most striking feature during the second half of the flight is the midlatitude jet at levels above 8000 m. The wind speed reaches more than $70\,\mathrm{m\,s^{-1}}$ with a predominantly westerly direction between drop #7 and #9 when the aircraft headed slightly further north into the jet streak. At low levels the directional wind change from easterly to westerly flow is striking. Considering that major differences in the model analysis and observational sections result from the interpolation in data-sparse regions, the overall qualitatively good agreement for wind between both data sets is outstanding.

Quantitatively this is manifested in the high correlation coefficients between the model analysis and lidar wind speed (0.99) and wind direction (0.93) (Figure 4.10a,b). The mean bias of wind speed is $1.2\,\mathrm{m\,s^{-1}}$ and the mean relative difference 4.2%. This indicates a weak overestimation of the wind speed in the model analysis. Despite the rather symmetric shape, the frequency distributions of absolute and relative differences (left panel of Figure 4.11a) show a weak shift to positive values, reflecting the overestimation of wind speed in the model analysis. The right panel of Figure 4.11a shows the wind lidar data coverage (availability) at each level. Although the mean wind lidar data coverage over all

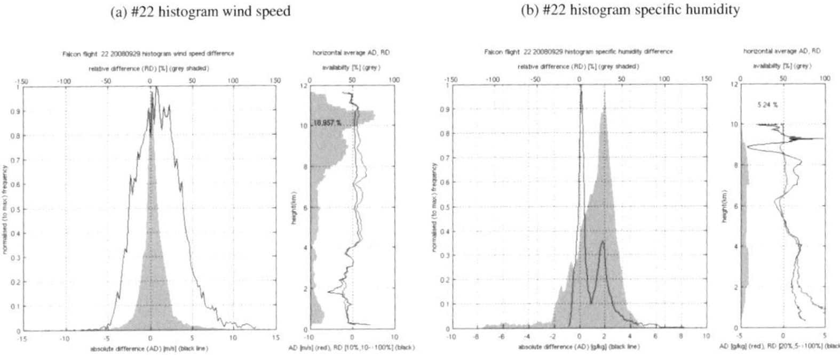


Figure 4.11: Statistical overview of the differences between model analysis and lidar data. (a): wind speed, (b) specific humidity. Left panels: frequency distribution of absolute difference (AD, black contour, bottom axis) and relative difference (RD, grey shaded, top axis) normalised to the maximum frequency. Right panels: lidar coverage (availability, grey shaded, in %, top axis) at each lidar level. The number indicates the total lidar coverage. Profiles of horizontal mean AD (red, bottom axis) and RD (black, bottom axis, range is from -100% to +100%)

levels was only 19.0%, the lidar data coverage at jet-level (8000-11000 m amsl) reached mostly 45% and up to 75% in the layer 10000-11000 m amsl (right panel of Figure 4.11a). Also, the data coverage at jet-level was higher during the second flight leg in the jet core region than during the first half of the flight (see white outlined region in Figure 4.9). Thus the jet was captured very well by the wind lidar. The profiles of absolute and relative difference at each level emphasise that the mean positive bias is mainly due to a weak positive bias above 4000 m amsl (right panel of Figure 4.11a).

The cross sections of specific humidity $q_v$ and equivalent potential temperature $\Theta_e$ (Figure 4.12) highlight the baroclinic zone. Water vapour lidar coverage was very poor with only a mean availability of 5.3% at all levels. The lidar mainly captured the very first part of the flight (see white outlined region in Figure 4.12c). The correlation coefficient of model analysis and lidar was very high (0.98), however the absolute mean bias of 1.0 gkg$^{-1}$ and mean relative difference of 17.2 % rather indicate a significant wet model bias (Figure 4.10c). The section of $q_v$ indicates a strong gradient at around profile 48 where measurements from drop#1 and the lidar data are available (Figure 4.12c). The DIAL data was captured during the ascend of the aircraft. Such data should have been filtered out by quality control due to problems with the height attribution (An-

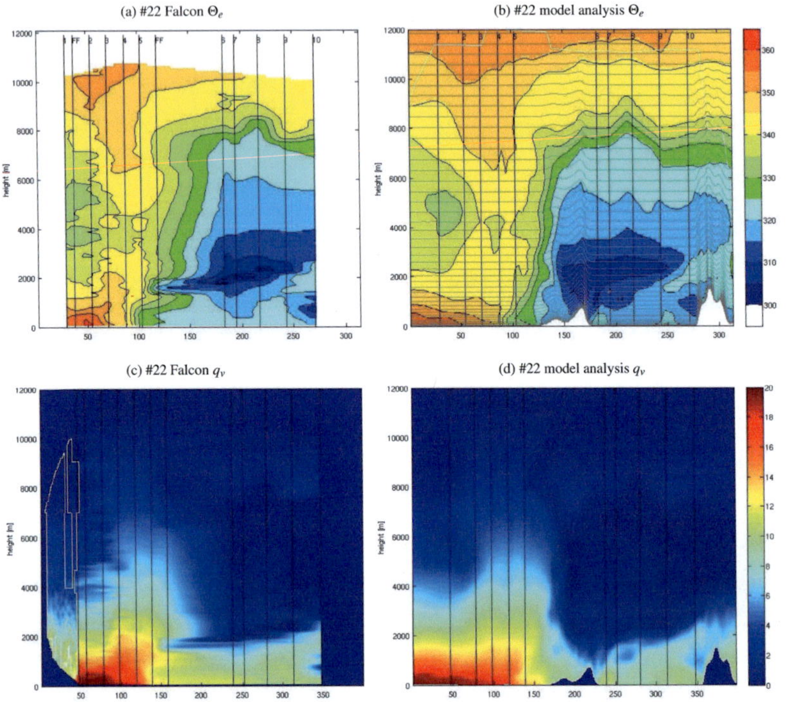


Figure 4.12: Same as Figure 4.9 but for equivalent potential temperature $\Theta_e$ (top, in K) and specific humidity $q_v$ (bottom, in $gkg^{-1}$). In addition to Figure 4.9 "fast-fall" sondes (labelled "FF") are included.

Table 4.1: CAPE (J), pressure (hPa) at level of free convection (Plfc) and at lifting condensation level (Plcl), temperature (°C) at lcl (Tlcl), and precipitable water (pwat in 10 mm increments) from drops #1-5 and two fast-fall sondes (ff) on the first flight leg of flight #22. The CAPE is calculated from lfc to 310 hPa only, to be comparable for all dropsondes. (*All variables calculated by the NCL function skewt_func.ncl available at* `http://ncl.ucar.edu/Applications/skewt.shtml`.)

| drop | 1 | ff | 2 | 3 | 4 | 5 | ff |
|---|---|---|---|---|---|---|---|
| CAPE | 1085 | 835 | 1376 | 873 | 0 | 0 | 10 |
| Plfc | 943 | 831 | 965 | 1010 | 488 | 987 | 938 |
| Plcl | 961 | 963 | 970 | 1010 | 1010 | 991 | 964 |
| Tlcl | 24 | 23 | 25 | 25 | 23 | 18 | 17 |
| pwat | 50 | 50 | 60 | 60 | 60 | 60 | 40 |

dreas Schäfler (DLR), personal communication). For the rest of the flight cloud coverage was too thick for DIAL measurements. Hence, a more detailed discussion of the lidar data is omitted. In the following we concentrate on the section interpolated from dropsonde data.

The section captured by sondes #1-5 and two "fast-fall" sondes (profiles 0-150) in the vicinity of the convective burst reveals a moist layer ($q_v$ partly exceeds 20 $gkg^{-1}$) up to about 2000 m amsl (Figure 4.12c,d). At the northern edge (sondes #3-5) this layer is elevated and moist air ($q_v$ up to 4 $gkg^{-1}$) reaches up to 8000 m amsl. This elevated moist layer coincides with a region of unstable stratified $\Theta_e$ south of and almost neutrally stratified $\Theta_e$ in a marked baroclinic zone starting north of dropsonde #4 (Figure 4.12a,b). The CAPE values of all dropsondes from the first flight leg are calculated from the level of free convection to 310 hPa, which is the layer in which data is available from all sondes (Table 4.1). The sondes south of the baroclinic zone (# 1,ff,2,3) exhibit very high values of CAPE while in the baroclinic zone CAPE is practically 0 (sondes #4,5,ff). Thus ahead of the baroclinic zone the environmental conditions are favourable for convective, upright ascent, whilst in the baroclinic zone the neutral stratification is favourable for dynamically forced slantwise ascent along the tilted (moist-)isentropes. The elevated moist layer consequently results partly from buoyant ascent of moist air ahead of the baroclinic zone and partly from dynamically forced ascent of moist air along the baroclinic zone. The baroclinic zone is seen in the analysis data around profile 125 (Figure 4.12b). In that region the $\Theta_e$ gradient at the baroclinic zone in observational data is weaker due to the lack of data (Figure 4.12a). Nevertheless, the sondes at the southern (#4,5,ff) and northern edge (#6,7) give enough information so that the interpolation captures the baroclinic zone in the data-sparse region. The low-level temperature gradient of more than 5 K at the baroclinic zone between drop #4 and #5 is particularly well represented in observational and analysis data (Figure 4.12a,b and Table 4.1). The major difference between model analysis and observational data is the dry layer from about 1700-2000 m amsl at profile 160 (2nd "fast-fall" sonde, Figure 4.12). This dry layer is not captured with a comparable vertical gradient in the model analysis data. Despite that, the qualitatively good agreement between model analysis and observational data gives confidence in the existence of the baroclinic zone separating warm humid air in the south from cooler dryer air in the midlatitudes and in the existence of ascent of this warm humid air layer by upright ascent within the convective burst ahead of the baroclinic zone and by slantwise ascent along the tilted isentropes.

Falcon flight #23 was a round trip from Atsugi on a diamond pattern through the jet core to the north of Atsugi (Figure 4.8). The flight started at 03:50 UTC 30 September 2008 heading northwestward to the Sea of Japan. Then the aircraft flew a profile through the jet core ascending from 8000 to 11000 m amsl. North of the jet streak the aircraft turned and headed back to Atsugi over land. The wind lidar coverage was 34.8% and more evenly distributed over the different levels (right panel in Figure 4.13a). In addition, dropsonde data is mainly available in data-sparse regions while regions without dropsonde data are captured by the wind lidar. Thus an excellent cross section of wind speed and wind direction was measured along the flight track (Figure 4.14). The sections of wind speed reveal the double penetration of the jet core during the two flight legs with wind speed reaching up to 80 m s$^{-1}$ (Figure 4.14a,b). The correlation coefficient between analysis and lidar wind speed is 0.99 and mean absolute bias as low as 0.62 m s$^{-1}$ (1.2%). This is reflected in the symmetrical frequency distributions of AD and RD, with AD only slightly emphasising positive values (left panel in Figure 4.13a). The profile of AD shows a maximum deviation of about 3.5 m s$^{-1}$ at jet level (right panel in Figure 4.13a). Similarly the section of wind direction from model analysis and observations qualitatively coincide well. It is remarkable that the directional wind change in the planetary boundary layer close to land in the first third of the flight (profiles 20-100) is comparable in model analysis and observations. However, details of the circulation in the

(a) #23 histogram wind speed

(b) #23 histogram specific humidity

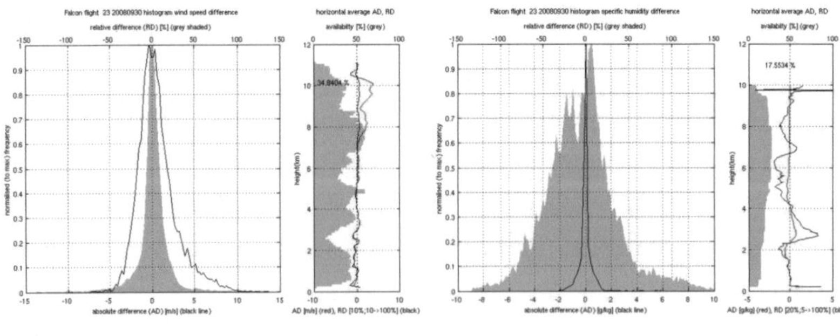

Figure 4.13: As Figure 4.11 but for Falcon flight #23 from 03:50 UTC 30 September 2008 to 06:55 UTC 30 September 2008.

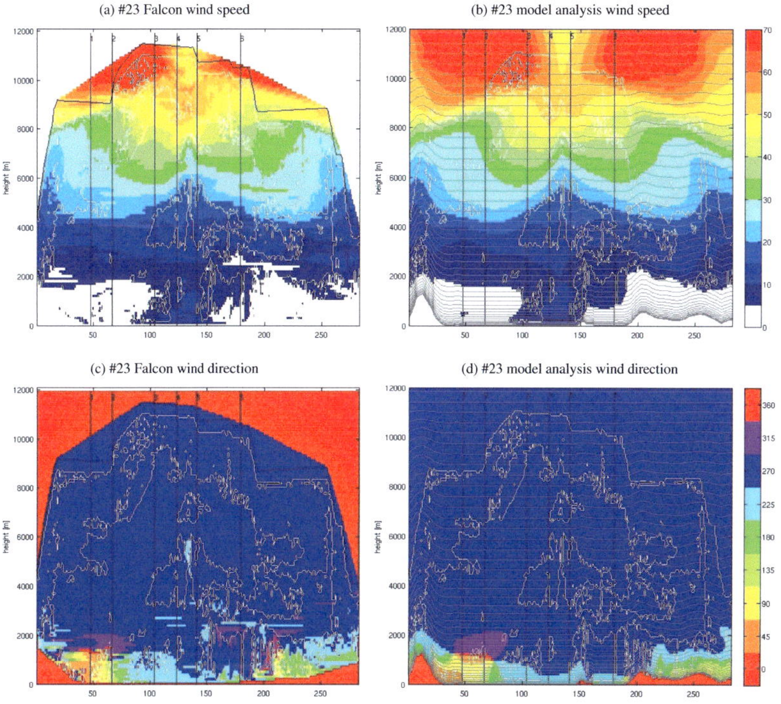


Figure 4.14: Same as Figure 4.9 but for Falcon flight #23 from 03:50 UTC 30 September 2008 to 06:55 UTC 30 September 2008.

planetary boundary layer over land near profile 210 are different in analysis and lidar data. The correlation coefficient for wind direction is 0.87.

The section of specific humidity (Figure 4.15c,d) reveals a shallow humid planetary boundary layer over sea during the first leg of the flight (until profile 70). The analysis data shows a more shallow layer than the observations especially between drops #3-5 (profile 40-70). A second feature is an elevated moist layer at a height between about 2000-5000 amsl. Here observational data indicates more humid conditions than the model analysis data. During the last third of the flight over land, no observations are available. The lidar coverage was 17.6% and evenly distributed over the levels (right panel of Figure 4.13b). The correlation coefficient of model analysis and lidar data was 0.95 and the absolute

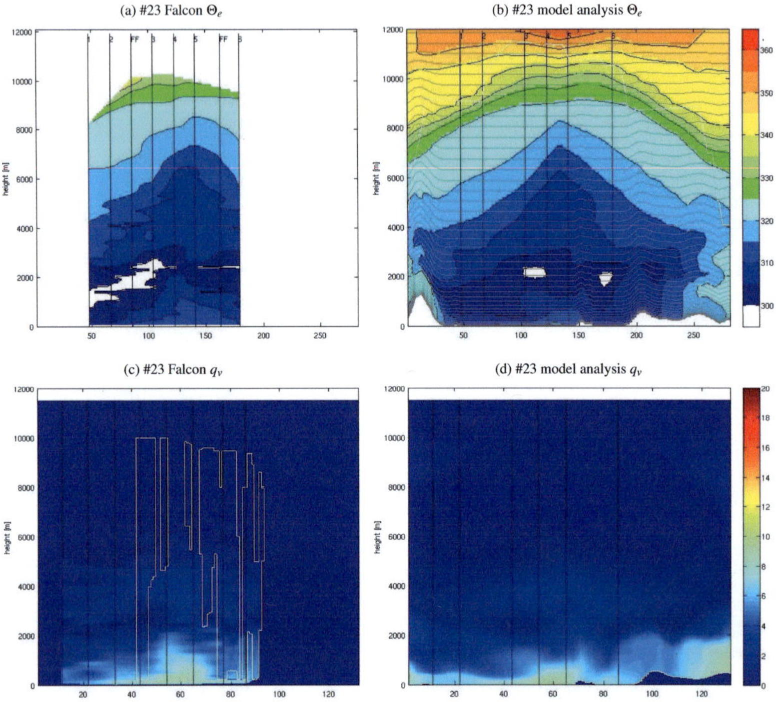


Figure 4.15: Same as Figure 4.12 but for Falcon flight #23 from 03:50 UTC 30 September 2008 to 06:55 UTC 30 September 2008.

mean difference -0.02gkg$^{-1}$, the mean relative difference -4.4%. This indicates that the model analysis was too dry on average. However, the frequency distribution reveals that very significant relative differences occur and due to the symmetric distribution only a weak negative total difference results (left panel of Figure 4.13b). The profiles of AD and RD show that major errors occur mainly in the elevated moist layer between 2000-5000 m amsl with the analysis being too dry in the lowest and upper part of this layer and too moist in between.

The section of $\Theta_e$ shows a baroclinic zone at mid and upper levels (2000-10000 m amsl) to the north of Atsugi (profiles 0-120 and 180-260 Figure 4.15b). The vertical $\Theta_e$ gradient is most pronounced at the location of the jet streak. This baroclinicity is partly captured by the $\Theta_e$ section interpolated from the

Table 4.2: Statistical numbers of the model-lidar comparison for Falcon flights #19-25: Flight number, wind lidar coverage in % of the grid defined by the lidar profiles and vertical levels, correlation coefficient of wind speed, absolute wind speed difference (AD) in m $s^{-1}$, relative wind speed difference (RD) in % and correlation coefficient of wind direction; dial water vapour lidar coverage in%, correlation coefficient of water vapour, AD in $gkg^{-1}$, RD in %.

| flight # | WIND lidar | | | | | DIAL $q_v$ lidar | | | |
|---|---|---|---|---|---|---|---|---|---|
| | cov. | $R_{wspd}$ | AD | RD | $R_{wdir}$ | cov. | $R_{q_v}$ | AD | RD |
| 19 | | 0.98 | | | | 30.3 | 0.98 | 0.30 | 8.4 |
| 20 | | 0.98 | | | | 8.4 | 0.98 | 0.27 | 11.4 |
| 21 | 19.9 | 0.93 | -0.29 | -2.1 | 0.92 | 3.4 | 0.91 | 0.96 | 15.4 |
| 22 | 19.0 | 0.99 | 1.16 | 4.2 | 0.93 | 5.2 | 0.98 | 0.99 | 16.8 |
| 23 | 34.8 | 0.99 | 0.61 | 1.1 | 0.87 | 17.6 | 0.95 | -0.02 | -4.3 |
| 24 | 15.8 | 0.97 | -0.16 | 0.2 | 0.98 | 51.6 | 0.96 | 0.45 | 19.5 |
| 25 | 35.5 | 0.98 | 0.61 | 2.8 | 1.00 | 72.7 | 0.96 | -0.07 | -2.2 |

drop sondes. At upper levels dropsonde #2 and the first "fast-fall" sonde also indicate the increased vertical $\Theta_e$ gradient at upper levels.

Table 4.2 summarises some statistical numbers for the model-lidar comparison of all Falcon flights during the lifetime of Jangmi. It is striking that correlation coefficients are generally very high and only weak differences occur for the wind lidar. However, the differences in humidity are more severe indicating that the model analysis is on average too moist. Nevertheless this model-aircraft data comparison proved that the ECMWF (pseudo-)analyses represent the jet, baroclinic zone and the convection at the baroclinic zone accurately and thus encourage to use the model analysis data for further investigations.

### 4.2.2 Satellite observations

#### 4.2.2.1 The interaction with the baroclinic zone

The synoptic discussion showed that convective bursts at the baroclinic zone are an important feature of the interaction of Typhoon Jangmi and the midlatitudes. In terms of observational data the convective bursts are best detectable in satellite data. For the ECMWF analysis cloud top temperature, calculated using the post-processing tool introduced in Section 3.3.4, is validated against MTSAT brightness temperature for IR channel 1 ($10.3 - 11.3\,\mu m$). The two variables are not directly quantitatively comparable. However, they indicate qualitatively the spatial extent of convective clouds. The period of investigation is 30 September 2008, capturing a first convective burst around 00 UTC 30 September 2008, its decay during the day and a new convective burst around 00 UTC 01 October 2008.

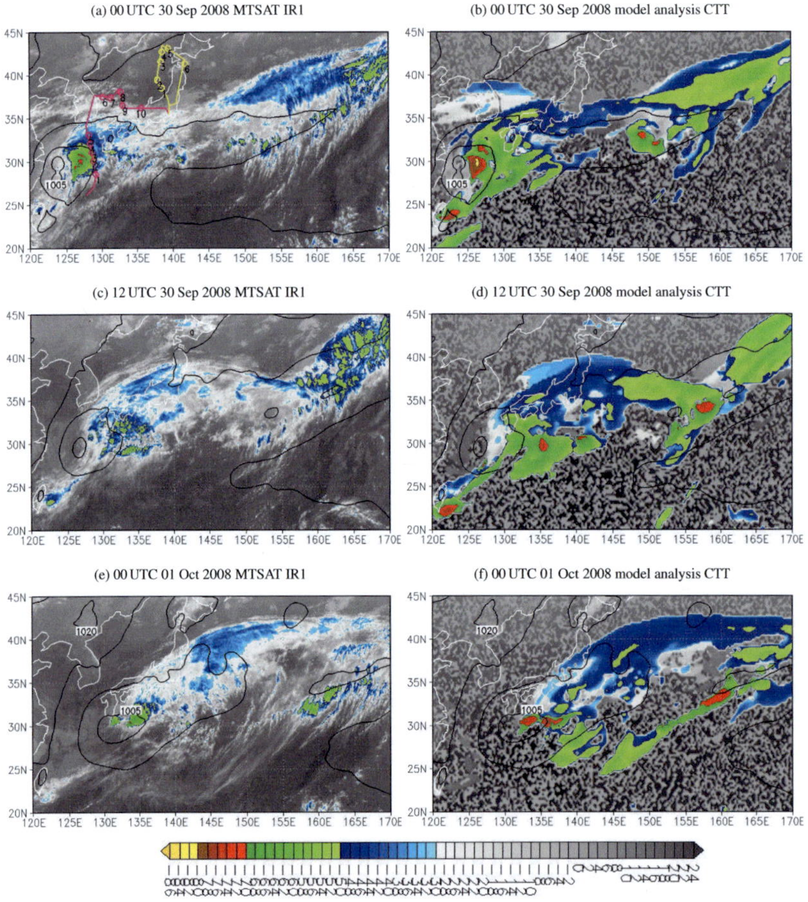


Figure 4.16: Cloud top temperature calculated from MTSAT IR channel 1 ($10.3 - 11.3 \mu m$) brightness temperature (left) and ECMWF analysis (right) both in °C. The time for the satellite data is 15 minutes prior to the times indicated. Additionally the pmsl from ECMWF analysis (black with a 5 hPa contour interval) is shown in all panels. The horizontal resolution of the model data is 0.25° and 0.05° for the satellite data. The DLR-Falcon flight track and locations of dropsondes for flight 22 centred 00Z/30 (magenta) and flight 23 centred 05Z/30 (yellow) are indicated in panel (a) .

At 00 UTC 30 September 2008 a first convective burst occurred at about 127°E, 30°N slightly south of the baroclinic zone to the northeast of Jangmi's centre (Figure 4.16a). At exactly the same position the ECMWF analysis shows a maximum of cloud top temperature (Figure 4.16b). However, in the model this convective burst is embedded in an area of high clouds which extends southwestward and includes another convective active area close to Taiwan (123°E, 24°N). This is not reflected in the observations which only show two smaller scattered cells (blue colours) close to Taiwan. The same overestimation of cloud coverage appears at about 150°E, 32°N where some larger cells are evident in the satellite data and an area of high clouds in the ECMWF analysis. The model data further tends to show homogenous regions of high clouds in the forming warm sector and downstream of Jangmi. Much more fine structure is visible and in the observations the brightness temperature is generally higher.

After twelve hours the first convective burst at the baroclinic zone decayed and only weaker scattered cells remain (Figure 4.16c). This is reflected in model analysis data with a maximum of cloud top temperature at about 135°E, 30°N (Figure 4.16d) which is the location of the largest cell in satellite data. The other convective active area close to Taiwan is now clearly evident in both model and satellite data. Furthermore, the model data shows homogenous regions with embedded convection downstream of Jangmi which are not present in the observations. Nevertheless, the northern edge of the cloud shield in the warm sector and downstream are captured accurately in the model.

At 00 UTC 01 October 2008 a second convective burst with two cells appears close to the centre of the remnants of Jangmi (Figure 4.16e, Figure 4.16f). The model analysis also shows an area of high cloud centred at about 140°E, 28°N, where only a few scattered cells are seen in the observations. A linear zone of high cloud extends from 145°E, 24°N to 165°E, 34°N in the analysis. Scattered cells along this feature are only observed in the eastern half of this band. At about 160°E, 32°N a larger convective cell is evident in both model analysis and satellite data.

The model-satellite data comparison with focus on convective bursts during ET showed, that the large convective cells ahead of the baroclinic zone are well captured in the analyses. Also the spatial extent of the cloud shield in the developing warm-sector of Jangmi resulting from the dynamically forced slantwise ascent along tilted isentropes is similar to the observations. However, analysis data lacks the fine structure and often indicates higher and more homogenous clouds than observed. In particular, the spatial extent of high clouds south of the baroclinic zone is much larger than in satellite data. This is partly due to the coarser resolution of model data but also reflects an erroneous cloud representation in the model.

#### 4.2.2.2 The outflow-jet interaction

The outflow-jet interaction during the ET of Jangmi was most distinct at 30 September 2008. For that day the ECMWF analysis of the jet-level wind (200 hPa) is compared against observational wind data from MTSAT derived atmospheric motion vectors (AMVs - e.g. Velden 1996; Velden et al. 1997, 2005). Hourly AMV data at irregular locations were provided by CIMSS. The AMVs are derived by tracking edges in the satellite imagery (at different wavelengths) for subsequent timesteps. The height assignment varies around ± 25-35 hPa (Chris Velden, personal communication). Having in mind that uncertainties exist in both the AMVs and model data, neither of the data source can be judged as representing the truth. However, both data sets can be given more confidence if they show quantitatively and qualitatively similar structures. In this section we compare AMVs from all MTSAT channels located in the layer from 225-175 hPa against the model fields at 200 hPa. This is a common configuration and was proposed by CIMSS.

At 00 UTC 30 September 2008 the ECMWF analysis shows a jet streak located over the northern part of the Japanese main island (Figure 4.17a). This jet streak extends southwestward towards Jangmi, which is located at 124°E, 30°N. The model streamlines indicate that the outflow of Jangmi is directed towards the jet streak. East of Jangmi the anticyclonic flow around the subtropical ridge is evident. Hardly any AMVs are available in the jet region at this time. The few AMVs at the western jet entrance region show the gradual increase of wind speed towards the jet core. South of 35°N the AMVs give observational

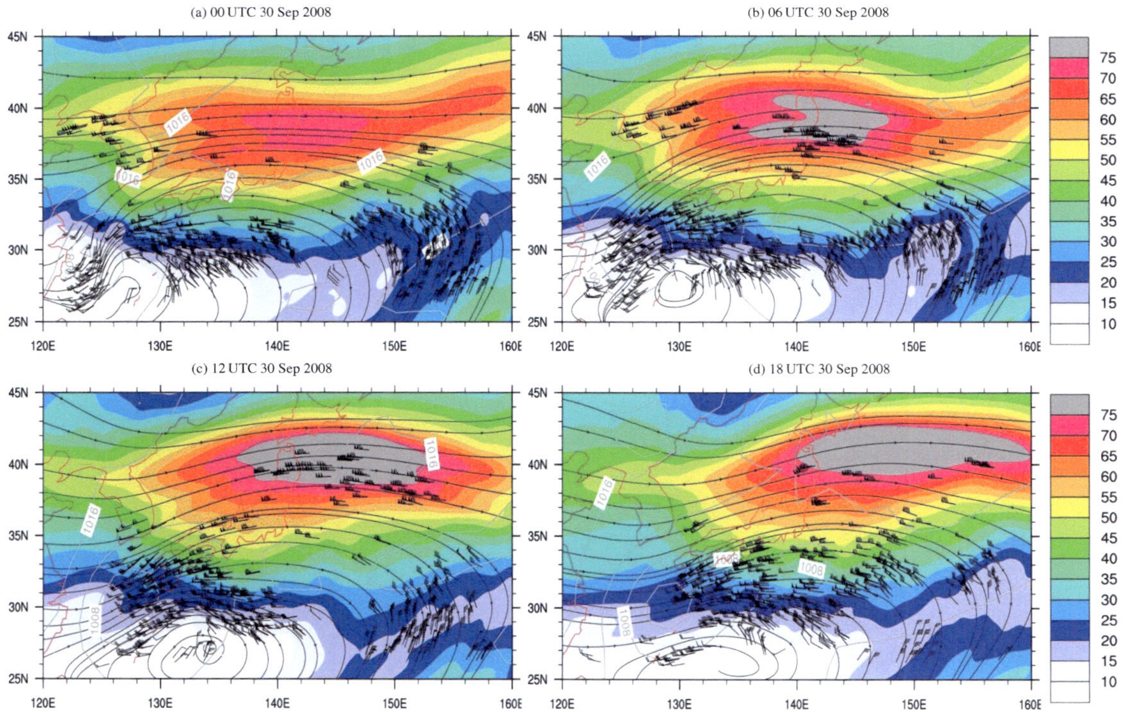


Figure 4.17: Wind barbs from MTSAT atmospheric motion vectors (AMV, data from all channels, provided by CIMSS) in the layer 175-225 hPa (black barbs, half flag 2.57 m $s^{-1}$, full flag 5.144 m $s^{-1}$, triangle 25.72 m $s^{-1}$). Wind speed (in m $s^{-1}$ shaded), streamlines (black lines) at 200 hPa, and pressure at mean sea level (grey with a 4 hPa contour interval) from ECMWF analysis. Times shown are 00, 06, 12, 18 UTC 30 September 2008 for the model data. The MTSAT data is time-labelled for HH:13, where HH is day-1 23, 05, 11, 17 UTC.

evidence of two important features: the anticyclonic outflow of Jangmi, directed towards the jet and the enhanced anticyclonic flow between the eastern edge of the subtropical ridge and the TUTT cell. In general the model wind speed and direction coincide remarkably with the AMVs.

Six hours later the jet streak became stationary over Japan and the jet core wind speed increased (Figure 4.17b). AMVs and model fields show a jet core wind speed of more than $75\,\mathrm{m\,s^{-1}}$. The AMVs south of 35°N capture the sustained outflow of Jangmi towards the jet and the flow around the subtropical ridge at about 155°E, 30°N. Despite the generally good coincidence between model and satellite winds, in the southwest corner of the domain the AMVs rather show a cyclonic flow around the southern part of Jangmi in contrast to the model streamlines.

At 12 UTC 30 September 2008 the jet streak remains stationary over northern Japan (Figure 4.17c). It expanded further and the region of high jet core wind speed ($|\vec{v}| > 75\,\mathrm{m\,s^{-1}}$) spans from the eastern Sea of Japan over northern Honshu to the West Pacific. Streamlines and AMVs continue to indicate the outflow of Jangmi directed towards the jet streak. AMVs also emphasise the flow around the centre of the subtropical ridge with an anticyclonic circulation centred at about 132°E, 28°N. This centre is slightly inconsistent with the model streamlines. As for the previous time steps the flow at the eastern edge of the ridge remarkably coincides in AMV and model data.

At 18 UTC 30 September 2008 the jet streak starts moving eastward and the southwesterly outflow of Jangmi weakened (Figure 4.17d). Nevertheless over Japan the jet streak still shows a strong extension towards Jangmi in both satellite and model analysis data. The majority of the AMVs highlight the circulation of the subtropical ridge in agreement with the analysis fields.

In summary, a good agreement of the irregular satellite AMVs and ECMWF analysis was found at jet-level on 30 September 2008. This enabled us to highlight an interesting evolution of the midlatitude jet streak over Japan. Starting with a strong southwesterly outflow from Jangmi into the jet core, the jet becomes stationary over Japan and at its southwestern edge extends towards Jangmi. As long as the outflow of Jangmi continues to be directed towards the jet, the jet streak intensifies over Japan. Once Jangmi's outflow decays, the jet starts moving eastward. Another interesting feature in that period is the enhanced flow at about 152°E, 28°N between the eastern edge of the subtropical anticyclone and the TUTT cell approaching from the east.

The validation of the ECMWF analysis against various observational data sets, demonstrated that the model analysis captures accurately the main features during the interaction of Typhoon Jangmi and the midlatitudes. This justifies our use of the ECMWF analysis for analysing the physical processes that occur during this interaction in more detail.

## 4.3 Details of the outflow-jet interaction

In the following, the physical processes governing the observed interactions between the outflow of Typhoon Jangmi and the midlatitude jet streak during the ET period (00 UTC 29 September 2008 and 00 UTC 02 October 2008) is investigated in more detail. After an overview of the temporal evolution in this section, the physical processes that occur at the baroclinic zone (subsection 4.3.1) and that impact the midlatitude jet (subsection 4.3.2) will be explained in more detail.

A daily sequence of three-dimensional views in the ET region is given (Figure 4.18) and discussed together with the same panels overlaid with the paths of representative trajectories (Figure 4.19). A detailed explanation of the Figures is given in the additional captions. Information on the trajectories are summarised in Table 4.3.

Table 4.3: Information on the four representative trajectory calculations: Number, start and end time (day of September,October 2008/time UTC), longitude, latitude and height of the region where trajectories originate from, total number of trajectories, and note on weather systems represented by the calculation.

| No | start | end | lon. °E | lat. °N | height | total # | note |
|---|---|---|---|---|---|---|---|
| 1 | 28/12Z | 30/12Z | 118-135 | 17-45 | 0-2000 | 13910 | Jangmi |
| 2 | 29/12Z | 01/12Z | 118-135 | 17-45 | 0-2000 | 16502 | Jangmi, ascent near baroclinic zone |
| 3 | 30/12Z | 02/12Z | 118-150 | 17-45 | 0-2000 | 10208 | baroclinic zone |
| 4 | 01/12Z | 03/12Z | 118-166 | 17-45 | 0-2000 | 9251 | barocl. zone, cyclogenesis downstream |

At 00 UTC 29 September 2008, the time when convection is decaying in the centre of Jangmi (cf. Figures 4.5e, f), the tropical cyclone is still present as a vertical PV tower extending through the entire troposphere (blue tower labelled "1" in Figure 4.18a). To the northwest of Jangmi the lowered 1.5 PVU surface at upper levels and the associated tropopause step (blue rather vertical surface in the upper left corner of Figure 4.18a, west of the green shading) indicates the approaching midlatitude trough. The first midlatitude jet streak is located to the northeast of Jangmi (green "tube" at upper levels) and is still distant from the TC. A steep baroclinic zone in the 320 K $\Theta_e$ surface (grey/white transparent shading in the middle of the panels, initially almost upright in the lower troposphere and sloping at upper levels to the north (back of the volume) is aligned underneath the tropopause step to the north of Jangmi. The tropopause step and the slope of the baroclinic zone are better evident when looking from west to east through the volume (Figure 4.20a). At the surface the baroclinic zone is reflected in a strong temperature gradient (orange to green shades). A wave developing on this gradient indicates that at its northern and eastern side Jangmi advects warm air polewards, while colder air penetrates equatorward west of Jangmi (see also Figure 4.20a). Most of the trajectories are rising in or in the vicinity of Jangmi's PV tower (Figure 4.19a). The trajectories start with low PV values (PV<0.2 PVU) in the low and mid troposphere. During ascent the PV increases strongly to values in excess of 2 PVU.

Thirty hours later Jangmi has reached the baroclinic zone (Figure 4.18b). A wave structure in the $\Theta_e$ surface northeast of Jangmi, also reflected in the surface temperature gradient, indicates a deflection of the baroclinic zone by the building of a warm sector east of Jangmi (Figure 4.18b). The surface temperature also indicates that west of Jangmi cooler air is advected cyclonically around the tropical cyclone undergoing ET (see also Figure 4.20b). A new narrow and upright PV tower (labelled "2") developed close to Jangmi at the baroclinic zone.

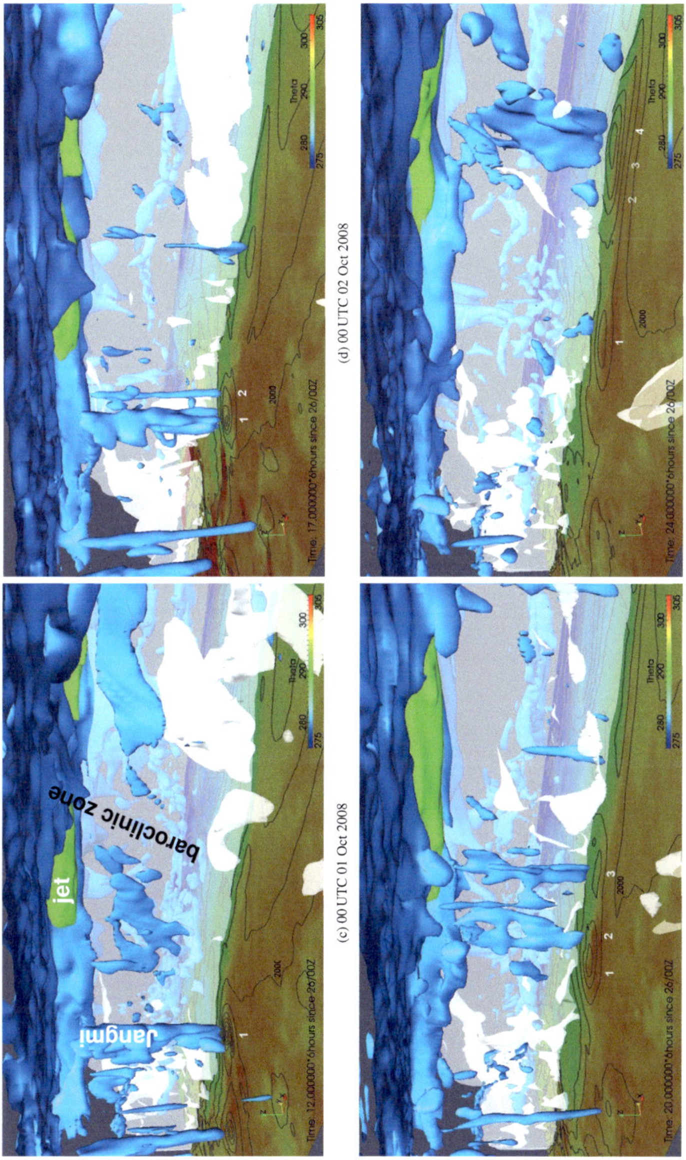


Figure 4.18: Three-dimensional view of Jangmi from the southeast. Shown are the 1.5 PVU PV surface (blue shading), the 320 K surface of equivalent potential temperature (transparent grey shading), the midlatitude jet in regions where $|\vec{v}| > 60\,\mathrm{m\,s^{-1}}$ (green shading), and at 990 hPa, potential temperature (in K, see legend) and geopotential (with a 250 gpm contour interval). Different PV features are labelled by white numbers. ... *See continued caption on page 98 for a more detailed explanation.*

Figure 4.18: ... The viewing point for all panels is static and from the southeast into the model data volume. The view is centred on the region where ET occurs (centred at about 145°E, 30°N, 500 hPa). The southern (lower) edge of the bottom plane is at 1°N, the northern (upper) edge at 69°N. On the left the western edge is at 81°E. The lower right corner of each panel is at about 180°E. At the bottom plane shading indicates potential temperature $\Theta$ (in K) and the grey contours geopotential at 990 hPa (250 gpm contour interval, 2000 gpm height contour is labelled). Both "surface fields" help to illustrate the surface conditions. E.g., the closed height contours in subfigure 4.18a, labelled "1", show the surface centre of Jangmi and the sharp gradient from red to green colours indicate the strong surface temperature gradient at the baroclinic zone. The blue shaded surface shows the 1.5 PVU PV surface. This PV surface represents the PV tower associated with Jangmi, convective bursts, and the dynamic tropopause. The grey/white transparent surface shows the 320 K surface of equivalent potential temperature $\Theta_e$ which points up the three-dimensional baroclinic zone. The green shaded "tube" in the upper levels is the surface of wind speed $|\vec{v}| = 60\mathrm{m\,s^{-1}}$ illustrating the midlatitude jet.

In the same region the high values for CAPE (Table 4.1), unstable stratification in $\Theta_e$ and an elevation of the moist air layer were observed by the Falcon dropsondes (Figure 4.12), indicating favourable conditions for convection. New convection was also observed in this region in the satellite imagery and a northeastward extension of the surface isobars occurred (cf. Figures Figure 4.5j,l, 4.16a,b and Figure 4.18b). The evolution of a new PV tower ahead of the baroclinic zone has been observed also during the ET of Hurricane Irene (1999 - Agustí-Panareda et al. 2004) and of Hurricane Floyd (1999 - Atallah and Bosart 2003). The trajectories have risen in that region during the preceding 30 hours (Figure 4.19b). PV along these trajectories increased during the rapid ascent in the mid troposphere. Reaching the upper troposphere, the slope of the trajectories indicates that ascent continues along the baroclinic zone and PV decreases. This is the typical path of an air parcel ascending through a region of steady latent heating (Wernli and Davies 1997, cf. Figure 3.5). The air parcels start with low PV in the lower troposphere. As they approach the level of maximum heating, PV increases due to diabatic PV production and reaches a maximum at the heating maximum. Above PV is diabatically reduced during the subsequent ascent. In the upper troposphere, the air parcel reach low PV values similar to the start values. The new PV produced in the mid troposphere is advected vertically so that the new PV tower extends through the entire troposphere. The PV profiles are discussed in more detail in Section 4.3.1. At upper levels a new midlatitude jet streak replaced the precedent jet over Japan. Now the jet streak is located close to the northeast of Jangmi. The trajectories which have been ascending since the previous time step have reached the jet level and are directed into the jet streak (Figure 4.19b). By transport of low PV air to upper levels

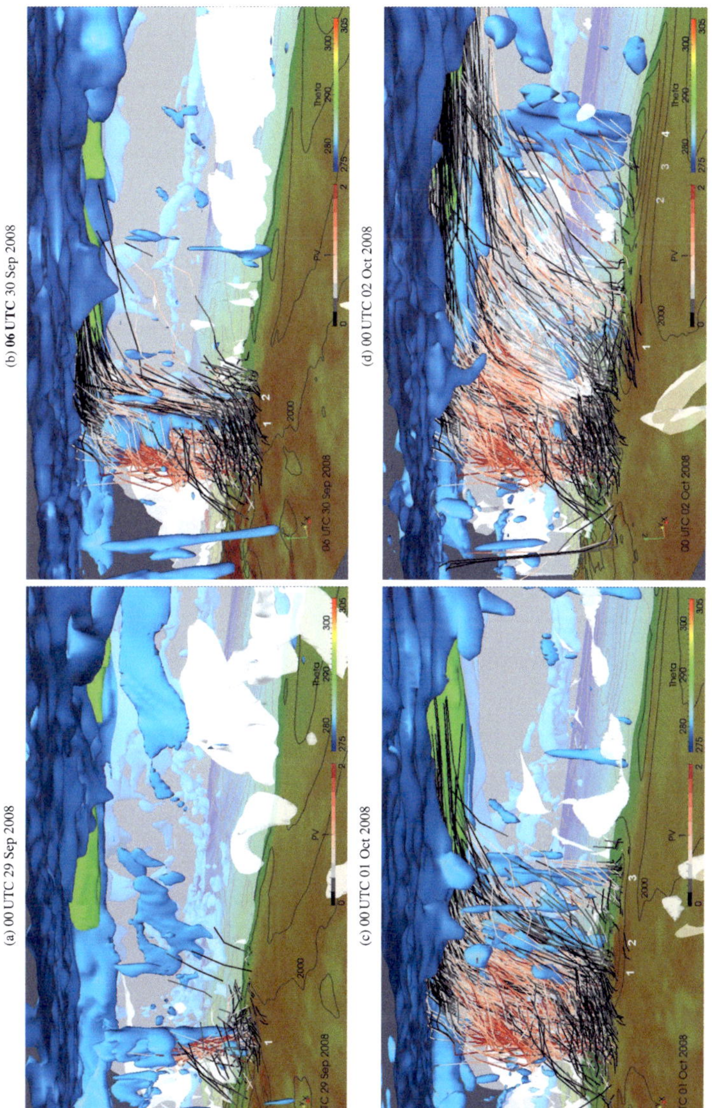


Figure 4.19: As Figure 4.18 but with trajectories from four representative trajectory calculations. The trajectories indicate the parcel paths until the time of each plot. *See continued caption on page 100 for details of the trajectory calculations.*

Figure 4.19: ... Shown are the same panels as in Figure 4.18 overlaid with the paths of representative trajectories coloured by the PV of an air parcel moving along this trajectory. Low-PV air (PV<0.6 PVU) is indicated by grey shades. Higher values of PV (PV>0.6 PVU) appear in red shades. All trajectories show the parcel paths until the time of each panel. Trajectories from four representative 48 hour forward LAGRANTO calculations are shown (Table 4.3). To highlight only the trajectories that are directly linked to the ET of Jangmi and the subsequent evolution at the baroclinic zone, the spatial origins of the trajectories are limited compared to the more general set of trajectory calculations introduced in Section 4.1.2. In the first calculation (shown in panels (a)-(d)) trajectories start in the lower troposphere (118-135°E, 17-45°E, 0-2000 m amsl) between 12 UTC 28 September 2008 and 12 UTC 30 September 2008. This calculation captures all trajectories that ascend in the vicinity of Jangmi. In the second calculation (shown in panels (b)-(d)) trajectories start from the same box but one day later between 12 UTC 29 September 2008 and 12 UTC 01 October 2008. As Jangmi is closer to the baroclinic zone in that time window, this calculation also captures the initial ascent near the baroclinic zone. The third calculation (shown in panels (c)-(d)) starts in a box extending further downstream (118-150°E, 17-45°E, 0-2000 m amsl) one day later (from 12 UTC 30 September 2008 to 12 UTC 02 October 2008), to account for the evolution and propagation along the baroclinic zone. For the last calculation computed from 12 UTC 01 October 2008 to 12 UTC 03 October 2008, the box is extended further downstream (118-166°E, 17-45°E, 0-2000 m amsl) to account for the cyclogenesis at the baroclinic zone. In each calculation the trajectories fulfil the criterion of ascent larger than than 8500 m (same as in Section 4.1.2) during the 48 h calculation period and only every 100th trajectory is shown.

they raise the dynamic tropopause and increase the PV-gradient at the tropopause step to the north (see also Figure 4.20b for the elevated tropopause). A steep tropopause fault developed similar to the "PV wall" reported by Atallah and Bosart (2003) in the case of the ET of Hurricane Floyd (1999). Through enhancing the upper-level horizontal temperature gradient this process sustains the jet streak and increases the jet core wind speed.

During the next day dynamically forced moist isentropic ascent along the baroclinic zone becomes dominant. At 00 UTC 01 October 2008 new positive PV anomalies (labelled "2" and "3" in Figure 4.18c) now are tilted along the baroclinic zone rather than forming a vertical PV tower. Jangmi's PV tower has decayed and only two blobs of PV remain in the mid and upper levels (above the closed 990 hPa geopotential height contours and the white label "1" in Figure 4.18c). However, a new scattered positive PV anomaly appears immediately to the northeast (labelled "2") where the extension of the height contours appear and new convection persists (cf. Figures 4.5n and Figure 4.18c). The circulation induced by the new PV tower that developed at 06 UTC 30 September 2008 advects tropical air on its eastern side against the baroclinic zone. There new PV is produced diabatically in the ascending air.

The PV anomaly labelled "2" in Figure 4.18c consists of the preexisting narrow PV tower and of the new PV produced at the baroclinic zone. The circulation induced by anomaly "2" likewise advects tropical moist air poleward at its eastern side and another positive PV anomaly appears downstream of Jangmi (labelled “3”). This PV anomaly consists partly of the diabatically produced PV when the air advected by anomaly “2” impinges on the baroclinic zone and ascends and partly of PV which appears in the region where QG forcing from upper and lower levels is favourable for cyclogenesis. The PV anomaly “3” is already located over a closed height contour indicating the incipient cyclogenesis. The closed height contour reaches into a weak wave in the surface temperature gradient. Thus a frontal wave-like new extratropical cyclone develops at Jangmi’s “warm front”.

In contrast to the previous time steps, when ascending trajectories were rather upright, the new trajectories follow the tilt of the $\Theta_e$ surface reflecting the dynamically forced ascending motion along the baroclinic zone (Figure 4.19c). Again higher values of PV in the mid troposphere indicate diabatic PV production during the moist isentropic ascent. It is striking that the end points of the new trajectories cluster mainly to the (south)east of the major PV anomalies (“2” and “3”) and that the trajectories are tilted eastward. This reflects the propagation along the baroclinic zone. It is also interesting, that trajectories ascending in the vicinity of the older anomaly “2” have already reached higher levels than trajectories ascending later close to the newer anomaly “3”. This reflects that the flow induced by the new PV anomalies continuously triggers new lifting further downstream and thus leads to an eastward propagation of PV anomalies along the baroclinic zone. However, some trajectories exhibit rather

(a) 00 UTC 29 Sep 2008 (b) **06 UTC** 30 Sep 2008

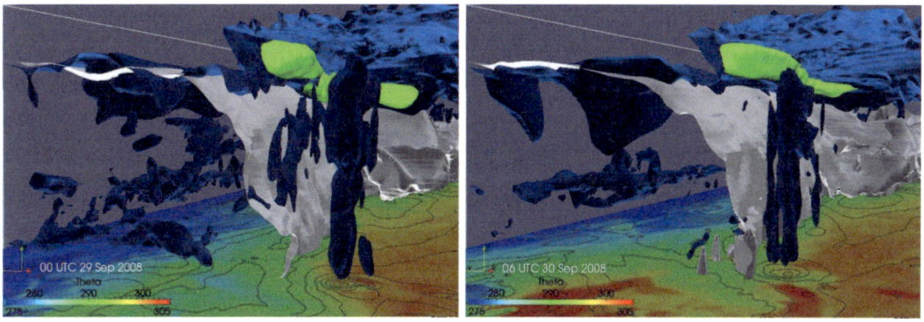


Figure 4.20: As Figure 4.18a, b but the view is from the west to the east through the data volume. In contrast to Figure 4.18 PV and $\Theta_e = 320$ K is shown starting at 120°E eastward, to highlight the baroclinic zone and the fault at the tropopause close to Jangmi.

low PV values during their entire ascent (e.g. black bundle in Figure 4.19c near label "2"). The air parcels here presumably move upward along the baroclinic zone without passing through a region of condensational heating.

At upper levels the trajectories which have risen in the vicinity of the centre of Jangmi during the previous two days (Figures 4.19a, b) end in the jet core. The jet is maintained by the continuous transport of low PV air to the jet level and to the southern edge of the tropopause step. The spatial and vertical extent of the jet core increased (cf. Figure 4.18b and Figure 4.18c).

During the next 24 hours cyclogenesis (10 hPa in 24 hours) and a rapid propagation along the baroclinic zone occurred linked to anomaly "3" (Figure 4.18d). Hardly any PV remnants of Jangmi are seen (at the location labelled "1"). Instead, a broad PV anomaly (labelled "4") extends through the low and mid troposphere. On the surface it is reflected in several closed height contours. This PV anomaly consists of remnants from anomaly "2" and "3" in its western part but mainly of new PV which was diabatically produced when air that was repeatedly advected by the flow induced by the PV anomalies "2" and "3" ascended at the baroclinic zone. The anomaly is located underneath the southwestern jet core (green) and ahead of a midlatitude trough, which is seen in the deep tropopause and steep tropopause step at upper levels to its northwest. The wave structure in the baroclinic zone and in the surface temperature gradient reveals that the new cyclone forms along a frontal wave. Warm air masses from Jangmi's warm sector became embedded in the new cyclone's warm sector. To the west a weak southward flow of cooler air marks the evolution of a cold front.

The trajectories show that major lifting occurred along the baroclinic zone during the past 24 hours (Figure 4.19d). They reflect the continuing process of propagation along the baroclinic zone by advection towards it, moist isentropic ascent, diabatic PV production and thus induction of advection further downstream. Similar to the trajectories which has ascended in Jangmi's inner core earlier (Figure 4.19a, b), the trajectories ascending at the baroclinic zone (Figure 4.19c, d) transport low PV air to jet level. In that way they sustain the jet streak and help to its eastward propagation.

### 4.3.1 Focus on the baroclinic zone

In this section the interaction of Jangmi with the baroclinic zone is analysed. Therefore it is investigated how air ascends and how PV is produced diabatically in the lower troposphere above the planetary boundary layer (Figure 4.22).

At 00 UTC 29 September 2008 (Figure 4.22a) high PV values and the closed isobars indicate Jangmi at 121°E, 26°N (labelled "1"). The trajectory intersection points show that most of the air parcels ascend in or in the vicinity of Jangmi's centre. Strong northward flow into the midlatitudes is evident to the east of Jangmi. To its north the baroclinic zone extends along 34°N, as seen in the dense isotherms of $\Theta_e$.

During the following 30 hours a transition occurs from convective ascent in the inner core of Jangmi to buoyant ascent ahead of the baroclinic zone in the region where unstable conditions were observed. To the northeast of Jangmi a deflection of the $\Theta_e$ isotherms (Figure 4.22b) results from the northward advection of tropical air in the evolving warm sector. West of Jangmi the baroclinic zone is perturbed by the southward advection of cooler air. The $\Theta_e$ gradient has increased significantly. The trajectory intersection points indicate continuing ascent in and around of Jangmi's inner core. At the southern edge of the baroclinic zone the new PV tower is reflected by a positive PV anomaly (at about 128°E, 32°N, labelled "2"). Slightly to the east of this region a cluster of trajectory intersection points indicates lifting in the new PV tower. There the southerly flow induced by the new PV tower advects air against the baroclinic zone resulting in ascent. Further to the east (129-135°E,29-34°N) southerly flow indicates the beginning of larger scale advection towards the the baroclinic zone.

A cross section through Jangmi, the new PV tower, the baroclinic zone and the midlatitude jet reveals the different paths of lifting to jet level (Figure 4.21). The high PV values in Figure 4.21a between 600-900 km belong to Jangmi's decaying PV tower. The wind vectors in the plane of the section show very strong buoyant ascent to the northeast of this tower at about 1000 km. The new narrow PV tower appears at 1100-1200 km with high PV values of more than 2.5 PVU (cf. Figure 4.18b, 4.19b). This is directly at the low-level edge of the baroclinic zone seen in the closely packed isotherms of $\Theta_e$ at the surface between 1200-1400 km (Figure 4.21a). As seen in the Falcon dropsondes (Figure 4.12) and reflected in the model data here the region ahead of the baroclinic zone is characterised by unstable stratification of $\Theta_e$ and thus exhibits favourable conditions for convective ascent. In the new PV tower the wind vectors show strong lifting to a level of about 300 hPa. Above, the vectors rather follow the (moist-)isentropes (which almost coincide with the dry isentropes at this level) and are directed into the midlatitude jet core. The jet core is located at about 2400 km, 200 hPa where a strong PV gradient occurs (Figure 4.21b). The isentropes become more closely packed towards the jet core. Assuming that the flow is stationary and along the (moist-)isentropes, mass continuity requires an acceleration where the isentropes are closer together. This is one mechanism explaining the increase in jet core wind speed. Along this path very moist low PV air (PV<-0.2 PVU) is transported to the tropopause (Figure 4.21a, b). The moist and low PV air follows the closely packed isentropes from about 1500 km, 500 hPa to the jet core. In this region the contours of wind speed across the plane of section (Figure 4.21b) show a southwest- and downward extension of high wind speeds, which indicates that the outflow of Jangmi and the new PV tower is directed into the jet core. The very low PV values indicate small stability or even instability to slantwise ascent so that the flow may further accelerate towards the jet due to symmetric instability (Thorpe et al. 1989). A dry anomaly is located within the baroclinic zone at around 1600 km, 850 hPa. Because of the low specific humidity, $\Theta_e$ also has a minimum. The reasons for the dry anomaly remain unclear. It is conspicuous that the anomaly appears where the baroclinic zone hits the coast of South Korea.

In summary, the cross section revealed three distinct paths of ascent to jet level. Initially buoyant convective ascent occurs in the TC inner core. When the storm approaches the midlatitudes air rises convectively in unstable stratified tropical air ahead of the baroclinic zone. Further the cyclonic TC circulation and the cyclonic flow induced from the new PV anomalies advects (sub-)tropical towards the baroclinic zone. Dynamically forced slantwise ascent occurs along the tilted (moist-)isentropes when the air impinges on the baroclinic zone.

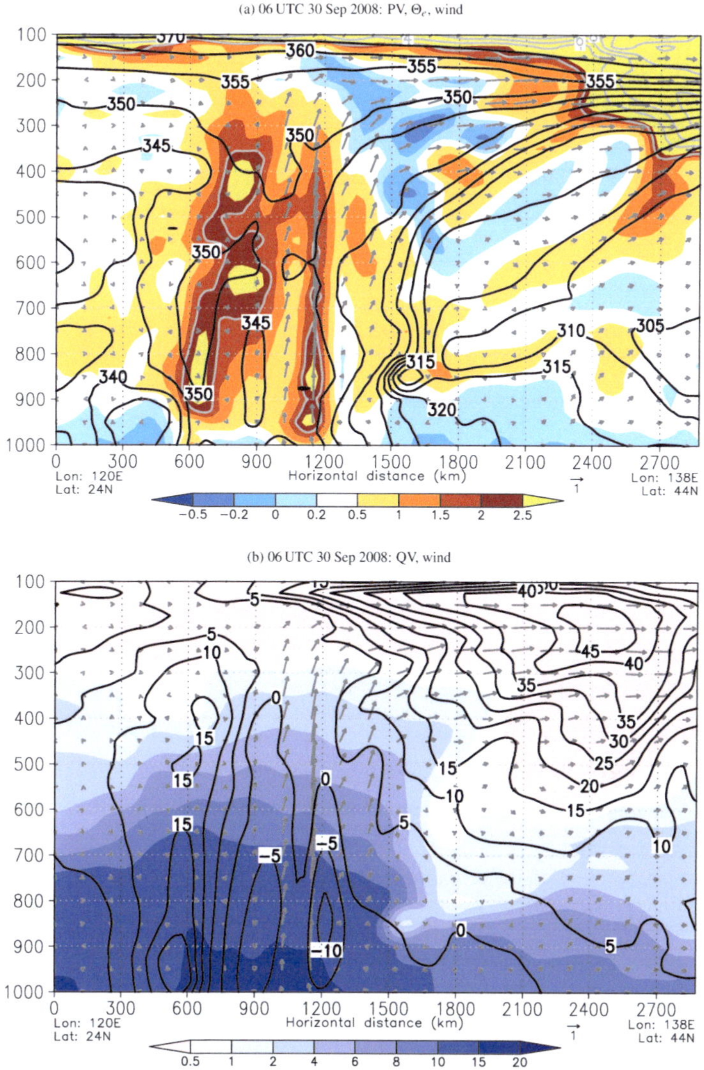


Figure 4.21: Vertical cross sections from 120°E, 24°N to 138°E, 44°N at 06 UTC 30 September 2008. Left: potential vorticity (shaded), equivalent potential temperature (black with a 3 K contour interval, 5 K for $\Theta_e > 345\,K$), and wind in the plane of section (grey arrows). Right: specific humidity (shaded in $\mathrm{g\,kg^{-1}}$), wind across the plane of section with positive values coming out of the section (black with a $5\ \mathrm{m\,s^{-1}}$ contour interval), and wind in the plane of section (grey arrows). The location of this cross section is marked in Figure 4.22b as a red line. Note different colours are used for PV. Data from ECMWF analysis.

Six hours later the new PV tower and Jangmi form one larger region of high PV values (Figure 4.22c). East of this region the southerly flow impinges on the baroclinic zone. This results in new upright ascent ahead of and slantwise ascent along the baroclinic zone in addition to the ongoing ascent in the region of Jangmi and the new PV tower (see intersection points in Figure 4.22c). The new ascent further east is more linearly aligned (130°E, 30°N to 135°E, 33°N), reflecting the more dynamically forced ascent along the baroclinic zone. As the air parcels rise below the level of maximum heating PV increases. A new positive PV anomaly appears along the baroclinic zone at about 136°E, 33°N (labelled "3"). The flow induced by this anomaly results likewise in poleward transport of tropical air towards the baroclinic zone further downstream.

The sustained transport of tropical air towards the baroclinic zone, by the flow induced by positive PV anomalies triggers a process of diabatic Rossby wave-like propagation along the baroclinic zone (Parker and Thorpe 1995; Böttcher and Wernli 2011). The cyclonic circulation induced by a positive PV anomaly advects air poleward on its eastern side towards the baroclinic zone. When the air parcels impinge on the baroclinic zone convective ascent and (moist-)isentropic upgliding occurs. Thereby new PV is produced diabatically at the baroclinic zone where condensation occurs in the ascending air. This new positive PV anomaly in turn induces its own circulation and the process is repeated. In this manner PV anomalies can propagate quickly along the baroclinic zone. This process is referred to as diabatic Rossby wave propagation (Parker and Thorpe 1995). In the case of Jangmi there is strong evidence of such a behaviour during the interaction with the baroclinic zone. In contrast to diabatic Rossby wave cases in the literature (e.g. Böttcher and Wernli 2011), the PV anomalies are not restricted to low levels, and cyclogenesis occurs at the same time as the propagation.

At 18 UTC 30 September the new PV anomaly has moved to about 139°E, 33°N which corresponds to a propagation speed of about 13 m s$^{-1}$ (Figure 4.22d). Ascent continues at the southern edge of the anomaly, but more important trajectory intersection points are concentrated at the baroclinic zone directly to the east of the PV anomaly (at about 142°E, 34°N) where it induces southerly flow

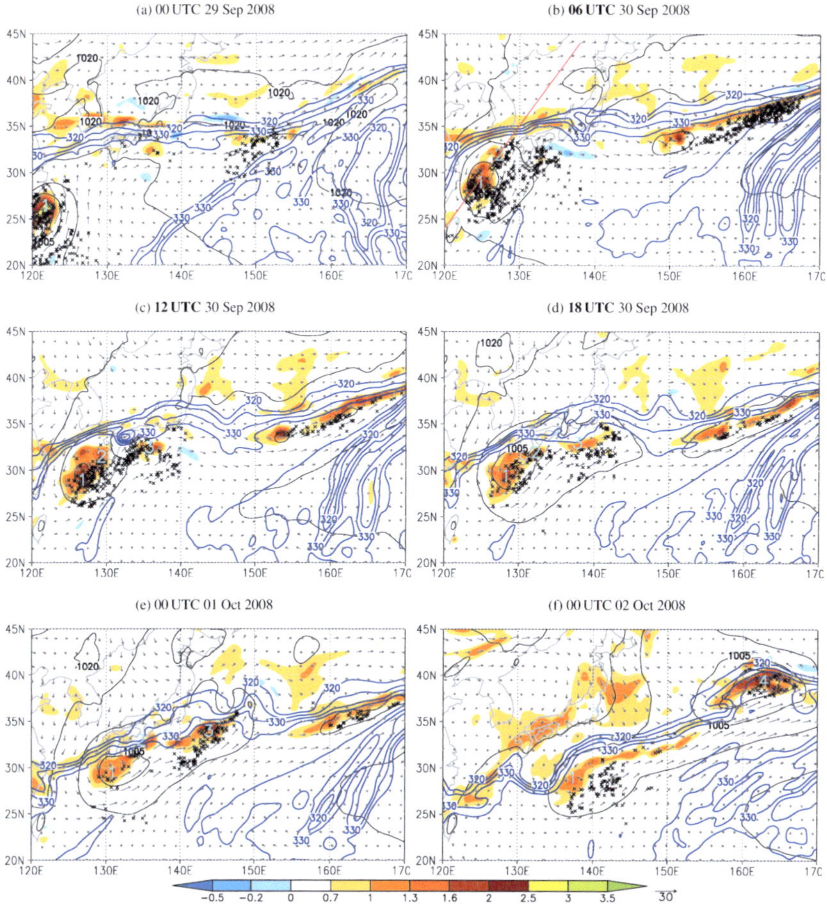


Figure 4.22: Vertically averaged fields in a layer from 850-750 hPa: low-level potential vorticity (shaded, in PVU), horizontal wind vectors (grey), equivalent potential temperature (blue with a 5 K contour interval for $320 \leq \Theta_e \leq 335$ K), and pmsl (black with a 5 hPa contour interval). The crosses mark every 50th location where a trajectory intersects the pressure layer 850-750 hPa. The trajectory data used is the set of 6 hourly 48 hour forward trajectory calculations introduced in Section 4.1.2. Data taken from ECMWF analysis. The number labels refer to the systems introduced in Figures 4.18, 4.19. The red line in (b) marks the location of the cross section in Figure 4.21. The region shown focuses on the baroclinic zone and is the same as in Figure 4.16.

towards the baroclinic zone and thus sustains the propagation process. Jangmi (labelled "1") continues to weaken. The smaller number of intersection points indicates the reduced ascent in its inner core. The new PV tower (labelled "2") now separates from Jangmi as it produces another convective burst (cf. Figure 4.5n or Figure 4.16e).

In the next 6 hours the new PV anomaly and associated ascent at the baroclinic zone intensify (Figure 4.22e). The anomaly has propagated to about 144°E, 34°N, with an increase of the propagation speed to about 21 $m\,s^{-1}$. The transport of tropical air towards the baroclinic zone by a southwesterly flow continues and subsequent ascent (as reflected by the intersection points) at the baroclinic zone remains concentrated east of the anomaly. Jangmi is weakening further and the remnant PV of the new PV tower (labelled "2") becomes aligned along the baroclinic zone. The almost complete absence of trajectory intersection points in the inner core of Jangmi show that the transition from predominant buoyant ascent in the TC inner core to predominant buoyant ascent ahead and dynamically forced slantwise ascent along the baroclinic zone is completed.

During the next 24 hours a midlatitude cyclogenesis occurs (labelled "4" in Figure 4.22f) in association with the PV anomaly at the baroclinic zone. The new cyclone has propagated fast along the baroclinic zone with a propagation speed of about 19 $m\,s^{-1}$ to 162°E, 39°N. The deflection of the $\Theta_e$ isotherms reveals the frontal wave character of this new cyclone. Ascent, indicated by the trajectory intersection points, continues predominantly in the eastern half of the PV anomaly ("4") at the baroclinic zone. Jangmi and the new PV tower almost have decayed, but east of the PV remnants trajectory intersection points show lifting linked to the ongoing convective bursts (cf. Figure 4.5r).

In summary, the foregoing analysis revealed the transition from buoyant ascent and associated diabatic PV production in the inner core of Jangmi to upright ascent ahead and dynamically forced slantwise along a midlatitude baroclinic zone. In the first phase of ET (until about 00 UTC 30 September 2008) Jangmi is located south of the baroclinic zone and convectively enhanced lifting is concentrated within its inner core. In the second phase (until about 18 UTC 30 September) Jangmi interacts with the baroclinic zone and gradually weakens. Upright, buoyant ascent occurs in the storm's inner core and in convection in an unstable air mass to the northeast of Jangmi ahead of the baroclinic zone. Dynamically forced slantwise motion along the tilted (moist-)isentropes occurs when tropical air advected by the low-level cyclonic TC circulation impinges on the baroclinic zone. In the last phase, new PV anomalies at the baroclinic zone sustain the poleward advection of tropical air towards it and thus trigger a propagation along the baroclinic zone. This results in the fast DRW-like propagation of a new midlatitude cyclone.

The representative trajectory calculations introduced in Section 4.3 (Figure 4.19, Table 4.3) enable us to analyse this transition process quantitatively. The three trajectory calculations started at 12 UTC 28 September 2008, 12 UTC 29 September 2008, and 12 UTC 30 September 2008 correspond to the lifting during the three phases of ET. The last trajectory calculation, started at 12 UTC 01 October 2008, mainly reflects the lifting in the new midlatitude cyclone and is therefore omitted.

The first set of trajectories (Figure 4.23a) shows that lifting occurs almost vertically in Jangmi's inner core (bundle of trajectories labelled "a") and in the new PV tower to its northeast (labelled "b"). The air parcels start with low PV values at low-levels. During ascent their PV increases strongly, especially in the low and mid troposphere within Jangmi's inner core and the new PV tower. When reaching the upper levels the PV of the air parcels decreases, reaching their values of the lower troposphere or below. At jet level only very weak further ascent occurs while the parcels are advected eastward with the jet. Thus low PV air out of Jangmi's environment in the lower troposphere has been transported to jet level. When Jangmi and the new PV tower are closer to the baroclinic zone (Figure 4.23b) tropical air advected east of the PV anomaly impinges directly on the baroclinic zone. The air parcels enter a region of dynamically forced ascent and partly move along the sloping isentropes. The ascent along the baroclinic zone is reflected by the more tilted trajectories following the moist isentropic surface especially at upper levels. Most of the air parcels start with low PV values at low levels, the PV increases below the level of maximum heating and decreases to the lower tropospheric values or below at jet level. New positive PV anomalies appear at the baroclinic zone at the location of these trajectories (the small PV tower in the centre of Figure 4.23b belongs to another system, which dissipates in the following 24 hours). However, some air parcels exhibit almost constant low PV during their ascent (e.g. black bundle slightly east of "b" in

Figure 4.23b). These air parcels do not move through a region of condensational heating and thus lack of diabatic PV production. During the last phase of ET, trajectories indicate that lifting is almost entirely slantwise along the baroclinic zone (Figure 4.23c). Several new PV anomalies appear at the baroclinic zone.

A more quantitative and statistical analysis of the three trajectory calculations is given by the temporal evolution of average, standard deviation, minimum, 25% percentile, median, 75% percentile, and maximum of PV, diabatic heating rate (DH), $\Theta_e$, and pressure along the trajectories (Figure 4.24). More than ten thousand trajectories in each calculation (Table 4.3) allow us to conduct this statistical analysis. By definition the trajectories represent ascending air masses. Thus the temporal sequences correspond to approximate profiles of the variables from the boundary layer to the upper troposphere. In the following these sequences or profiles are described by means of the temporal evolution of the average as this is qualitatively similar to the evolution of the percentiles, extremes and standard deviation.

In the first 12 hours of trajectory calculation 1 (representing the ascent in and in the vicinity of Jangmi cf. Figure 4.23a) average PV increases strongly from about 0.3 PVU to 1.0 PVU (Figure 4.24a). The diabatic heating rate increases from 2 to 10 K (6 hours)$^{-1}$ (Figure 4.24d), thus the increase in PV occurs below the level of maximum diabatic heating.

In the same period, the air parcels ascend moderately from an average level of 940 hPa to 800 hPa (Figure 4.24j). During the next 12 hours the moderate average ascent continues to about 620 hPa while PV remains almost constant at 1.0 PVU, although diabatic heating slightly decreases. However, diabatic heating has a second maximum of about 14 K(6 hours)$^{-1}$ at 18 UTC 29 September (hour 30) at an average pressure height of 500 hPa. This peak reflects the diabatic PV production in the new PV tower that develops at the baroclinic zone. Thus some trajectories are below and some are above their level of maximum diabatic heating and consequently the average PV remains constant. Also higher PV air from lower levels is partly advected vertically. While at 12 UTC 29 September (hour 24) the 25% and 75% percentiles of pressure show a wide range from 850 hPa to 420 hPa this range decreases to 460 hPa to 240 hPa in

(a) 12 UTC 29 September 2008, trajectories 28/12Z → 30/12Z

(b) 12 UTC 30 September 2008, trajectories 29/12Z → 01/12Z

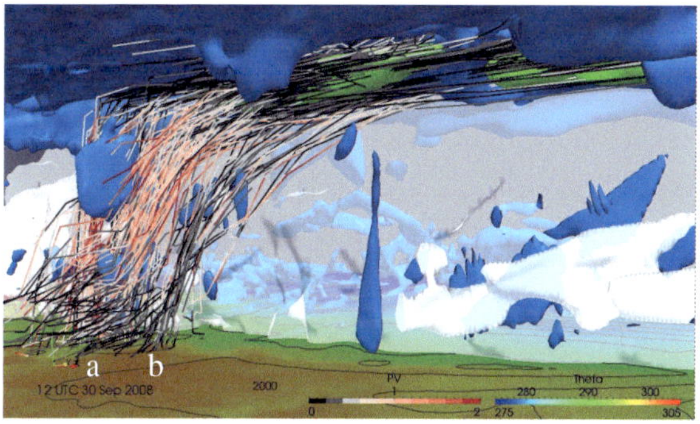


(c) 12 UTC 01 October 2008, trajectories 30/12Z → 02/12Z

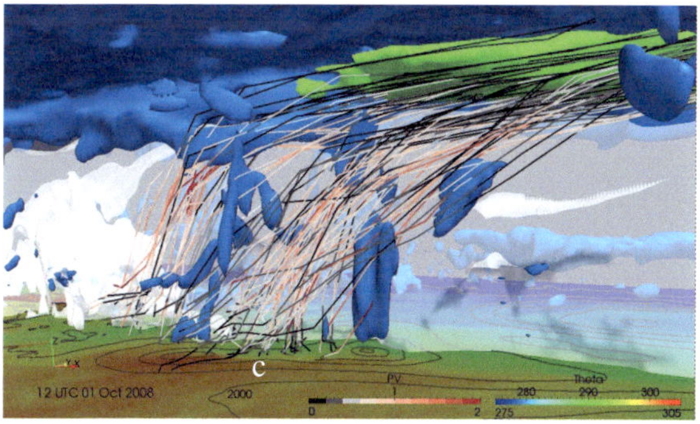


Figure 4.23: As Figure 4.18 but zoom on the baroclinic zone at the times indicated above each panel. The times are centred around three of the four representative 48 hour forward trajectory calculations (starting → end times of the calculation also indicated above each panel) introduced in Section 4.3. Every 100d trajectory is shown for the full 48 hour calculation period. See caption of Figure 4.18 and Figure 4.19 for more details of the plot and trajectory calculations.

the following 12 hours indicating a rapid ascent from an average level of 620 hPa to 360 hPa. This reflects partly the ascent in the upper part of Jangmi's PV tower and mainly the strong ascent from low to upper levels in the new PV tower which has developed in exactly that period (see the two bundles of trajectories close to Jangmi's and the new PV tower in Figure 4.23a). At the beginning of this period rather high average PV of 1.0 PVU, reflects the diabatic PV production at lower levels in the new PV tower, after 18 UTC 29 September (hour 30) the mean PV starts to decrease. PV then continues to decrease above the region of maximum diabatic heating to low values of 0.2 PVU. All air parcels reach the jet level at an average pressure level of 220 hPa (maximum 350 hPa, minimum 80 hPa). In the last six hours of the calculation almost moist isentropic upgliding (average $\Theta_e \approx 351$ K) of on average 50 hPa occurs (Figure 4.24g, j) reflected in the trajectories that are slanted in the upper levels rather than upright (Figure 4.23a).

Interesting changes occur when Jangmi approaches the baroclinic zone (Figure 4.24e-h cf. Figure 4.23a). The increase of average PV is less marked, increasing from initial values of 0.3 PVU at 930 hPa to 0.7 PVU at 750 hPa after 12 hours (Figure 4.24b). Also ascent occurs more rapidly. The air parcels reach an average height of 530 hPa after 24 hours (Figure 4.24k, compared to 620 hPa in the previous calculation). Once reaching this level, the air parcels ascend on average moist isentropically ($\Theta_e \approx$345 K) to jet level (Figure 4.24h). More trajectories exhibit a slantwise ascent along the baroclinic zone (Figure 4.23b) rather than an upright ascent in the PV towers as in calculation 1 (Figure 4.23a). In the last six hours of the calculation $\Theta_e \approx 345$ K is almost equal to $\Theta \approx 344$ K. Thus the air moves almost adiabatically. This is consistent with the almost constant value of PV which has decreased to low values of on average 0.25 PVU at jet level. Similar to the previous calculation the diabatic heating shows a layer of high heating rates between 870 hPa and 420 hPa (Figure 4.24e). Highest values of PV occur in that layer while above PV decreases.

The statistics for trajectory calculation 3, representing the ascent along the baroclinic zone (Figure 4.23c), reveal that ascent occurs much faster in the lower troposphere with the air parcels reaching 490 hPa after the first 24 hours (Fig-

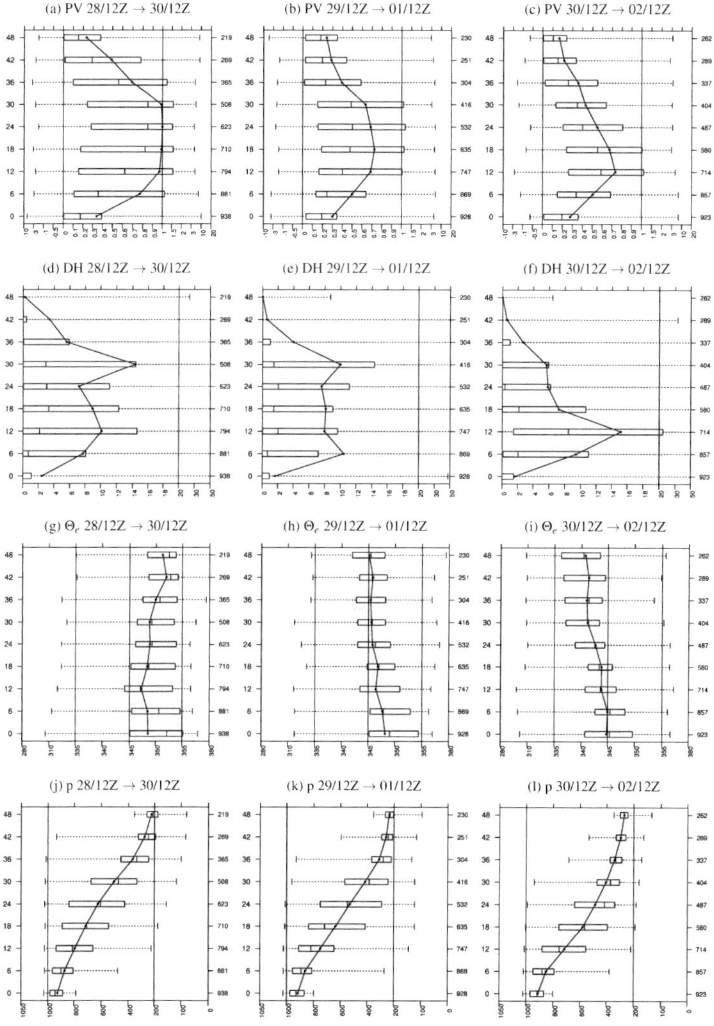


Figure 4.24: Temporal evolution of statistical properties of specific variables along the first three representative trajectory calculations (Table 4.3). Start → end times of the calculations are shown above each panel. PV (in PVU), (a),(b),(c), diabatic heating rate (in K/(6 hours)), (d),(e),(f), $\Theta_e$ (in K) (g),(h),(i), and pressure (in hPa) (j),(k),(l) traced along the trajectories. The boxplots show minimum, 25%, 50% (Median), 75% percentiles, and maximum. The black solid line marks the average and the grey shading the average ± one standard deviation. Note that the abscissa is expanded between the two vertical lines and becomes irregular outside. The ordinate shows time after the initialisation of the calculation. On the right of the figure the ordinate is labelled with the pressure averaged over the trajectories at a given time.

ure 4.24l, compared to 530 hPa in calculation 2 and 630 hPa in calculation 1). This is possibly due to a more coherent and synchronous slantwise ascent along the baroclinic zone reflected in the reduced standard deviation for pressure compared to calculation 1 and 2 (grey shading in Figure 4.24l cf. Figure 4.24j and Figure 4.24k). The average PV maximum of 0.7 PVU occurs earlier (Figure 4.24c after 12 hours) and at a lower average level of 710 hPa (see the new low to mid-level PV anomaly in Figure 4.23c). This is also the level of a maximum in diabatic heating (Figure 4.24f, 14 K(6 hours)$^{-1}$). Above this peak, PV starts to decrease. Periods of almost moist isentropic ascent (constant average $\Theta_e$) occur in the first 24 hours of the calculation (Figure 4.24i, $0 \rightarrow 6$ and $12 \rightarrow 18$ hours). After 30 hours (18 UTC 01 October 2008) air ascends moist isentropically at average $\Theta_e = 341$ K. In the last 6 hours PV remains almost constant as the air is ascending slowly and adiabatically ($\Theta_e \approx 341\ \mathrm{K} \approx \Theta$) from an average pressure of 290 to 260 hPa. The air parcels reach a lower height compared to the previous calculations, but nevertheless they reach the lower jet level.

In summary, the statistical analysis of PV and $\Theta_e$ along the representative trajectories proved two important physical processes: Firstly, diabatic PV production occurs at low and mid levels in ascending air masses in the PV towers linked to Jangmi and at the baroclinic zone. This is followed by diabatic PV reduction in upper levels above the layer of maximum diabatic heating. And secondly, a transition occurs from buoyant upright ascent to jet level in the PV towers (Figure 4.23a) to dynamically forced moist isentropic slantwise ascent to jet level along the baroclinic zone during the later stages of ET (Figure 4.23c).

### 4.3.2 Focus on the midlatitude jet

During the ET of Jangmi a distinct outflow-jet interaction was observed which resulted in a strong acceleration of the jet core wind speed to more than 80m s$^{-1}$ at 00 UTC 01 October 2008 (cf. Section 4.1.2, Figure 4.5m). The analysis so far showed that physical processes at the baroclinic zone are crucial for that interaction of Jangmi and the midlatitudes. In this section the analysis is concluded by having a closer look at the evolution of the midlatitude jet.

The origin of the air in the jet core is investigated by backward trajectory calculations that end in a box at jet level (100-179°E, 20-60°N, 8000-15000 m above mean sea level) at 00 UTC 01 October 2008 and started 48 hours earlier. The trajectories fulfil the criterion of ascent in excess of 8500 m during the 48 hour period. Almost all air parcels, that ascend in that time period start in the environment of Jangmi, are either lifted in Jangmi or at the baroclinic zone, and end in the vicinity of the jet core (Figure 4.25). Thus this backward calculation is comparable to the forward calculation 2, which captured the trajectories that ascend from low levels in the vicinity of Jangmi and at the baroclinic zone. The PV traced along the trajectories indicates that low PV air is transported to jet level (black shades in Figure 4.25). The northernmost trajectories end directly in the jet core ($|\vec{v}| > 50\,\mathrm{m\,s^{-1}}$ marked by green shading) while the southernmost trajectories move around the subtropical anticyclone to the east of Jangmi.

A more detailed look at the transport of low PV air to the jet is given in Figure 4.26. The trajectories end with low PV air in and south of the jet core (green

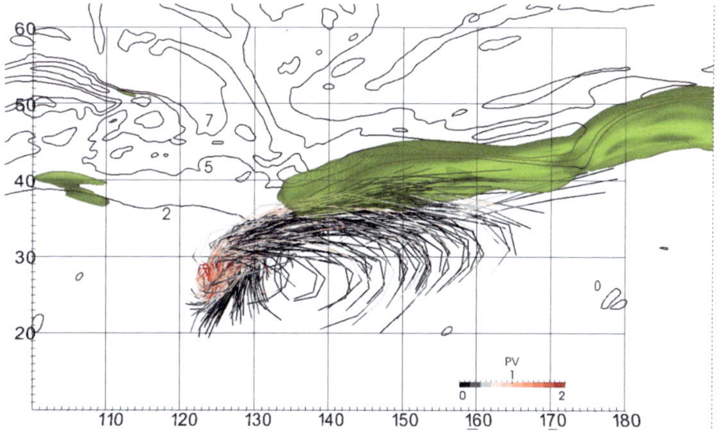


Figure 4.25: Trajectories ending in the box 100-179°E, 20-60°N, 8000-15000 m above mean sea level at 00 UTC 01 October 2008, started at 00 UTC 29 September 2008 and calculated backward for a 48 hour period. The air parcels fulfil the criterion to ascent more than 8500 m in the 48 hour period. The colours indicate PV along the trajectory. PV below 0.6 PVU is shown in grey colours, above in red colours. Every 200d trajectory is shown. The green transparent shading shows the threedimensional surface of horizontal wind speed $|\vec{v}| > 50\,\mathrm{m\,s^{-1}}$. The black contours show the 2, 5, and 7 PVU contour at the isentropic level of $\Theta = 340\,\mathrm{K}$. View is from top and centred on the domain in the threedimensional cube. Please note, that the grid refers to the level of about 400 hPa and is due to optical distortion only an estimate for the meteorological fields shown. Date shown is 00 UTC 01 October 2008. To be compared with Figure 4.5m.

transparent surface in Figure 4.26a). The air mostly ascended vertically in the centre of Jangmi and the new PV tower, and partly slantwise along the baroclinic zone (cf. trajectory calculation 2, discussed before). The jet is located at the strong tropopause step to the north of Jangmi (reflected in the blue 1.5 PVU surface in Figure 4.26b). The trajectories reveal that the low PV air is transpor-

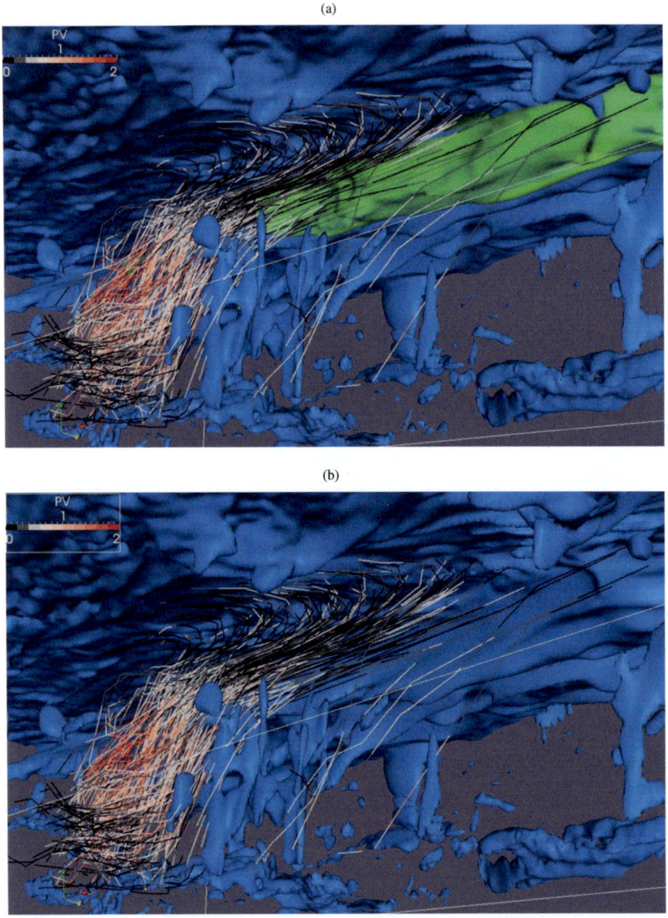


Figure 4.26: Threedimensional view from the southeast corner in the jet core region. Trajectories are the same as in Figure 4.25 and shaded by PV (in PVU). The 1.5 PVU PV surface is shown in blue shading. The green transparent shading (trajectories ending in the jet appear pale) in (a) illustrates the midlatitude jet ($|\vec{v}| > 50\,\mathrm{m\,s^{-1}}$). In (b) the jet is omitted to give a clearer view of the dynamic tropopause and the tropopause step. ECMWF analysis shown at 00 UTC 01 October 2008.

ted directly beneath the dynamic tropopause (also represented by the 1.5 PVU surface) and to the south of the tropopause step, resulting in a lifting of the tropopause. This is consistent with the occurrence of the low PV region in the cross section discussed earlier (Figure 4.21) and similar to the "PV wall" and tropopause fault described by Atallah and Bosart (2003) for the ET of Hurricane Floyd (1999). The statistical analysis of the evolution of PV along the representative trajectories confirms that low PV air is transported to jet level (Figure 4.27). As in calculation 2 the highest PV values occur in the lower half of the troposphere in the layer of maximum diabatic heating (-42 to -24 hours). At jet level (timestep 0, average pressure 210 hPa, minimum 130 hPa, maximum 350 hPa) average PV is very low at about 0.2 PVU. As a consequence of the transport of low PV air the PV and temperature gradient at the tropopause and tropopause step increases, resulting in an acceleration of the jet. In the next section, the impact of Jangmi on this acceleration will be quantified using PV inversion techniques.

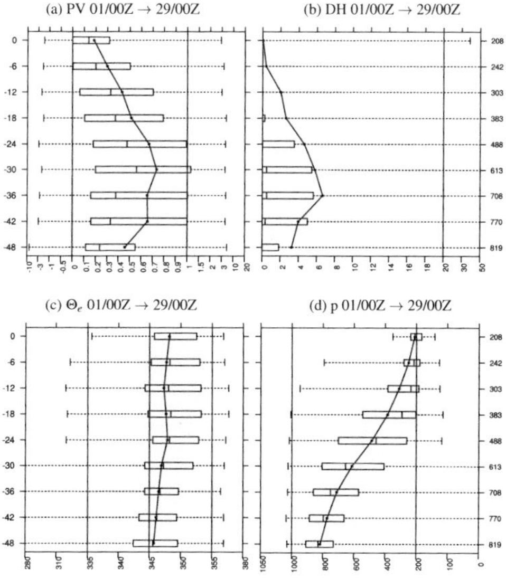


Figure 4.27: As Figure 4.24 but for the backward trajectories used in Section 4.3.2.

## 4.4 Quantification of the impact of Jangmi on the midlatitudes

The foregoing detailed investigation based on analysis and observational data revealed important interactions between Typhoon Jangmi undergoing ET and the midlatitude flow. This is most prominently manifested in the advection of warm moist air at low levels by the cyclonic circulation of the TC towards a baroclinic zone, subsequent ascent, transport of low PV air to upper levels, and acceleration of the midlatitude jet. However, the quantitative contribution of Jangmi to this midlatitude evolution remains unclear. In this section, the impact of Jangmi on the midlatitude jet stream and on the downstream flow evolution will be quantified with the help of a PV surgery experiment. The technique outlined in Section 3.2.4 is designed to remove a mature TC cleanly from an analysis field, while leaving the larger-scale environment unchanged. We apply this method to remove Jangmi from the initial conditions of a COSMO simulation. The role of Jangmi in the evolution of the midlatitude flow is quantified by comparing the synoptic evolution in this model run against a control simulation.

### 4.4.1 Setup of the control and no-TC experiments

Two COSMO simulations are computed, both with the same configuration except for the initial conditions. Six-hourly ECWMF analysis (with 0.25° horizontal resolution, experiment rd f7m4 see Section 3.1.1, named ECMWF analysis in the remainder) serve as boundary conditions. The model runs start at 00 UTC 28 September 2008, which is a few hours prior to landfall in Taiwan. Jangmi is located at about 123°W, 23°N and had reached peak intensity in the previous 24 hours (cf. Section 4.1.2). At this time an impact of Jangmi on the midlatitudes is not yet evident (cf. discussion in Section 3.2.5). The simulations run for 150 hours until 06 UTC 04 October 2008. The horizontal domain covers the entire North Pacific. Information on the COSMO model runs are summarised in the Appendix A in Table A.1. The control simulation (CNTRL=S20) is initialised from the unmodified ECMWF analysis. The positive PV anomaly linked with Typhoon Jangmi and the associated anomalies in the meteorological fields have been removed in the initial conditions for the no-TC simulation (NOTC=N22) using the PV surgery technique (Section 3.2.4). Details of the PV inversion and the anomaly definition are summarised in the Appendix, Table A.2 (cf. steps in Section 3.2.4).

### 4.4.2 Verification of the control run

It is difficult to disentangle the flow contributions from the TC and from the midlatitudes during the interaction of the TC undergoing ET with the midlatitude flow. Therefore in this study, the extratropical flow contribution is represented by a COSMO simulation (NOTC) in which the storm had been removed from the initial conditions using the PV surgery method Section 3.2.4 prior to ET.

The temporal evolution of the meteorological fields in a COSMO simulation is determined by the COSMO model physics, dynamics and numerics. Thus, a quantitative comparison of the COSMO simulation with modified initial conditions (e.g. the no-TC run) against the total flow evolution requires a reference simulation (CNTRL), which uses the same model configuration except that it is initialised from the unmodified analysis. Nevertheless, for this study it is very important that the control simulation represents the direct impact of the ET on the midlatitudes close to the analysis. The most prominent features during the ET of Jangmi are:

- Following landfall in Taiwan, Jangmi started to recurve, transitioned into an extratropical system and weakened.
- During recurvature, an important interaction of the outflow of Typhoon Jangmi and the midlatitude jet occurred.
- An enhanced flow between the subtropical anticyclone east of Jangmi and a TUTT cell was evident.
- After recurvature, Jangmi underwent the transformation stage of ET then decayed and lacked reintensification.
- Slightly downstream of Jangmi an extratropical cyclone developed. With respect to the direct impact of Jangmi on the midlatitudes, this is of secondary interest.

Obviously, due to the extended time period (>6 days) in which these events occur, it can not be expected that a COSMO simulation represents the entire evolution accurately. However, choosing the initial time 00 UTC 28 September 2008 and using six hourly analysis data as boundary conditions allows the direct impact of Jangmi on the midlatitude flow to be simulated accurately. As will be shown in the following, the main limitation in the representativeness of the CNTRL run is in the evolution of the downstream cyclone. This cyclone develops in the forecast time range beyond 96 hours (4 days) and may show sensitivity to other factors beside the ET system. Since the downstream cyclone is only of secondary interest, the CNTRL run is found to be suitable as a reference simulation for the investigation of the impact of Jangmi on the midlatitudes and on the outflow jet interaction.

The control simulation starts at 00 UTC 28 September 2008 shortly before landfall in Taiwan. In the first 24 hours Jangmi follows a track similar to the track in the analysis data (Figure 4.28) and the pressure at mean sea level in the centre of Jangmi increases from 974 hPa to 993 hPa (Figure 4.29) which is almost identical to the increase in the analysis (994 hPa). During the subsequent recurvature Jangmi lags slightly behind the analysed track. This may be explained by a little different background flow in the control simulation. At 00 UTC 01 October 2008 the track of Jangmi in the control simulation is slightly north of the analysed track. Due to the slower propagation, Jangmi is located closer to the midlatitude trough that approaches from the west (cf. Figures 4.5k,m and Figures 4.37a,c) and a weak reintensification occurs (from 1001 hPa at 12 UTC 30 September 2008 to 996 hPa at 00 UTC 01 October 2008, Figure 4.29). In the subsequent days Jangmi decays in both the CNTRL run and the analysis. The minimum pmsl increases to more than 1005 hPa (Figure 4.29) and Jangmi heads somewhat back to the tropics (Figure 4.28). In summary, the comparison of the track and minimum pmsl in the control simulation and the analysis showed that Jangmi's recurvature and decay after landfall in Taiwan are represented accurately in CNTRL.

The track (Figure 4.28) and minimum central pressure (Figure 4.29) of the extratropical cyclone downstream of Jangmi indicate that the cyclogenesis is fore-

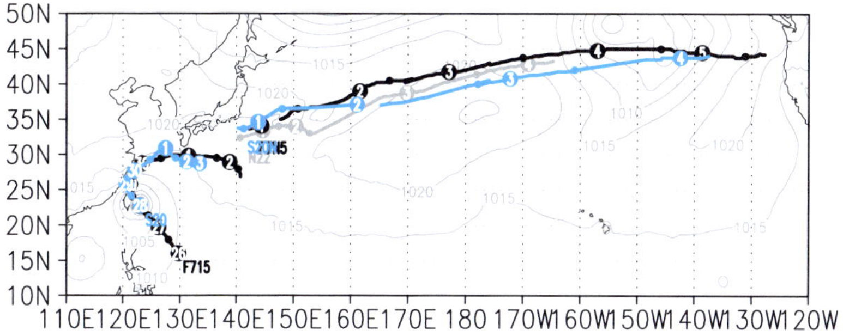


Figure 4.28: Track of Jangmi/the downstream cyclone in CNTRL simulation (blue, labelled S20/S20N), starting at 00 UTC 28 September 2008 and in the ECMWF analysis (black, labelled F715/F7N5) starting at 00 UTC 26 September 2008. Track of the downstream cyclone in the NOTC simulation starting at 06 UTC 30 September (grey, labelled N22). Thick points, labelled with the day of September/October 2008 mark 00 UTC times, thin points mark 12 UTC times. pmsl from ECMWF analysis shown at 00 UTC 28 September 2008 (grey contours).

cast to occur at about the correct location and time in CNTRL. However, the control simulation fails to predict the deepening after 00 UTC 01 October 2008 and the downstream cyclone in CNTRL propagates much faster than in the analysis. Hence the control simulation provides only a limited representation of the impact of Jangmi on the downstream cyclogenesis.

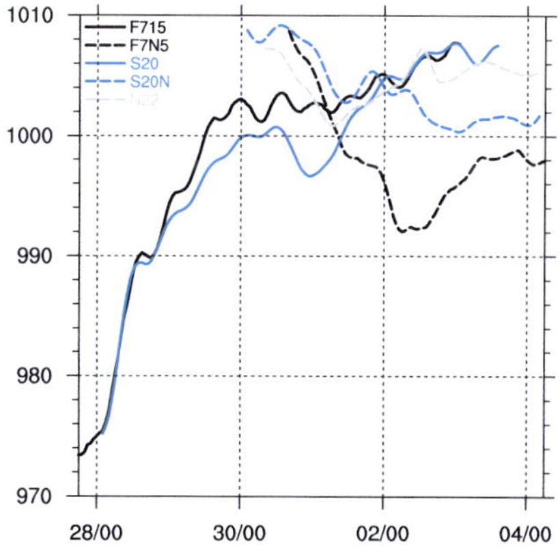


Figure 4.29: Minimum pmsl (in hPa) in the centre of Jangmi (solid) / of the downstream cyclone (dashed) in ECMWF analysis (black, labelled F715/F7N5), CNTRL (blue, labelled S20/S20N) and NOTC (grey, labelled -/N22).

A good representation of Jangmi's ET and direct impact on the midlatitudes in the control simulation is reflected in the high anomaly correlation coefficient of geopotential at 500 hPa (*ACC*, blue in Figure 4.30). The *ACC* is well above 0.6 - the threshold for a useful simulation (Persson and Grazzini 2007) - through the entire calculation period. During the first 48 hours of simulation the *ACC* is above 0.99 and during the outflow-jet interaction (on 30 September 2008) it remains above 0.95. The control simulation degrades to $ACC = 0.85$ until 00 UTC 4 October during the cyclogenesis of the

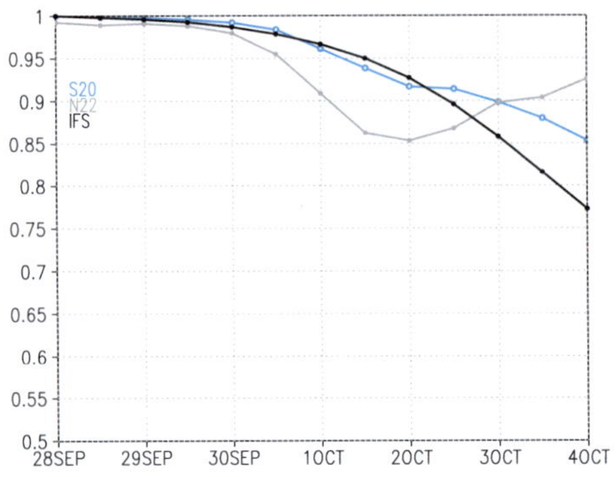


Figure 4.30: Anomaly correlation coefficient (ACC) of 500 hPa geopotential calculated on the subdomain 100-245°E, 20-60°N every 12 hours. Data shown for the control simulation (blue) and the no-TC simulation (grey). The ECMWF analysis is used as the analysis data and the daily mean values of ERA-INTERIM interpolated on a 0.25°x0.25° horizontal grid as the climatology (cf. Section 3.3.3). Additionally the mean ACC for all deterministic ECMWF IFS forecasts during 1 August 2008 to 1 November 2008 calculated for the domain 110°E-300°E, 20°N-60°N is shown in black.

downstream cyclone. Until 2 October (+120 h) the *ACC* of the control run is almost identical to the mean ACC in the deterministic IFS forecast. After 3 October 2008 (+144 h) the COSMO CNTRL $ACC \geq 0.85$ characterises a very good simulation compared to the mean IFS-ACC.

The representation of the outflow-jet interaction in the control simulation is verified using the jet-level wind speed (Figure 4.31). After 12 hours, the wind speed at 200 hPa differs only slightly in the region of the midlatitude jet (Figure 4.31a,c). Most of the differences occur in the region of the TC outflow at around 128°E, 30°N to the north and east of Jangmi (Figure 4.31c). Although the control simulation captures the outflow at around 128°E, 30°N (Figure 4.31a), it is weaker than in the analysis (cf. 15 m s$^{-1}$ contour from analysis (dashed) and CNTRL). At 12 UTC 30 September 2008, when the outflow-jet interaction occurs, differences in the region of Jangmi are mainly confined to the southern side of the jet streak over Japan (Figure 4.31d). The largest differences are found near the second jet streak over the Pacific and near the flow around the subtropical anticyclone south of Japan. In the analysis (dashed contours in Figure 4.31b), the second jet streak is more distinctly separated from the jet streak over Japan by a trough near 160°E than in CNTRL (Figure 4.31b). The flow around the subtropical anticyclone is weaker in the control simulation (cf. 15 m s$^{-1}$ contour from analysis (dashed) and CNTRL in Figure 4.31b) but it captures the enhanced flow between the anticyclone and the TUTT cell, centred at about 153°E, 23°N. Thus there is an overall good representation of

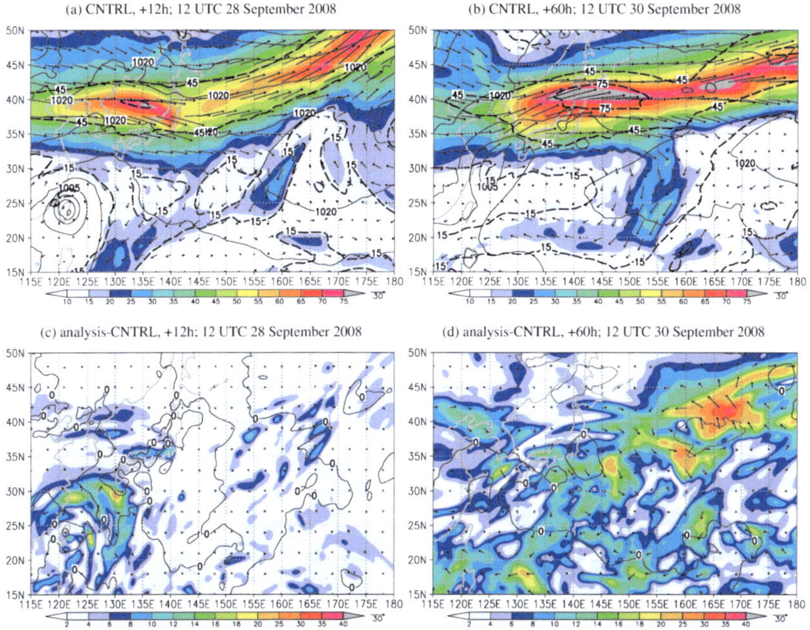


Figure 4.31: (a),(b) wind vectors (black), wind speed (shaded, in m s$^{-1}$) at 200hPa, pmsl (black contours, every 5 hPa) from CNTRL and contours of 200 hPa wind speed (black dashed, 15,45,75 m s$^{-1}$) in ECMWF analysis. (c),(d) difference wind ECMWF analysis minus CNTRL simulation is shown. The valid time is indicated above each panel. Note the different colour bars in a,b and c,d.

the outflow-jet interaction during the ET of Jangmi in the control simulation and of the enhanced flow at the eastern side of the subtropical anticyclone.

In summary, the verification of the CNTRL simulation against the analysis showed that this COSMO run is well suited as a reference simulation for the ET of Jangmi. The most prominent ET features, e.g. the recurvature and decay, the outflow-jet interaction, and the enhanced flow between the subtropical anticyclone and the TUTT cell are represented accurately. However, the extratropical cyclone downstream of Jangmi beyond day 4 of the simulation is different. Therefore after 96 simulation hours the CNTRL run is only a limited representation of the analysis. Nevertheless, the control simulation is suited to investigate the evolution of the downstream *scenario* constituted by this simulation.

### 4.4.3 The interaction with the jet

#### 4.4.3.1 The effect of PV surgery at initialisation time

The interaction of Typhoon Jangmi and the midlatitude jet is quantified at the 200 hPa pressure level since the jet streak and the outflow of Jangmi were most prominent at this level. At 00 UTC 28 September 2008, the initialisation time of the CNTRL and NOTC simulations, a distinct midlatitude jet is evident at 200 hPa in the northwest Pacific region (Figure 4.32a). The zonally oriented jet streak which will interact with Jangmi's outflow is located over China (at about 115°E, 40°N) and approaches Japan (cf. with discussion of Figures 4.5a, c, e in Section 4.1.2). Another zonally oriented jet streak is located over Japan. East

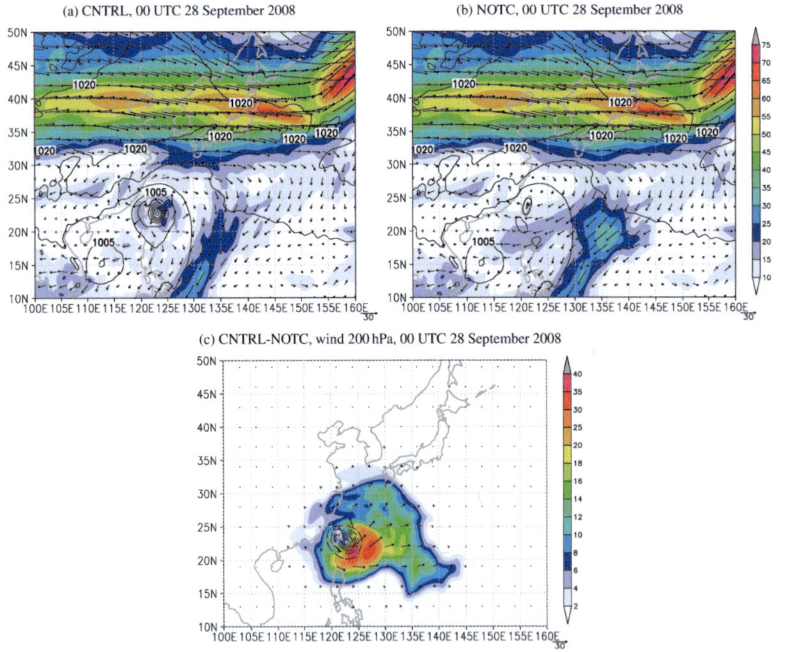


Figure 4.32: (a),(b) wind vectors (black), wind speed (shaded, in m s$^{-1}$) at 200hPa, and pmsl (black contours, every 5 hPa) at the initialisation time of the CNTRL/NOTC simulations (00 UTC 28 September 2008). For (a) CNTRL and (b) NOTC simulations. (c) 200 hPa difference wind of CNTRL minus NOTC and difference pmsl (every 5 hPa; values >5 hPa solid, <-5 hPa dashed).

of 155°W a southwesterly jet streak is obvious. The effect of the PV surgery is confined to the region around Jangmi. At the surface a difference pmsl of less than -34 hPa reflects the TC removal (Figure 4.32c). At 200 hPa in CNTRL (Figure 4.32a), a divergent cyclonic flow in the close vicinity of Jangmi (with a radius of less than 5° longitude/latitude) curves anticyclonically into a larger-scale anticyclonic circulation representing the TC outflow (with a radius of up to 20° longitude/latitude to the east). In the NOTC run (Figure 4.32b), the upper-level anticyclone to the east of Jangmi extends westwards to the Chinese coast (as indicated by the wind vectors in Figure 4.32b). In the region of Jangmi (north)easterly flow occurs at 200 hPa. The complex transition from cyclonic circulation close to Jangmi to the anticyclonic outflow is not present. This is reflected in the difference wind, which is most pronounced to the southeast of Jangmi where the upper-level cyclonic circulation close to Jangmi opposes directly the northeasterly flow without the storm (Figure 4.32c). Despite this, the difference wind at 200 hPa in the larger scale is pronounced to the north and east of Jangmi and can be attributed to its anticyclonic outflow. South of Japan (centred at about 128°E, 29°N) the outflow leads to an enhanced westerly flow and east of Jangmi (from about 133°E, 27°N to about 133°E, 15°N) to a more northerly flow (cf. Figure 4.32a-c).

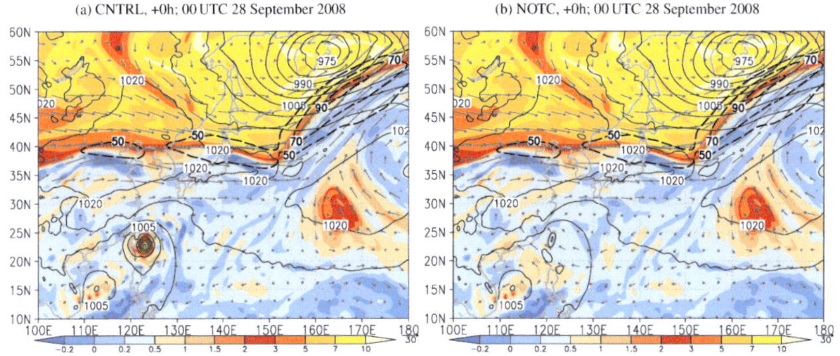


Figure 4.33: PV (in PVU, shaded), wind (grey vectors and black dashed with a 20 m s$^{-1}$ contour interval for $|\vec{v}| > 50$ m s$^{-1}$) at the isentropic level of $\Theta$=340K, and pmsl (black contours every 5 hPa). On the left data from the CNTRL simulation is plotted and on the right data from the NOTC simulation.

The westward extension of the subtropical anticyclone is clearer on the 340 K isentropic surface (at about 350 hPa in the region of Jangmi, Figure 4.33). Without Jangmi, a closed anticyclonic circulation extends from 115-155°E, 15-35°N at 340 K (Figure 4.33b), while in the CNTRL run the circulation is deflected sharply northward at the eastern edge of Jangmi at about 130°E (Figure 4.33a) due to the cyclonic circulation associated with Jangmi's positive PV anomaly at 340 K (cf. Figure 4.35d).

#### 4.4.3.2 The evolution of the upper-level flow

The interaction of Jangmi and the midlatitudes during ET resulted in marked changes of the jet structure. In the following, the impact of Jangmi on these changes is quantified with a 24 hourly sequence of 200 hPa wind in the CNTRL and NOTC simulations (Figures 4.34 and 4.35).

At 12 UTC 29 September 2008 in the CNTRL simulation (Figure 4.34a), the jet streak, detected over China 36 hours earlier (Figure 4.32), has become stationary over Japan. This is accompanied by a pronounced southwesterly outflow centred at 128°E, 30°N extending from Jangmi over the northern East China Sea, Korea, and the Sea of Japan towards the midlatitude jet streak. Without Jangmi (Figure 4.34b) the southwesterly flow west of Japan is not present, the jet streak is weaker and displaced eastward to about 152°E, 40°N. Another clear difference is the enhanced flow at the eastern (centred at about 152°E, 27°N) and southern flank (127-143°E, 20°N) of the subtropical anticyclone when Jangmi is present (cf. Figures 4.34a,b). The anticyclone is also located slightly further north in CNTRL (with a ridge axis from about 126°E, 27°N to 142°E,27°N) compared to NOTC (with a ridge axis from about 112°E, 25°N to 142°E, 25°N). These differences are reflected in the difference wind at 200 hPa (Figure 4.35a). The dipole in pmsl anomaly at about 111°E, 17°N in Figure 4.35a also reveals that Typhoon Mekkhala in the South China Sea has a different track which leads to some differences in the 200 hPa wind in that region.

The interaction of the outflow of Typhoon Jangmi and the midlatitude jet in the control simulation becomes apparent at 12 UTC 30 September 2008 (Figure 4.34c). As Jangmi approaches the midlatitudes, the southwesterly outflow

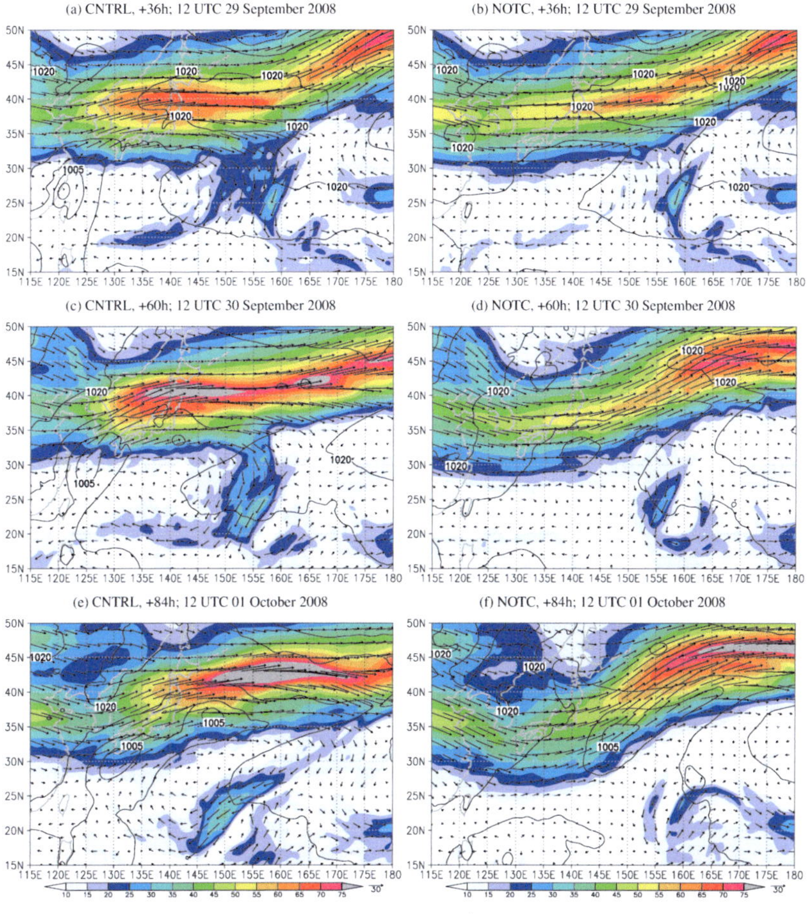


Figure 4.34: Wind vectors (black), wind speed (shaded, in $m\,s^{-1}$) at 200hPa, and pmsl (black contours, every 5 hPa). Data in the left/right column is from the CNTRL/NOTC run. The forecast hour and valid time are indicated above each panel.

(centred at 130°E, 35°N) towards the jet streak over Japan persists (Figure 4.34c) and the jet streak expands southwestward towards Jangmi (see the extension of the green shades in the southwestern corner of the jet streak). The jet core wind speed has increased to more than $75\,m\,s^{-1}$. In contrast, without Jangmi (Figure 4.34d), the jet wind speed over Japan reaches values of only about $50\,m\,s^{-1}$. In the control simulation (Figure 4.34c), a weak ridgebuilding begins down-

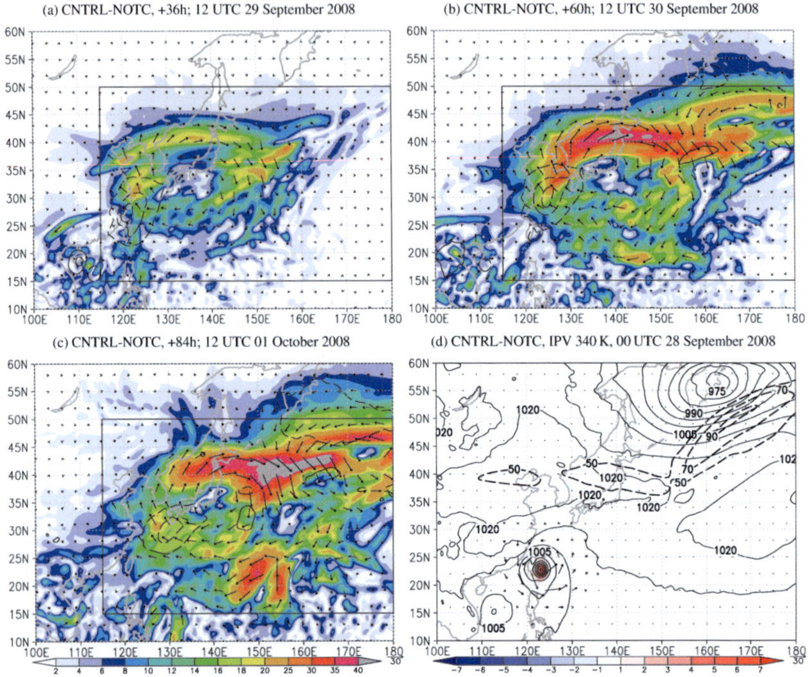


Figure 4.35: Wind vectors (black), wind speed (shaded, in $m\,s^{-1}$) at 200hPa, and pmsl (black contours, every 5 hPa). Data shown is the difference of CNTRL minus NOTC simulation. The box marks the subsets shown in Figure 4.34. (d) difference of PV (shaded) and wind vectors (grey) on the 340 K isentropic surface CNTRL minus NOTC, pmsl and $|\vec{v}| > 50\,m\,s^{-1}$ on 340 K (dashed contours every $20\,m\,s^{-1}$) of the CNTRL simulation. The forecast hour and valid time are indicated above each panel.

stream of Jangmi and south of the jet streak. Further eastward a rather zonally oriented strong upper-level flow is present ($|\vec{v}_h| > 70\,m\,s^{-1}$). In the NOTC experiment (Figure 4.34d), a broad trough is present over Japan, followed by a broad ridge to the east over the Pacific. A jet streak with wind speed of more than $70\,m\,s^{-1}$ is present at about 175°E, 45°N near the ridge axis. As on the previous day the flow of the subtropical anticyclone is enhanced in CNTRL, in particular between its eastern flank and the upper-level TUTT cell. The enhancement and the location further north of the jet streak are reflected in an absolute wind speed difference of more than $40\,m\,s^{-1}$ between the two model runs (4.35b). Over the Pacific the 200 hPa difference wind reflects the different amplification of the midlatitude wave guide in CNTRL and NOTC. The flow around the subtropical anticyclone is more than $20\,\mathrm{m\,s^{-1}}$ faster at the eastern (150°E, 34°N) and southern flank (145°E, 19°N) in the presence of the enhancement due to Jangmi's outflow.

One day later, at 12 UTC 01 October 2008, the jet streak has started to propagate eastward and an elongated region of very high wind speed ($|\vec{v}_h| > 75\,\mathrm{m\,s^{-1}}$) extends east of Japan (Figure 4.34e) in the control simulation. Jangmi continued to decay and the southwesterly outflow is less distinct. The extratropical cyclone downstream of Jangmi develops in the southwestern entrance region of the jet *core* (at about 148°E, 36°N). In the NOTC simulation (Figure 4.34f), the extratropical cyclogenesis occurs also but slightly further south ahead of the trough and in a southwestern jet entrance region at about 148°E, 34°N. The evolution of the downstream cyclone in both scenarios will be compared in Section 4.4.5. The TUTT cell moved westward in the control simulation and enhanced flow between its northwestern flank and the subtropical anticyclone persists centred at about 150°E, 23°N (Figure 4.34e). The difference wind reflects the differences in the deflection of the midlatitude wave guide and the associated jet streaks (4.35c). It is most pronounced at about 155°E, 40°N between the trough and ridge in the NOTC experiment. Here wind speed differences over a wide area exceed $40\,\mathrm{m\,s^{-1}}$. The different position of the TUTT cell in the simulations results in a second region of high difference winds. A dipole in the pmsl anomaly at about 148°E, 35°N reflects the different locations of the downstream cyclogenesis. The broad region of pmsl anomaly $< -5\,\mathrm{hPa}$ in the northeastern edge of the panel indicates that the central midlatitude low is weaker and located further northeastwards when Jangmi is not present (cf. Figures 4.34e,f and 4.35c).

The comparison of the CNTRL and NOTC simulations enabled us to quantify the impact of Jangmi on the midlatitude jet. The main results are:

- The interaction of the outflow of Jangmi and the midlatitude jet during ET (29 September to 01 October 2008) resulted in an acceleration of the jet core wind speed of about $25\,\mathrm{m\,s^{-1}}$. In conjunction with the northward shift of the jet streak over Japan an absolute wind speed difference of more than $40\,\mathrm{m\,s^{-1}}$ at 200 hPa occurred.

- At the same time, downstream of Jangmi and south of the jet streak over Japan, a weak upper-level ridgebuilding occurred when Jangmi was present. Without Jangmi a broad trough developed which transitions into a broad ridge over the Pacific.
- As a consequence, the midlatitude jet is rather zonally oriented when Jangmi is present while a wave pattern with a long wave length of about 100° longitude is present without Jangmi. Thus the outflow of Jangmi acts to zonalise the midlatitude flow.
- The outflow of Jangmi initially enhanced the flow around the upper-level subtropical anticyclone with a magnitude of up to 20 $m\,s^{-1}$ at its northern side (close to the jet). Later during ET (after 12 UTC 30 September 2008) the flow on the eastern (close to the TUTT cell) and southern flank was enhanced.
- In the presence of Jangmi the TUTT cell was more pronounced at the end of ET and propagated westward, in accordance with the analysis. The flow between the subtropical anticyclone and the TUTT cell was significantly accelerated at 12 UTC 01 October 2008. The impact of Jangmi on this TUTT cell may constitute a feedback from ET to the tropics.
- Since the ET of Jangmi alters the upper-level midlatitude flow, the forcing for the cyclogenesis downstream is altered resulting in a different location and development of this cyclone in the subsequent days (see also Section 4.4.5).

### 4.4.4 The interaction with the baroclinic zone

The acceleration of the midlatitude jet was investigated in Section 4.3 using analysis data. With the help of trajectory calculations it was found that the increase of the upper-level PV gradient in the jet region was due to warm moist tropical air advected by Jangmi's low-level circulation impinging on the baroclinic zone. This leads to buoyant and dynamically forced ascent ahead of and along the baroclinic zone accompanied by enhanced diabatic heating at mid levels. Diabatic PV destruction occurred above the level of maximum heating. Thus low

PV air was partly transported from lower levels to the upper troposphere and advected by the TC outflow enhancing the upper-level midlatitude PV gradient. This resulted in an acceleration of the midlatitude upper-level jet. In the following, the representation of this physical mechanism in the CNTRL and NOTC simulations will be investigated.

Representative 72 hour forward trajectories are computed that originate from low-levels in the region of the baroclinic zone (Table 4.4). In the NOTC simulation 55% fewer trajectories are detected (18069) compared to the CNTRL scenario (40542). As in the analysis data (Figure 4.19), the trajectories based

Table 4.4: Details of the 72 hour forward trajectories based on 3 hourly CNTRL and NOTC data

| | |
|---|---|
| **origin lon, lat, z** | 118°E - 179°E, 17°N - 45°N, 0-2000 m amsl |
| **calculation period** | 72 hour *forward* from 00 UTC 28 Sep 2008 to 00 UTC 01 Oct 2008 |
| **criterion** | ascent in 72 hours > 8500 m |

(a) CNTRL

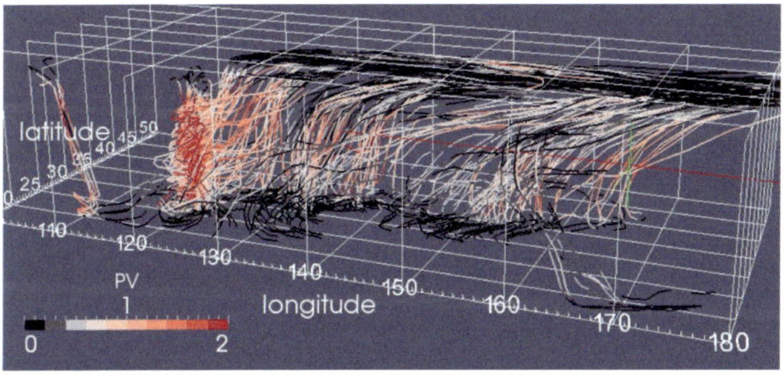


(b) NOTC

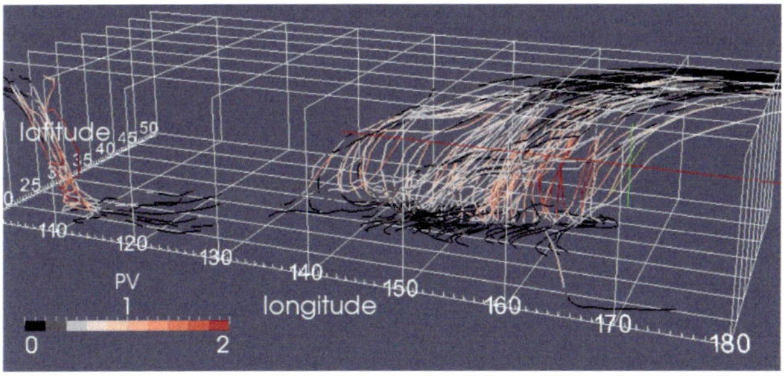


Figure 4.36: Representative 72 hour forward trajectories based on CNTRL (a) and NOTC (b) simulation. Calculation details in Table 4.4. Shading indicates PV along the trajectories. Every 200th trajectory at the end of the calculation period shown. The view is from the south.

on the control simulation reveal strong upright motion in Jangmi, ahead of the baroclinic zone directly downstream of Jangmi, and along the baroclinic zone further downstream (Figure 4.36a). This reflects the upright, buoyant lifting in the TC inner core, the advection by Jangmi impinging on the baroclinic zone and subsequent slantwise, dynamically forced ascent along tilted isentropes, and the advection towards the baroclinic zone by the newly generated PV anomalies (cf. discussion in Section 4.3.1). The trajectories based on NOTC are mainly associated with slantwise, dynamically forced ascent along the baroclinic zone during the extratropical cyclogenesis east of Japan (Figure 4.36b). The trajectories in the region of Jangmi are completely missing. Thus the strong ascent at the baroclinic zone south of Japan and the net transport of low PV air to upper levels is not present.

As presented in Section 4.1.2 for the analysis, 48 hour forward trajectories were calculated for the CNTRL and NOTC runs. The calculation was restarted every

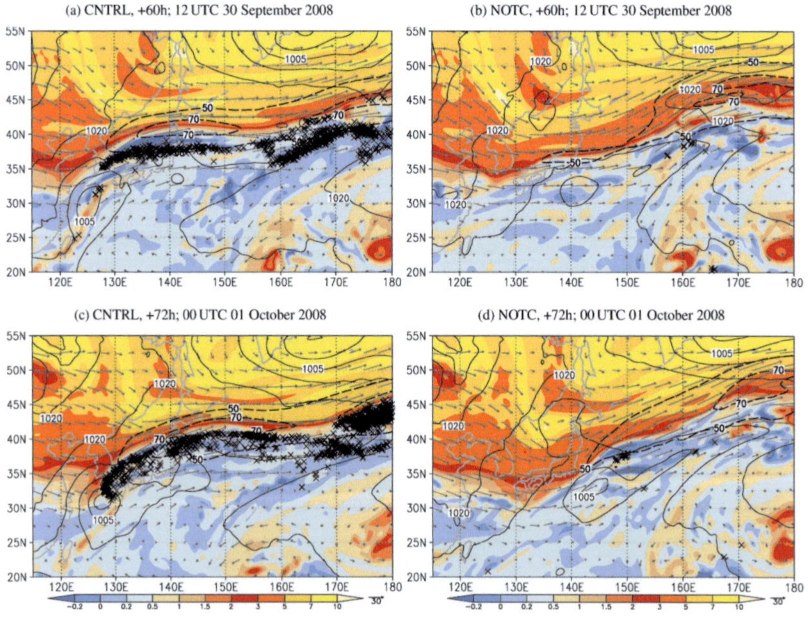


Figure 4.37: As Figure 4.33 but for 12 UTC 30 September 2008 and 00 UTC 01 October 2008. Crosses mark every 10th trajectory intersection point of the series of 48 hour forward trajectories (Table 4.5) with the 340K±0.5K isentropic layer at the time shown. Cf. with analysis in Figures 4.5k, m.

6 hours during the period 00 UTC 28 September 2008 to 18 UTC 30 September 2008 . The setup detailed in Table 4.5 is suited for the detection of the ascent along the baroclinic zone. Intersection points of these trajectories are shown in the following together with PV on the 340 K isentropic surface (Figure 4.37). The features on the 340 K isentropic surface in the control simulation constitute a very good representation of those in the analysis (cf. Figures 4.37a,c and 4.5k,m). Only a small difference exists in the trough downstream of the jet streak (at about 160°E at 12 UTC 30 September 2008) This trough is more pronounced in the analysis.

At 12 UTC 30 September 2008, the 340 K isentropic PV in the control simulation (Figure 4.37a) highlights a weak midlatitude trough over the Yellow Sea approaching Jangmi (as indicated by the curvature of the 2 PVU contour at approximately 123°E, 36°N). Downstream of Jangmi weak ridgebuilding occurs over Japan with very low PV air (PV$< -0.2$ PVU) to the southern side of an enhanced midlatitude PV gradient. As a consequence, a strong jet streak ($|\vec{v}_h| > 80\,\mathrm{m\,s^{-1}}$) is located over Japan. The trajectory intersection points are aligned along the southern edge of the low PV region. This indicates that ascent and the transport of low PV air occurs in the region of the jet streak. A second concentration of intersection points aligns along the midlatitude wave guide further downstream. In contrast, the NOTC experiment (Figure 4.37b) indicates a broad trough in the region of Japan and only a weak jet streak ($|\vec{v}_h| > 50\,\mathrm{m\,s^{-1}}$). Hardly any trajectory intersection points are present along the midlatitude wave guide. However, the first set of trajectories (Table 4.4, Figure 4.36b) indicates that some trajectories occur that ascend more than 8500 m in 72 hours. Ascent is much slower without Jangmi as indicated by the rarity of trajectory intersection points in NOTC in the second set of trajectories with a more strict criterion to ascend more than 8500 m in 48 hours (Table 4.5,Figure 4.37b).

The ridgebuilding due to Jangmi's outflow is reflected in the difference in PV at 340 K between the experiments (Figure 4.38a). South of the jet streak a broad region of negative difference PV of less than -6 PVU occurs, associated with a strong anticyclonic circulation of the difference wind. The lowest values of dif-

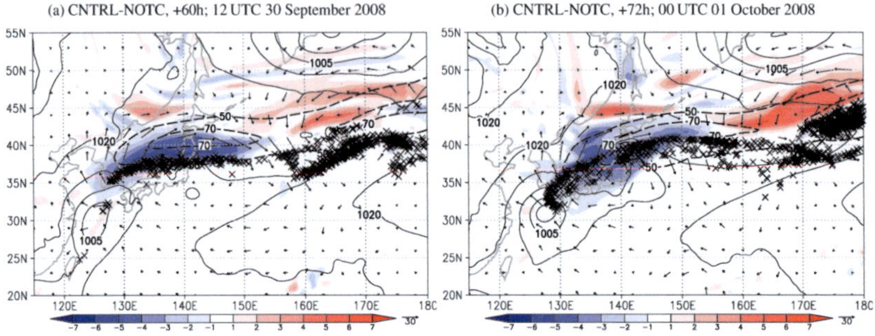


Figure 4.38: As Figure 4.35d: difference PV (shaded) and difference wind vectors (grey) on the 340 K isentropic surface (CNTRL minus NOTC). Pmsl, $|\vec{v}| > 50\,\mathrm{m\,s^{-1}}$ on 340 K (black dashed every $20\,\mathrm{m\,s^{-1}}$), and every 10th trajectory intersection point (as in Figure 4.37 (a),(c)) from the CNTRL simulation. Forecast hour and valid times indicated above the panels.

ference PV are slightly north of the location of the intersection point alignment (Figure 4.38a). The very low values result from the superposition of an adiabatic poleward advection of low PV air from the tropics at upper-levels and of the diabatic PV reduction by the dynamically forced slantwise ascent along tilted isentropes and the associated transport of low PV air from lower-levels along the baroclinic zone to the tropopause. This results in a significant lifting of the tropopause (cf. Atallah and Bosart 2003) and very low upper-level PV. The enhancement of the horizontal PV gradient $\nabla_h PV$ due to the diabatic PV reduction can be assessed by the magnitude of $\nabla_h PV$. In the CNTRL run the 340 K $\nabla_h PV$ is about 1 PVU $(100\,\mathrm{km})^{-1}$ higher (up to 5 PVU $(100\,\mathrm{km})^{-1}$) than in the NOTC run (up to 4 PVU $(100\,\mathrm{km})^{-1}$; not shown). The difference PV reveals some other features in the scenarios. The dipole at about 131°E, 44°N of negative difference PV to the south and positive values to the north reflects the reduced amplitude of the short-wave trough in CNTRL, that approaches Jangmi. The region of positive difference PV centred at about 165°E, 45°N shows that the trough downstream of the ridgebuilding is more pronounced in CNTRL than in the NOTC simulation. Significantly more ascent occurs ahead of this weak downstream trough, as indicated by the band of intersection points centred at about 165°E, 37°N. In the NOTC simulation a broad ridge develops in that region.

At 00 UTC 01 October 2008 the intersection points in the control simulation

Table 4.5: Details of the 48 hour forward trajectories based on 3 hourly CNTRL and NOTC data

| | |
|---|---|
| **origin lon, lat, z** | 100°E - 179°E, 10°N - 60°N, 0-2000 m amsl |
| **calculation period** | 48 hour *forward* trajectories starting every 6 hours from 00 UTC 28 Sep 2008 to 16 UTC 30 Sep 2008 |
| **criterion** | ascent in 48 hours $>$ 8500 m |

align directly along the midlatitude wave guide (4.37c). It is striking that the ridge over Japan as well as the alignment of intersection points along the wave guide are virtually not present in the NOTC experiment (4.37d). In both scenarios surface cyclogenesis occurs in a southwestern jet (core) entrance region at about 144°E. In NOTC this is about 2° further south (at about 33°N) than in CNTRL (at about 35°N). The ridge downstream of Jangmi and the less amplified trough in CNTRL result in a broad region of negative difference PV ($< -6$ PVU, Figure 4.38b). The alignment of intersection points from CNTRL occurs at the same location as the minimum in difference PV. The magnitude of 340 K $\nabla_h PV$ increased about 1 PVU $(100\,\mathrm{km})^{-1}$ in both the CNTRL and NOTC experiments thus $\nabla_h PV$ remains about 1 PVU $(100\,\mathrm{km})^{-1}$ higher in the presence of the ET system (not shown). The downstream trough, that is more pronounced in CNTRL, and the broad ridge in NOTC result in a region of positive difference PV ($>$ 5 PVU) at about 170°E 42°N. Ahead of the downstream trough intersection points indicate ascent in CNTRL.

The investigation of trajectories and 340 K isentropic PV presented here revealed the existence of the physical process governing the modification of the midlatitude jet in the control simulation and the lack of modification in the no-TC simulation. A crucial component of this process is the tropical air ascending within the TC inner core and ascending slantwise along the baroclinic zone. Both result in the diabatically enhanced transport of low PV air to upper levels and a lifting of the tropopause.

The low-level baroclinic zone is apparent in the vertically averaged (975-800 hPa) equivalent potential temperature $\Theta_e$ (Figure 4.39). In both the CNTRL and NOTC scenarios a baroclinic zone separates cooler dryer air in the north from warmer more moist air in the south. However, Jangmi significantly alters this baroclinic zone in the CNTRL scenario (Figure 4.39a,c). At 12 UTC 30 September 2008 the low-level circulation associated with Jangmi (located at about

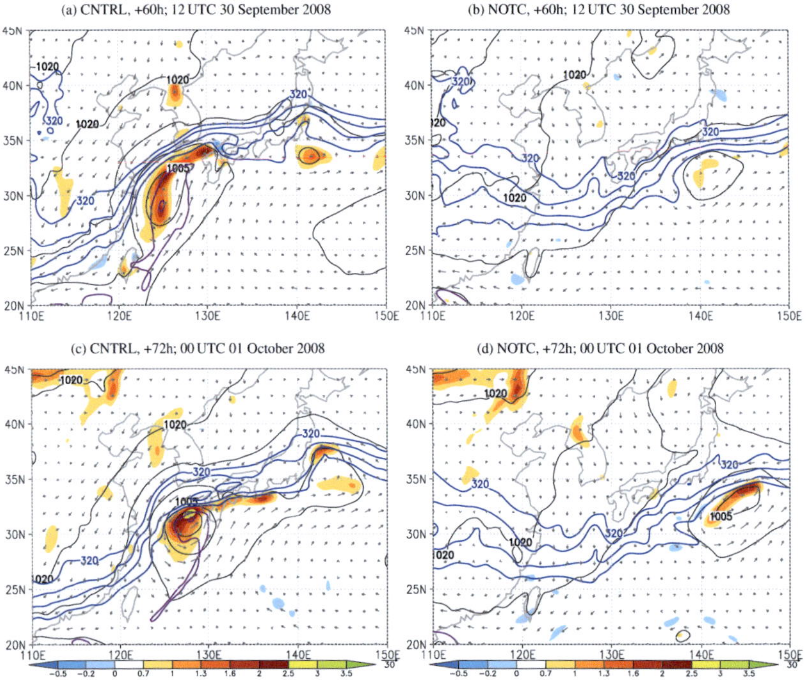


Figure 4.39: Vertically averaged fields in a layer from 975-800 hPa: low-level PV (shaded, in [PVU]), horizontal wind vector (grey), equivalent potential temperature (blue contours from 320 to 335 K every 5 K and purple contours from 350 to 355 K every 5 K), and pressure at mean sea level (black contours with a 5 hPa contour interval). (a),(c) data from CNTRL, (b,d) data from NOTC. Forecast hour and valid times indicated above the panels.

125°E, 30°N, Figure 4.39a) deflects the baroclinic zone and leads to an increase of the $\Theta_e$ gradient. West of Jangmi the northerly flow leads to a weak cold frontogenesis. It is striking that Jangmi still exhibits a warm core anomaly of $\Theta_e > 355$ K. To the northeast of Jangmi strong southwesterly flow against the baroclinic zone leads to a significant warm frontogenesis resulting in a band of diabatically produced PV along the baroclinic zone (Figure 4.39a, to be compared with the analysis data in Figure 4.22c). None of these features are present in the NOTC experiment (Figure 4.39b).

At 00 UTC 01 October 2008 Jangmi, still exhibiting a warm anomaly of $\Theta_e >$ 350 K, reached the baroclinic zone in the CNTRL scenario (Figure 4.39c). An elongated band of low-level PV extends along the baroclinic zone where warm frontogenesis continues (Figure 4.39c, cf. Figure 4.22e). In the NOTC experiment the baroclinic zone shifted southward (Figure 4.39d). The extratropical cyclogenesis at 144°E is associated with the evolution of a low-level PV anomaly at the baroclinic zone in both scenarios. The latitudinal position of this PV anomaly in the NOTC scenario (33°N) differs from that in the CNTRL scenario (35°N).

It was shown that the physical processes governing the interaction of Jangmi with the midlatitudes (explored in detail in Section 4.3) are not present when the tropical cyclone has been removed from the initial conditions. This confirms the crucial impact of the diabatic low-level PV production and upper-level PV reduction at the baroclinic zone in the presence of the ET system. The interaction with the baroclinic zone acts to modify the midlatitude flow in particular at upper levels.

### 4.4.5 Downstream development

At 00 UTC 01 October 2008 an extratropical cyclone is developing in both the CNTRL and NOTC simulations at about 144°E, 34°N (Figure 4.28). The minimum pressure of this cyclone in the scenarios dropped from 1010 hPa at 00 UTC 30 September 2008 to about 1002 hPa at 06 UTC 01 October 2008 (Figure 4.29). This is consistent with the analysis. However, the analysis data indicates a further pressure drop to 992 hPa while in the CNTRL simulation the minimum pressure reaches 1000 hPa during the next days and in the NOTC simulation the central pressure increases (Figure 4.29). The extratropical cyclone in the NOTC scenario moves generally more slowly than in the analysis (Figure 4.28). In the CNTRL simulation, the track of the downstream cyclone is similar to the track in the analysis until 00 UTC 02 October 2008, but the cyclone moves faster afterwards.

The *ACC* reflects some of the differences between the scenarios and analysis (Figure 4.30). The *ACC* for 500 hPa geopotential indicated a decrease in the CNTRL simulation skill after 00 UTC 01 October 2008. The *ACC* of the NOTC

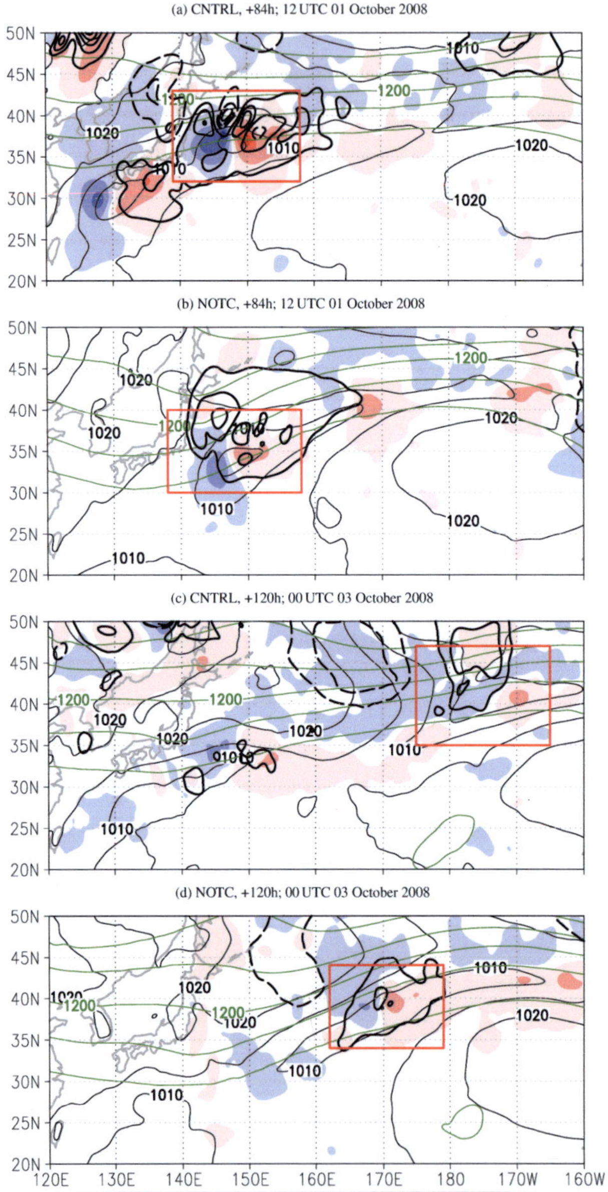


Figure 4.40: As Figure 4.6 but based on CNTRL (a,c) and NOTC (b,d). Experiment, forecast hour and valid times indicated above each panel. Red boxes highlight location of downstream cyclone. To be compared with Figures 4.6c,e

run (grey in Figure 4.30) shows a strong decrease when the extratropical cyclogenesis begins and reaches 0.85 at 00 UTC 2 October 2008, which is significantly lower than in CNTRL (0.92) or the typical IFS-ACC. The NOTC *ACC* remains lower than in CNTRL until 00 UTC 03 October 2008 indicating, that the downstream development in NOTC differs more from the analysis than CNTRL during the period of interest. However, the ACC for NOTC remains relatively high ($\leq 0.85$) indicating that this scenario represents a useful background flow. In the remainder of this section the reasons for the significantly different propagation and evolution of the downstream cyclone in the CNTRL and NOTC scenarios are explored.

At 12 UTC 01 October 2008, as in the analysis (Figure 4.6c), the QG forcing for vertical velocity highlights two important regions in the CNTRL scenario. A dipole of weak (up to $\pm$ 0.01 m s$^{-1}$) low-level forcing for descent to the west and for ascent to the east reflects the dipole of cold and warm air advection centred on decaying Jangmi (at about 130°E, 30°N, Figure 4.40a). In the region of low-level forcing for ascent, weak upper-level forcing for ascent (up to 0.01 m s$^{-1}$) exists. This forcing is associated with the weak upper-level trough over Korea and the Sea of Japan. Stronger QG forcing exists in the region where the downstream extratropical cyclogenesis occurs, centred at about 148°E, 36°N. Here upper- and low-level forcing reach values of more/less than $\pm$ 0.015 m s$^{-1}$, albeit weaker than in the analysis data ($\gtrless \pm$ 0.02 m s$^{-1}$, Figure 4.6c). This forcing occurs in the southwestern jet *core* entrance region (Figure 4.40a; the jet is located where the green contours of geopotential at 200 hPa in Figure 4.40 are most densely packed, cf. Figure 4.34g). In the NOTC scenario, QG forcing only occurs in the region of the extratropical cyclogenesis (Figure 4.40b) centred at about 148°E, 34°N. The QG forcing is located ahead of a broader upper-level trough over Japan and in the southwestern entrance region of an elongated jet (see also Figure 4.34h). The forcing is significantly weaker ($\gtrless \pm$ 0.01 m s$^{-1}$) than in the CNTRL scenario and analysis.

With only weak further intensification to about 1000 hPa at 00 UTC 03 October 2008, the downstream cyclone in the CNTRL scenario propagates to about 188°E, 41°N (Figure 4.40c, Figure 4.28). It moves very fast along a strong baroclinic zone that shows a wave-like structure at the location of the cyclone associated with a clear low-level PV band (Figure 4.41c). The propagation along the baroclinic zone without further strong intensification can be classified as a DRW (cf. discussion in Section 4.3.1). As in the analysis, the cyclone is continuously located in the southwestern jet *core* entrance region, where upper-level QG forcing for ascent of up to 0.01 $\mathrm{m\,s^{-1}}$ persists (Figure 4.40c). The forcing is weaker than during the previous period and significantly weaker than in the analysis data (Figure 4.6e). Also the propagation of the jet streak over the Pacific differs from the analysis, which along with the DRW characteristics explains the faster progression of the cyclone. In the NOTC scenario, the cyclone propagates slower, weakens and becomes an elongated band-like structure without a distinct centre (Figure 4.40d, Figure 4.41d). Only weak QG forcing for ascent exists.

In the CNTRL scenario, the baroclinic zone is zonally oriented downstream of the extratropical cyclone during its eastward propagation (Figure 4.41a,c). East of the extratropical cyclone the strong southwesterly flow induced by the associated low-level PV anomaly thus advects warm moist air against the baroclinic zone and by means of diabatic PV production inaugurates a DRW-like eastward propagation. In contrast, in the NOTC scenario, the baroclinic zone east of the extratropical cyclone is oriented from the southwest to the northeast. The circulation induced by the low-level PV anomaly is consequently rather parallel to the baroclinic zone east of the extratropical cyclone. Diabatic PV production is reduced and the DRW-like propagation is less efficient. This may explain the rather slow propagation in the NOTC scenario.

The investigation of the downstream development in the CNTRL and NOTC scenarios showed that in the presence of the interaction of the outflow of Jangmi and the midlatitude jet (CNTRL) a weak extratropical cyclone developed east of Japan, that crosses the Pacific as a DRW being continuously located in the southwestern entrance region of the modified jet *core*. This scenario is similar to the analysis, albeit the cyclone propagates slightly faster due to the faster propagation speed of the jet and the DRW characteristics. The intensity of the cyclone is weaker. In contrast, in the NOTC scenario, a downstream cyclone

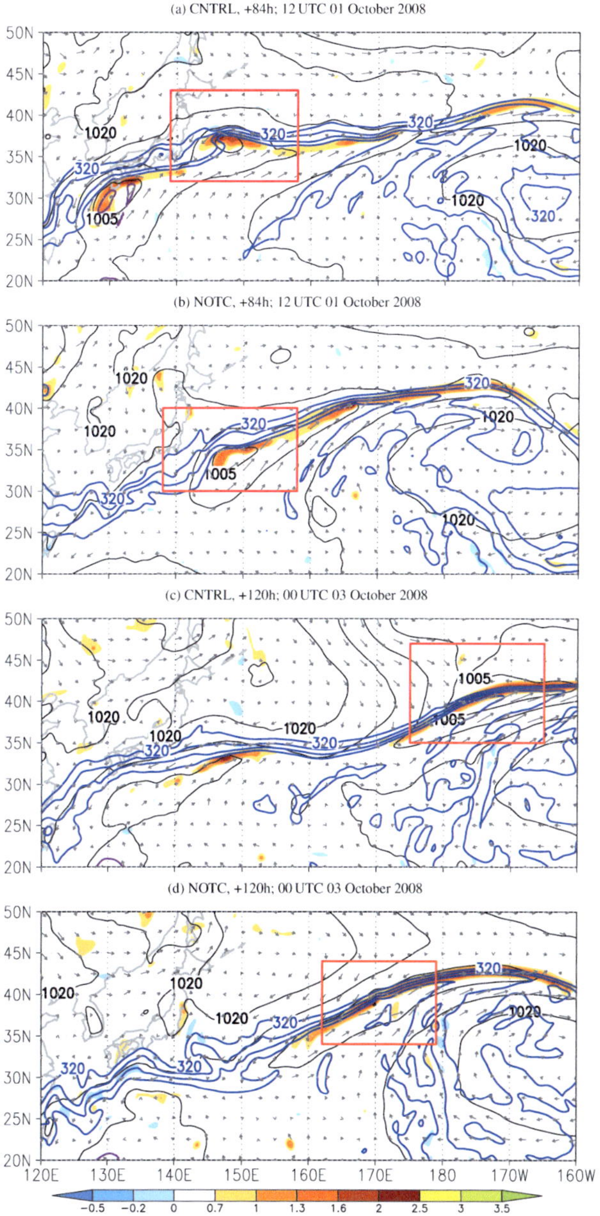


Figure 4.41: As Figure 4.39 but zoomed on downstream cyclone as in Figure 4.40. Data from CNTRL (a,c) and NOTC (b,d). Experiment, forecast hour and valid times indicated above each panel. Red boxes highlight location of downstream cyclone as in Figure 4.40.

initially develops at about the same location ahead of the broader midlatitude trough over Japan. However, this cyclone is located in the southwestern entrance region of an elongated jet whose jet *core* is less distinct. Consequently, the QG forcing is weaker and the cyclone becomes a band-like structure.

# 5 The midlatitude flow evolution during ET in different scenarios

The detailed analysis of Jangmi in the foregoing chapter revealed that the direct impact of the ET was manifested in an important outflow-jet interaction. However, the downstream impact in the midlatitudes was relatively weak. For a better understanding of this behaviour the TC removal technique was employed to study the flow evolution in scenarios in which the relative phasing of Jangmi and the midlatitude wave pattern has been altered. Furthermore, the downstream impact of ET in different initial midlatitude flow configurations has been quantified, by applying the PV surgery in other case studies. This enables us to elucidate more general conclusions about the downstream impact of ET than in the individual case study of Jangmi. In this chapter five ET cases including Jangmi are presented, each differing in the midlatitude environment into which the TCs recurve. The cases have been selected because of their different actual midlatitude flow evolution downstream of the ET system.

In the case of Typhoon Jangmi (2008) in the West Pacific, the TC recurves in phase with a weak midlatitude trough in a rather zonal midlatitude flow. The downstream impact is very weak. Therefore, the role of the relative position of the TC and the midlatitude trough in an unaltered midlatitude flow environment is studied (Section 5.1). Hurricane Hanna (2008) in the North Atlantic recurves into a preexistent ridge, which is part of a midlatitude flow that is characterised by pronounced Rossby wave activity. A chain of weather events occurs during and following ET including the triggering of a Mediterranean cyclogenesis (Grams et al. 2011). The impact of Hanna on this evolution is investigated in a run with and without the ET system (Section 5.2). A pronounced Rossby wave train exists also during the ET of Typhoon Choi-wan (2009). When Choi-wan reintensifies ahead of a short-wave trough Rossby wave breaking occurs and the downstream trough amplifies. The midlatitude flow evolution is investigated in a model run with and without Choi-wan (Section 5.3). During and after the ET of Typhoon Lupit (2009) a strongly amplified Rossby wave train over the Pacific results in an early outbreak of cold air in North America east of the Rocky Mountains. Lupit recurved into a rather long-wave trough-ridge couplet over the West Pacific. The role of Lupit on the Rossby wave triggering is investigated in sensitivity model simulations with and without the storm as well as in scenarios in which the storm has been relocated (Section 5.4). Typhoon Malakas (2010) moves ahead of a midlatitude trough in a flow that is characterised by Rossby wave activity upstream of the ET event. Following the ET, a Rossby wave train occurs in the eastern Pacific with a ridge over the western USA and a PV streamer over the eastern USA. At the same time, a heat wave was observed in southern California and heavy precipitation affected the US east coast. Simulations with and without Malakas in the initial conditions allow us to investigate the impact of the ET on these high impact weather events (Section 5.5).

## 5.1 Downstream development scenarios for Typhoon Jangmi (2008)

In the previous chapter, the interactions of Jangmi and the midlatitude flow during ET were explained as captured by observational and model analysis data. The impact of Jangmi on the flow evolution was quantified with the help of a TC removal experiment. It was found that apart from the strong outflow-jet interaction the ET of Jangmi had only weak midlatitude impact. The question arises as to why Jangmi decayed and did not reintensify as an extratropical cyclone such as in the ETs of Hurricanes David (1979), Iris (1995), Lili (1996) and Irene (1999) (Bosart and Lackmann 1995; Browning et al. 1998; Thorncroft and Jones 2000; Agustí-Panareda et al. 2005, 2004) or trigger a significant deflection of the midlatitude wave guide downstream (e.g. Hoskins and Berrisford 1988; Pinto et al. 2001; Agustí-Panareda et al. 2004; Riemer et al. 2008). The foregoing analysis also revealed that Jangmi's outflow and subsequent ridgebuilding downstream competes with an approaching midlatitude trough resulting in a zonalisation of the midlatitude flow. Earlier studies on western North Pacific typhoons (e.g. Klein et al. 2002) and in idealised frameworks (e.g. Ritchie and Elsberry 2007; Riemer and Jones 2010; Scheck et al. 2011b) showed that the phasing of the recurving TC and the midlatitude flow is crucial for the downstream impact of ET. An apparent sensitivity experiment is to investigate the relevance for the subsequent evolution of the position of Jangmi relative to the midlatitude trough. Therefore Jangmi is relocated in the initial conditions of a COSMO simulation using the PV surgery technique. The setup of the experiments is described in the following (Section 5.1.1). In Section 5.1.2 an overview about the scenarios is given. The evolution of the midlatitude flow is presented for all scenarios in Section 5.1.3 and in more detail for two selected scenarios in Section 5.1.4. Finally, an explanation for the different evolution amongst the scenarios is given in Section 5.1.5.

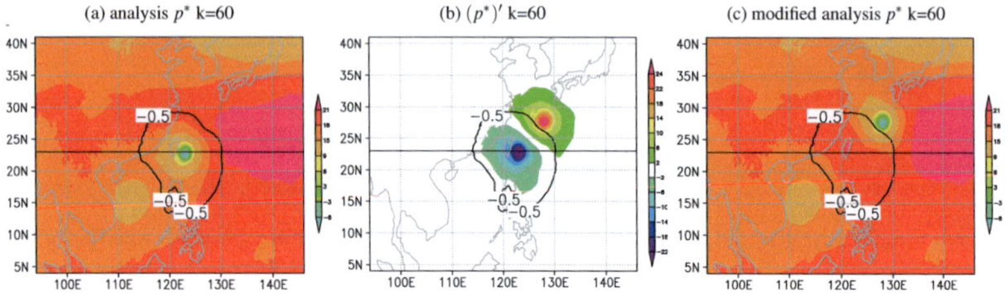


Figure 5.1: As Figure 3.9 but without the zonal profile and only illustrating the modification of the $p^*$ field for an anomaly that is moved by 5° to the east and north (scenario IW21).

## 5.1.1 Design of the scenarios

The PV surgery method defines the anomaly in meteorological fields associated with the storm (see Section 3.2.4 and Table A.2). This anomaly is subtracted from the analysis to provide an analysis without the storm. The anomaly can also be added at another location which will provide an analysis with the storm relocated. The relocation of Jangmi from a position at 123°E, 23°N to 128°E, 28°N is illustrated in Figure 5.1 cf. Figure 3.9. In contrast to only removing Jangmi (Figure 3.9b), the $p^*$ anomaly field shows a dipole of negative $(p^*)'$ at the position of Jangmi and positive $(p^*)'$ at the new position (Figure 5.1b). Subtracting this field from the analysis (Figure 5.1a) provides the modified analysis (Figure 5.1c) in which Jangmi is erased at its original position and relocated northeastward. The relocation procedure is summarised in the Appendix, Table A.3.

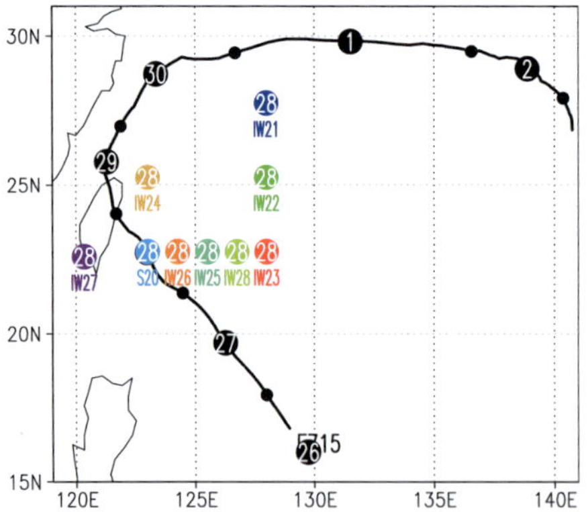


Figure 5.2: Initial position of Jangmi in the different scenarios (coloured and labelled, S20=CNTRL) and track of Jangmi in ECMWF analysis (black).

The modified analysis fields are used as the initial data for a COSMO simulation with a relocated storm. The setup of these COSMO runs is exactly the same as for the CNTRL and NOTC simulations (cf. Table A.1).

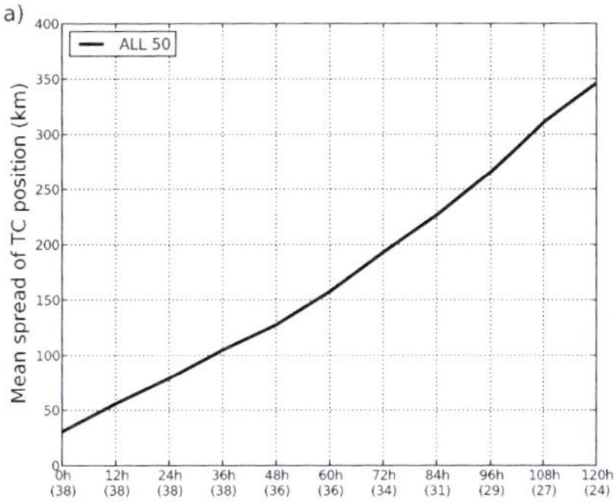


Figure 5.3: Example for the temporal evolution of the TC track error in the ECMWF ensemble prediction system (new operational (November 2010) configuration with 50 members, TL639 resolution (approximately 30 km)). Mean spread of TC position (solid black in km, mean difference of TC position of each ensemble member from ensemble mean) plotted over forecast time. Composite for 14 northern hemispheric TCs during the 2008 season. The numbers at the abscissa indicate how many ensemble forecasts (containing a TC) were used at each forecast time step. Figure from Lang (2011).

In particular, the initial time is 00 UTC 28 September 2008, the 6 hourly (unmodified) ECMWF analyses are used as boundary data and the simulations are computed for 150 hours until 06 UTC 04 October 2008.

### 5.1.2 Overview of the different scenarios

The impact of the phasing between Jangmi and the weak midlatitude trough has been addressed using a series of COSMO simulations with a relocated initial

Table 5.1: Details of the different experiments: label as used in the text, longitude (in °E) and latitude (in °N) of Jangmi's centre, zonal (in °E) and meridional (in °N) shift of the storm centre, zonal/meridional shift in number of grid points to the east/north, brief description of the scenario, and label used for the downstream cyclone (DS).

| label | lon. | lat. | $\Delta\lambda$ | $\Delta\varphi$ | $\Delta x$ | $\Delta y$ | remarks | DS | colour |
|---|---|---|---|---|---|---|---|---|---|
| F715 | 123.00 | 22.75 | 0 | 0 | 0 | 0 | ECMWF f7m4 analysis | F7N5 | **black** |
| S20 | 123.00 | 22.75 | 0 | 0 | 0 | 0 | reference run (CNTRL) | S20N | light blue |
| N22 | - | - | - | - | - | - | Jangmi removed from initial conditions (NOTC) | N22 | grey |
| IW21 | 128.00 | 27.75 | 5.00 | 5.00 | 20 | 20 | very strong reintensification (< 965 hPa) and ridgebuilding. RWT, central low in NE Pacific. | - | blue |
| IW22 | 128.00 | 25.25 | 5.00 | 2.50 | 20 | 10 | two periods of strong reintensification (< 965 hPa). RWT, central low in NE Pacific. | - | green |
| IW23 | 128.00 | 22.75 | 5.00 | 0.00 | 20 | 0 | two periods of reintensification (< 970 hPa), RWT triggering over NE Pacific | - | red |
| IW24 | 123.00 | 25.25 | 0.00 | 2.50 | 0 | 10 | Jangmi transitions into a moderate extratropical cyclone similar to the downstream cyclone in the analysis. | - | dark yellow |
| IW25 | 125.50 | 22.75 | 2.50 | 0.00 | 10 | 0 | Jangmi reintensifies, decays and transitions into downstream cyclone | I25N | turquoise |
| IW26 | 124.25 | 22.75 | 1.25 | 0.00 | 5 | 0 | Jangmi recurves back to tropics, downstream cyclone similar to CNTRL | I26N | orange |
| IW27 | 120.50 | 22.75 | -2.50 | 0.00 | -10 | 0 | Jangmi relocated west of Taiwan, decays at Chinese coast. Extratropical cyclogenesis 122°E, 30°N: I27a | I27N | violet |
| IW28 | 126.75 | 22.75 | 3.75 | 0.00 | 15 | 0 | two periods of strong reintensification, RWT triggering over NE Pacific | - | light green |

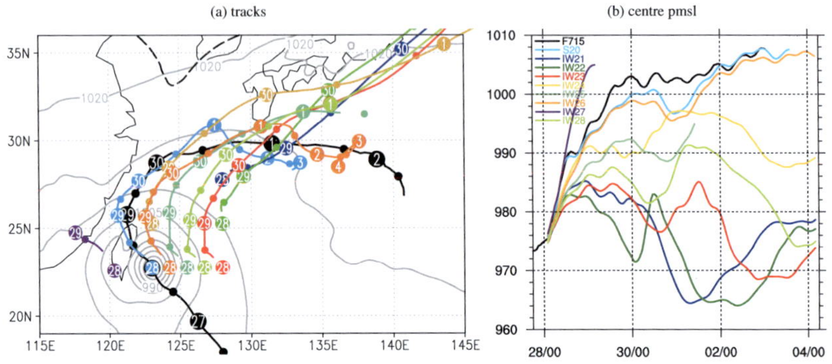


Figure 5.4: (a) Track of Jangmi in the scenarios (colours as in Figure 5.2 and (b)) and in the ECMWF analysis (black). Thick points, labelled with the day of September/October 2008 mark 00 UTC times, thin points mark 12 UTC times. Pmsl from ECMWF analysis (grey contours every 5 hPa) at 00 UTC 28 September 2008. PV on the 340 K isentropic level (black dashed 2 PVU contour) from the NOTC run at 12 UTC 29 September 2008. (b) Minimum pmsl in the centre of Jangmi in the different scenarios and in the ECMWF analysis.

position of Jangmi. For the relocation distances we followed typical track errors in the current ECMWF ensemble prediction system (Figure 5.3). The track error typically reaches more than 120 km after 48 hour forecast time and an error of around 350 km after 120 hours. This also underlines the relevance of knowing better the physical mechanisms during ET that determine this track error.

One set of experiments was designed so that Jangmi recurves ahead of the midlatitude trough, another set investigated the sensitivity on Jangmi's longitudinal position relative to the midlatitude trough. In all these scenarios we name the relocated storm "Jangmi" also. An overview of the scenarios, including their labelling conventions, is given in Table 5.1 and Figure 5.2. The labels enumerate the experiments, without indicating the parameters of the relocation.

In the first set of experiments, Jangmi was relocated in a way that it recurves *ahead* of the midlatitude trough that approaches Japan in the period 29-30 September 2008. Therefore scenarios IW21, IW22, IW23 (Figure 5.2) have a marked eastward shift of Jangmi by 5° and a varying meridional shift from 5° north (IW21, blue), 2.5° north (IW22, green), to 0.0° (IW23, red). In these scenarios Jangmi reintensifies as a strong extratropical cyclone (Figure 5.4b) and crosses the Pacific significantly further to the north (6°-10°) than the downstream cyclone in the analysis data (Figure 5.5). The initial decay of the storms

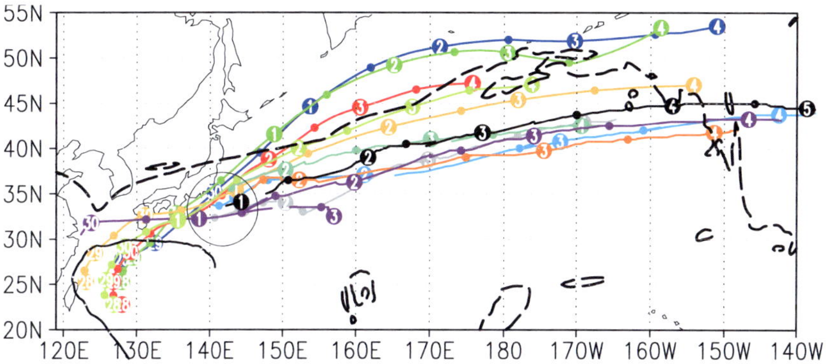


Figure 5.5: As Figure 5.4a but only tracks of reintensifying Jangmi or of the downstream cyclone in scenarios in which Jangmi decays. Track of Jangmi and the downstream cyclone in ECWMF analysis (black, solid). The black circle highlights the region of reintensification/extratropical cyclogenesis.

by 10-12 hPa in 24 hours stops at around 00 UTC 29 September 2008. A period of reintensification follows. In scenario IW21, Jangmi becomes a very intense extratropical cyclone. Its minimum pressure drops to 964 hPa at 06 UTC 01 October 2008. Jangmi moves very quickly into the Northern Pacific and has the northernmost track (Figure 5.5). In scenario IW22, the reintensification is interrupted from 00-12 UTC 30 September 2008, when the storm moves very close to the coast of Honshu (Figure 5.4b). However, after 12 UTC Jangmi reintensifies reaching a minimum pressure of 964 hPa at 18 UTC 01 October 2008. It follows a similar track to IW21, though the track shows a temporal shift of about 12 hours (Figure 5.5). A two-fold period of reintensification is also evident in scenario IW23 (Figure 5.4b). The central pressure minimum reaches 968 hPa at around 00 UTC 03 October 2008 and Jangmi moves significantly more slowly and slightly more southward than in IW21 and IW22 into the Pacific (Figure 5.5).

In the next sensitivity experiment Jangmi was shifted 2.5° to the north (IW24, dark yellow in Figure 5.2). Its northeastward motion is much faster than in the analysis or in CNTRL (Figure 5.4). It becomes located ahead of the secondary short-wave trough on 29 September 2008 (Figure 5.5) which halts the rapid decay. After 00 UTC 02 October 2008 it reintensifies moderately to a central pressure of 987 hPa at 12 UTC 03 October 2008 (Figure 5.4b). Jangmi in IW24

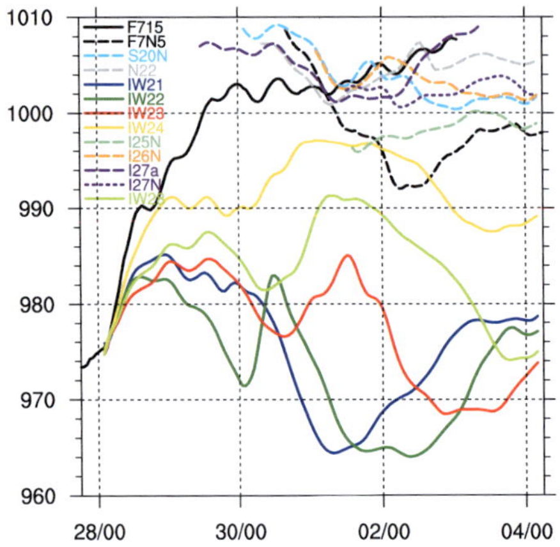


Figure 5.6: As Figure 5.4b but minimum pmsl in the centre of reintensifying Jangmi (solid) and of the downstream cyclones (dashed).

crosses the Pacific as a moderate wave-like cyclone, similar to the downstream cyclone in the analysis, but slightly stronger and about 3° further north (Figure 5.5, 5.6).

In the next series of sensitivity experiments (IW25-IW28, Figure 5.2) the role of the zonal position of Jangmi in the track and possible reintensification is investigated. When Jangmi is shifted 3.75° eastward (IW28, light green in Figure 5.4a) it develops as in the scenario with a 5.0° eastward shift (IW23, red), although its intensity is about 5 hPa weaker (Figure 5.4b). It also misses the secondary short-wave trough but reintensifies after 00 UTC 01 October 2008 to a central pressure of 974 hPa at 18 UTC 03 October 2008.

A critical relocation point is reached when Jangmi is shifted 2.5° eastward or less (IW25, turquoise in Figure 5.4a). Jangmi misses the secondary short-wave trough but as for IW28 and IW23 the decay stagnates ahead of the broader main trough (Figure 5.4b) before the cyclone dissipates on 01 October 2008 (138°E, 32°N). A new cyclone centre develops to the northeast (143°E, 35°N, I25N) in the region favourable for cyclogenesis (Figure 5.5). This cyclone moves into the Central Pacific like the downstream cyclone in the NOTC (N22) experiment but is slightly further north and more intense (centre pressure 996-1000 hPa, Figure 5.6). When Jangmi is shifted eastward by only 1.25° (IW26, orange in Figure 5.4), it recurves as in CNTRL (S20) and the analysis and decays on a southeastward track at the end of the calculation period. A weak wave-like downstream cyclone (I26N) develops and propagates across the Pacific following the same track and having the same intensity as the downstream cyclone in CNTRL (S20N) (Figure 5.5, 5.6). Finally, Jangmi is relocated by

2.5°westward to the west of Taiwan (IW27, violet in Figure 5.4). Jangmi decays when it makes landfall in China. However, due to its strong and long-lasting outflow and subsequent ridgebuilding a weak cyclone (I27a) develops at about 122°E, 30°N and moves south of Japan (Figure 5.5). It reaches a minimum pressure of 1001 hPa at about 144°E, 33°N at 12 UTC 01 October 2008 (5.6). Then to its northeast, a new cyclone emerges (I27N) and moves across the Pacific with similar intensity and track as the downstream cyclone in CNTRL (S20N).

From the foregoing overview of the scenarios we see that when Jangmi is relocated northward or far eastward so that it recurves ahead of the approaching midlatitude trough, it reintensifies and moves into the Central Pacific as an extratropical cyclone. Details of this reintensification depend crucially on the phasing with a secondary trough and will be explored in more detail in Section 5.1.3.

The experiments on the sensitivity of Jangmi's zonal position reveal that a critical relocation point exists. Storms relocated by more than 2.5° eastward reintensify and move across the Pacific. Storms relocated by less than 2.5° eastward decay. This is consistent with the existence of a critical zonal storm position found in an idealised model study by Scheck et al. (2011a,b). Storms located east of that position were attracted by the frontal zone, constituted by the midlatitude wave guide. Storms located west of that position were repelled back to the tropics. Furthermore, Scheck et al. (2011b) found the theoretical existence of a bifurcation point that governs this behaviour. It also shall be noted that between 30 September 2008 and 01 October 2008 all reintensifying cyclones pass the region favourable for extratropical cyclogenesis east of Japan (around 142°E, 34°N, marked by the black circle in Figure 5.5), whereas in the remnant scenarios with a decaying Jangmi the downstream cyclone develops in that region at the same time.

### 5.1.3 The downstream evolution in the ET scenarios

In the foregoing overview two main scenarios emerged: 1. reintensification of Jangmi and motion across the Pacific (IW21, IW22, IW24, IW23, IW28), 2. decay of Jangmi and weak downstream development (IW27, CNTRL / analysis / NOTW / NOTC, IW26, IW25). It was argued that the position of Jangmi relative to a short-wave trough was crucial for the subsequent evolution. In the following, the interaction of Jangmi with this secondary short-wave trough at 12 UTC 29 September 2008 is discussed for all scenarios. At this time an axis of the midlatitude main trough approaching Japan can be seen in the analysis data at around 125°E (Figure 5.7a). The secondary short-wave trough is evident at about 128°E, 38°N, reflected in the 3 PVU contour. In all scenarios the secondary trough is significantly sharper and located at about 122°E, 35°N (Figure 5.8, 5.9). The amplitude varies between the scenarios. In the NOTC scenario (Figure 5.8a) the short-wave trough is located slightly further west at 124°E compared to about 122.5°E in CNTRL (Figure 5.8e). Also the trough is slightly less amplified in NOTC due to the lack of the downstream ridge-building by Jangmi's outflow. However, the broader main trough approaching Japan is similar in all scenarios with a marked signature in the 7 PVU contour at about 120°E, 47°N and an axis from around 121°E, 45°N to 132°E, 51°N (e.g. in NOTC, Figure 5.8a).

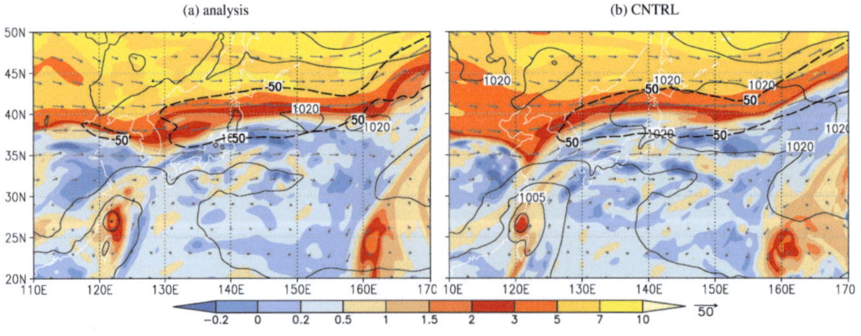


Figure 5.7: PV (in PVU, shaded), wind (grey vectors and black dashed with a $20\,\mathrm{m\,s^{-1}}$ contour interval for $|\vec{v}| > 50\,\mathrm{m\,s^{-1}}$) at the isentropic level of $\Theta$=340K, and pmsl (black contours every 5 hPa) for ECMWF analysis (a) and CNTRL (b) at 12 UTC 29 September 2008 (+36h).

In the scenarios with a 5° eastward shift and varying northward shift (IW21, IW22, IW23; Figure 5.8b,d,f) Jangmi is located ahead of the secondary short-wave trough at 12 UTC 29 September 2008 and begins a period of reintensification. When Jangmi is initially relocated 2.5° north (IW24; Figure 5.8c) it also moves ahead of the short-wave trough and stops decaying. In all these reintensification scenarios the rather strong downstream ridgebuilding leads to an amplification of the trough and an acceleration of the jet streak to a wind

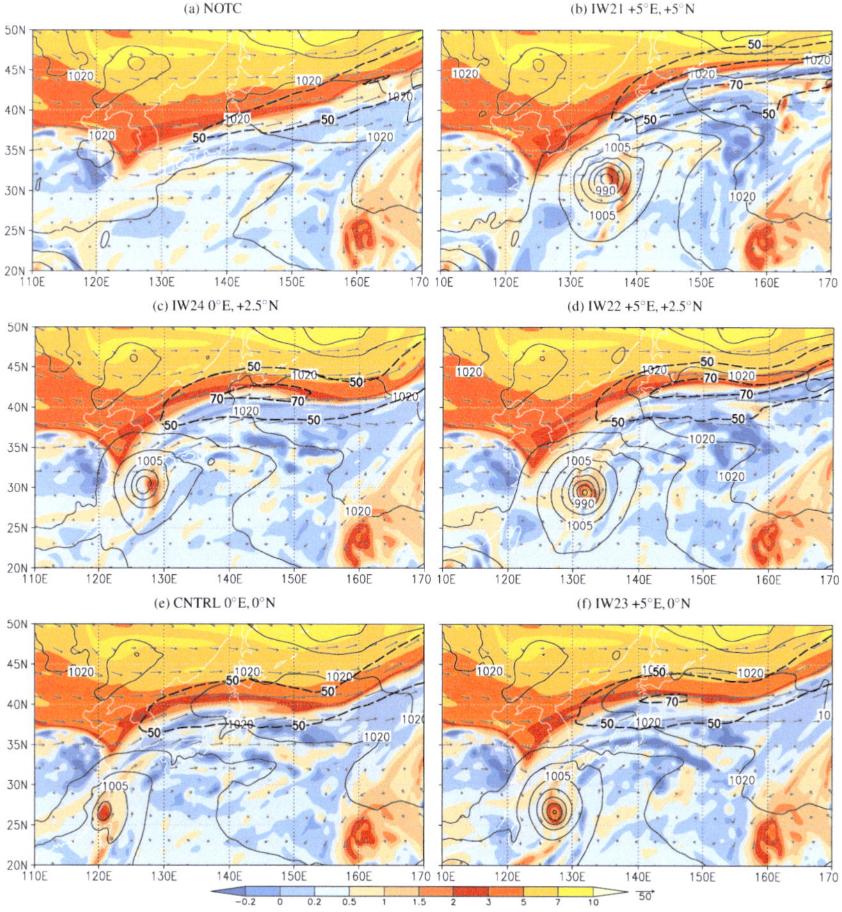


Figure 5.8: As Figure 5.7 but for NOTC, CNTRL and IW21-IW24 scenarios.

speed of more than 70 $ms^{-1}$. Jangmi being located further west- and southward in scenario IW23 initially reintensifies only slightly ahead of the short-wave trough and starts a second, stronger reintensification when the broader main trough passes Japan. This accounts for the temporal delay of the reintensification in IW23. It is also remarkable that a region of very low PV air is present south of the jet streak and the extent of this region is larger and the PV lower the stronger Jangmi reintensifies in the scenarios.

The longitudinal variation of the initial position of Jangmi reveals the crucial impact of the phasing with the short wave trough (Figure 5.9). When Jangmi is initially shifted 2.5° westward (IW27; Figure 5.9a) it is located upstream of the short-wave trough axis and almost dissipated. In CNTRL and IW26 (1.25° eastward shift, Figure 5.9c) Jangmi is located south of and in phase with the short-wave trough axis (Figure 5.9b,c) and has significantly decayed. The jet core wind speed is clearly below 70 $ms^{-1}$ over Japan. Although Jangmi is located slightly ahead of the trough axis in scenario IW25 (2.5° eastward shift, Figure 5.9d) it still decays. When Jangmi is shifted further to the east (IW28 and IW23 (3.75° and 5° eastward shift), Figure 5.9e,f) it moves ahead of the secondary trough, reintensifies and moves across the Pacific. In the latter three scenarios the jet core wind speed exceeds 70 $ms^{-1}$. It can be concluded that Jangmi decays less, the central pressure is lower, and the jet over Japan stronger, the further east Jangmi is located relative to its initial position. A relocation of more than 2.5° eastward yields in a reintensification. These results are in agreement with the relocation experiments of Ritchie and Elsberry (2007) for an idealised TC south and ahead of a midlatitude trough. Ritchie and Elsberry (2007), however, imposed a much farther shift to the storm by varying the latitudinal position between 0-20° in 5° steps and the longitudinal position between 0-40° in 10° steps. Here we showed a crucial sensitivity on a much smaller scale.

The downstream development differs substantially between the different scenarios (Figure 5.10). At 12 UTC 29 September 2008 the short-wave trough is evident in the scenarios at about 122°E and weaker in the analysis at about 128°E (Figure 5.10a, cf. Figure 5.7-5.9). The scenarios in which Jangmi is

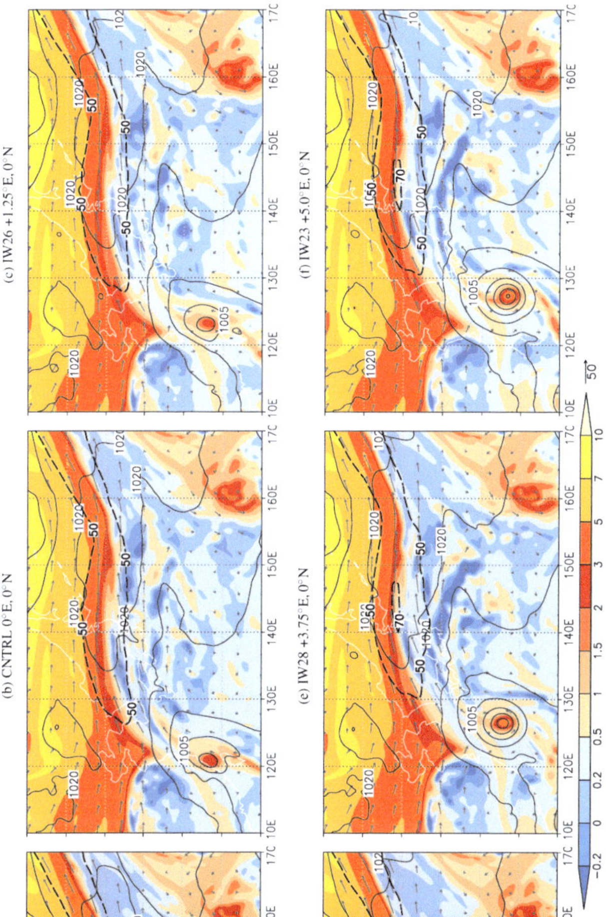


Figure 5.9: As Figure 5.7 but for CNTRL and IW23, IW25-IW28 scenarios.

initially shifted northward and reintensifies (IW21 (blue), IW22 (green), IW24 (dark yellow)) exhibit a stronger ridgebuilding over and downstream of Japan (Figure 5.10a). In theses scenarios Jangmi is located ahead of the short-wave trough.

One day later dramatic changes have occurred (Figure 5.10b). In scenario IW21 (blue) a very strong ridge has built downstream of Jangmi from 140°E to 180°E. In IW22 (green) and IW24 (dark yellow) this is also evident but with a smaller amplitude and about 5° further to the west. The other scenarios exhibit weaker ridgebuilding, except for NOTC (grey). It is notable that the scenarios with decaying Jangmi (IW27 (violet), IW26 (orange), IW25 (turquoise)) are closer to the CNTRL (light blue) run and the analysis (black). The main trough is evident at about 125°E in all scenarios and in the analysis. Jangmi is located ahead of it in IW23 (red) and IW28 (light green).

At 12 UTC 01 October 2008, a broad ridge is evident over the Central Pacific in IW21 (blue) with an axis at about 175°E (Figure 5.10c). In scenario IW22 (green) this ridge has a smaller amplitude and remains 5° further west. A broad downstream trough has developed over the Eastern Pacific in both scenarios (IW21 at about 155°W; IW22 at about 160°W). Further downstream, a second, narrower downstream ridge (axis about 132°W in IW21 and at 137°W in IW22) developed and the trough in the Eastern Pacific has narrowed into PV streamer at around 125°W. Thus in both scenarios a pronounced Rossby wave train was triggered downstream of Jangmi's ET. In IW24 (dark yellow) the midlatitude wave guide exhibits much less amplification although the development is qualitatively similar. In scenarios IW23 (red) and IW28 (light green) Jangmi reintensifies ahead of the main trough (almost stationary in all scenarios at about 130°E). The scenarios develop as in IW22 (green) but with 24 hour delay, with a ridgebuilding between 140°E-180°E. The other scenarios (decaying Jangmi) exhibit much weaker deflections of the midlatitude wave guide over the Pacific and are characterised by a broad ridge.

The next day, the downstream trough in IW21 (blue) at about 147°W and in IW22 (green) at about has 152°W amplified and is transitioning into a PV streamer. The trough together with the much broadened second downstream

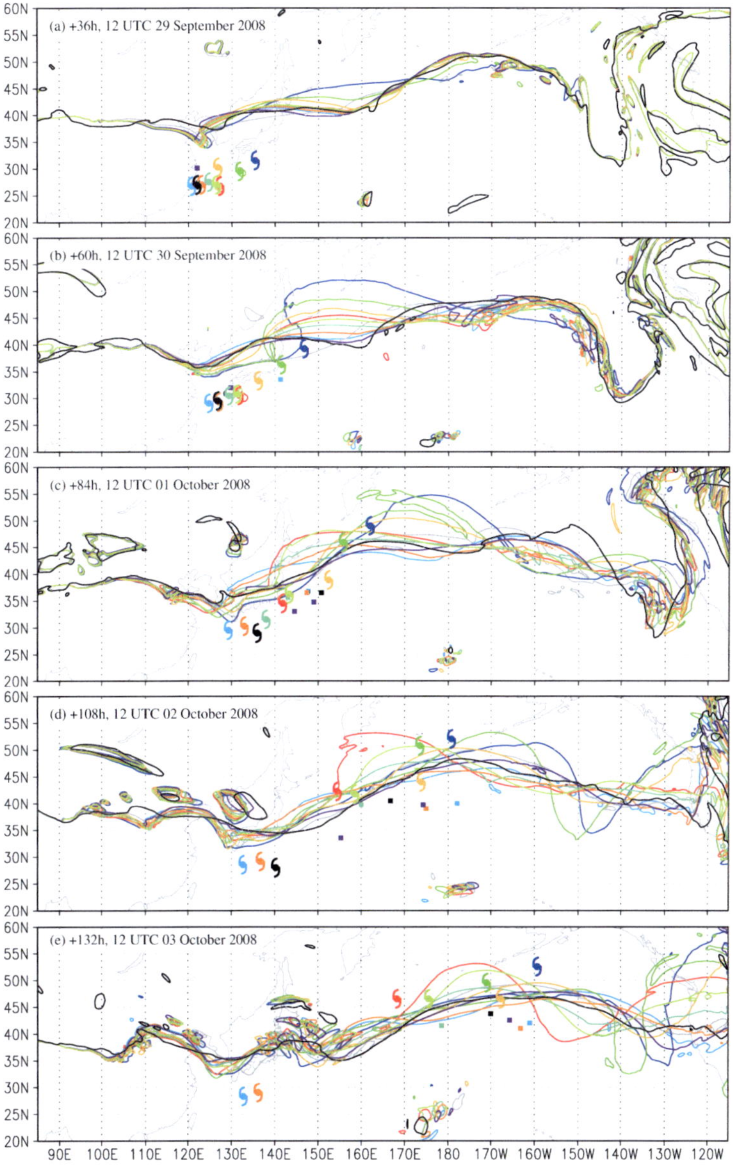


Figure 5.10: 340 K isentropic PV (3PVU contour) from the different scenarios (Table 5.1) coloured as in Figure 5.4. The times are indicated above each panel. The location - if present - of Jangmi is indicated with the TC symbol, and of the downstream cyclone (only in b-e) with a small rectangle.

ridge and the second downstream PV streamer (which moved out of the domain shown) form a marked Rossby wave train over the Eastern Pacific (Figure 5.10d). Again, a phase shift of about 5° is evident between these two scenarios. IW24 (dark yellow) has a much less amplified and broader upper-level wave guide. The other scenarios with reintensifying Jangmi (IW23, red and IW28, light green) show a strong ridging and initial downstream trough development similar to IW22 the day before. All other scenarios have a rather broad ridge across the Central Pacific with only minor deviations from the analysis.

At 12 UTC 3 October 2008 in IW21 (blue) and in IW22 (green) the downstream trough has transformed into a narrow PV streamer off the US and Canadian west coast (at about 132°W and 137°W respectively). In IW23 (red) and IW28 (light green) the downstream trough has amplified, a second downstream ridge developed so that a Rossby wave train is present over the Central and East Pacific. A broad ridge over the Pacific remains in the analysis and similarly in the other scenarios with decaying Jangmi.

It can be concluded that the midlatitude flow during the ET of Jangmi has the potential for strong Rossby wave activity. When Jangmi recurves so that it moves ahead of the midlatitude trough it reintensifies and in all of these scenarios a Rossby wave train develops. The amplitude and timing of this Rossby wave train varies amongst the scenarios. In the north- and eastward relocation scenarios (IW21, blue and IW22, green) an early Rossby wave train triggering occurs starting with a pronounced ridgebuilding on 30 September 2008 as Jangmi reintensifies ahead of the short-wave trough. This behaviour is evident also when Jangmi is relocated northward (IW24, yellow), but the Rossby wave train has a much smaller amplitude. When only relocated eastward (IW23, red and IW28, light green) the reintensification occurs ahead of the main trough with a 24 hour delay. Nevertheless, a pronounced Rossby wave train develops. The amplitude is smaller the closer Jangmi is to its original position. In the other scenarios, in which Jangmi decays as in the analysis, no pronounced Rossby wave activity occurs and a broad ridge is present over the Pacific.

The modification of the downstream flow by the relocated storms is striking when looking at the temporal evolution of the latitudinally averaged meridional wind component (Hovmöller diagram). We discuss the meridional wind in NOTC (Figure 5.11d) along with difference between the experiments and NOTC (Figure 5.11a-c, 5.12). In a Hovmöller diagram troughs approximately appear as a dipole of northerly flow (negative $v$) to the west and southerly (positive $v$) to the east, ridges are marked by southerly flow (positive $v$) to the east and northerly flow (negative $v$) to the west. The NOTC scenario is considered as the background flow without the modification of Jangmi. This is justified as this model run shows an initially strong decrease of the anomaly correlation coefficient during the ET of Jangmi (Figure 4.30). Afterwards the ACC remains almost constant suggesting that the skill reduction is primarily due to the lack of the impact of Jangmi in this scenario.

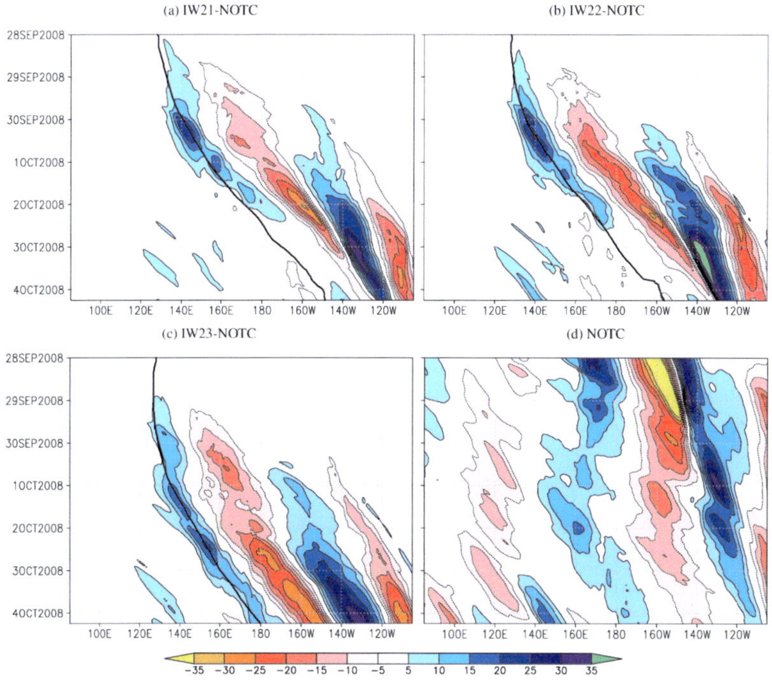


Figure 5.11: Hovmöller diagrams of the 250 hPa meridional wind component difference (scenario minus NOTC) (a-c) and of the 250 hPa meridional wind component in the NOTC scenario (d), meridionally averaged from 25°N to 55°N (with a 5 m s$^{-1}$ contour interval). The thick black line in (a-c) marks the location of Jangmi. The scenarios are indicated above each panel. Plots are sorted from a northern (a) to southern (c) position of the relocated storm.

The midlatitude flow in NOTC is dominated by an almost stationary broad ridge in the central North Pacific with an axis at about 170°W during the entire simulation period (Figure 5.11d). Initially a strong Rossby wave train propagates downstream of this ridge into the eastern North Pacific, but a ridge axis remains at 170°W. The downstream propagation of the weak midlatitude trough approaching Japan is reflected by weak northerly flow to the east of an axis at about 125°E on 29 September 2008. The Hovmöller diagrams for the strong reintensification scenarios IW21-IW23 with a 5° eastward and varied northward relocation, reveal the spatial and temporal shift of the Rossby wave train amongst these scenarios (Figure 5.11). The further northward Jangmi was initially relocated, the earlier the Rossby wave train occurs (see the gradual shift to later times of the Rossby wave train in Figure 5.11a-c). The maximum in northward difference wind at about 145°E in scenario IW21 and 140°E in scenario IW22 on 30 September 2008 (Figure 5.11a,b) reflects the strong ridgebuilding due to the outflow immediately downstream and north of reintensifying Jangmi. The 5° phasing difference is due to the initially faster eastward motion of Jangmi in IW21. Starting at around 00 UTC 1 October 2008 the amplification of the

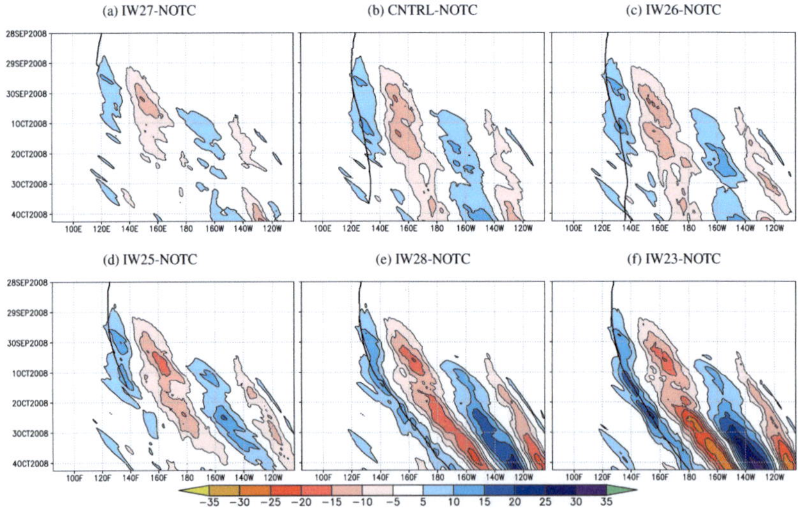


Figure 5.12: As Figure 5.11 but for the scenarios with a shifted initial longitudinal position of Jangmi. The plots are sorted from the easternmost (a) to the westernmost (f) position of the relocated storm.

downstream trough is seen in the couplet of southward and northward difference wind centred at about 165°W in IW21 and 170°W in IW22. Around 00 UTC 2 October 2008 the Rossby wave train is reflected in the difference wind in both scenarios with a pronounced downstream ridge and trough. With time, the ridge becomes less pronounced on its western side. However, the gradient in difference wind for the downstream trough axis becomes stronger, reflecting the transition into a PV streamer and the evolution of the second downstream ridge. The PV streamer is located at about 125°W in IW21 and at about 130°W in IW22 at the end of the calculation period. In scenario IW23 the strong re-intensification is delayed to 12 UTC 1 October 2008. At this time, a maximum in northward difference wind at 140°E reflects the ridgebuilding downstream of Jangmi (Figure 5.11c). The evolution of the downstream trough is reflected in an increasing gradient of meridional difference wind starting at about 170°W at 12 UTC 2 October 2008. Thus a pronounced Rossby wave train exists east of 180°E at the end of the calculation period.

The meridional difference wind in the scenarios with a variation of the initial longitudinal position reveals the evolution of a gradually more amplified Rossby wave train (Figure 5.12). Ongoing weak downstream ridgebuilding during the ET and decay of Jangmi is the most striking feature in CNTRL. It is reflected in the dipole of northward and southward difference wind centred around 140°E from 29 September 2008 to 2 October 2008 (Figure 5.12b). This leads to minor differences compared to NOTC that propagate downstream as a weak Rossby wave train. A similar but much weaker structure is evident when Jangmi is relocated westward (IW27, Figure 5.12a). The 1.25° initial eastward relocation of Jangmi in scenario IW26 (Figure 5.12c) has a weak effect on the downstream evolution as in CNTRL. However, when Jangmi is shifted 2.5° eastward (IW25, Figure 5.12d) a weak Rossby wave train occurs. It emerges east of 170°E downstream of the extratropical cyclogenesis after 00 UTC 2 October 2008. The amplitude significantly increases in scenario IW28 (Figure 5.12e) in which Jangmi reintensifies. A further eastward initial relocation (IW23, Figure 5.12f) results in an even more amplified Rossby wave train along with the stronger reintensification.

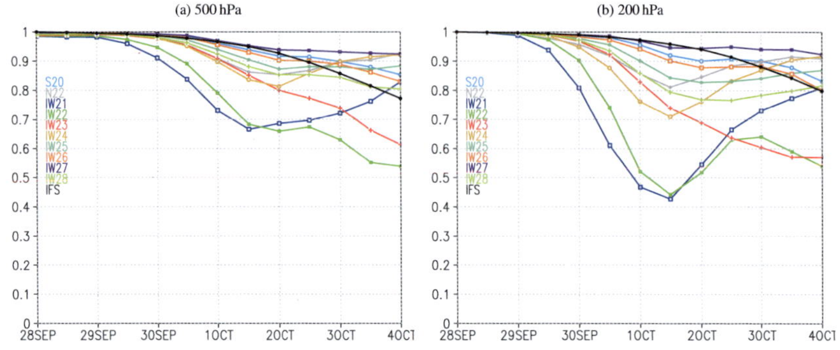


Figure 5.13: Anomaly correlation coefficient (ACC) of geopotential at 500 hPa (a) and 200 hPa (b). Data is from the different scenarios. The scenarios are coloured as in 5.4. The ACC is calculated on the subdomain 100-245°E, 20-60°N every 12 hours. The ECMWF analysis interpolated on the COSMO grid are used as the analysis data and the daily mean values of ERA-INTERIM interpolated on a 0.25° x 0.25° horizontal grid is used as the climatology. Additionally the mean ACC for all deterministic ECMWF IFS forecasts during 1 August 2008 to 1 November 2008 calculated for the domain 110°E-300°E, 20°N-60°N is shown in black.

The gradually stronger downstream development between the different scenarios is reflected in the anomaly correlation coefficient (ACC) for mid- and upper-level geopotential (Figure 5.13). Generally, the 200 hPa ACC characterising the upper level flow is lower than the ACC at 500 hPa. The lowest ACC occurs in most of the scenarios at 12 UTC 1 October 2008 when Jangmi either reintensifies or the extratropical cyclone develops. At this time, the ACC values reflect perfectly the decreasing intensity of the upper-level downstream impact amongst the different scenarios (from lowest to highest: IW21, IW22, IW24, IW23, IW28, NOTC, IW25, IW26, CNTRL, IW27; Figure 5.13b). Strongest deviations from the analysis and lowest ACC occur in IW21. Interestingly, scenario IW27 in which Jangmi decays and two weak extratropical cyclones develop exhibits the highest ACC. All scenarios with a decaying TC have a similar ($+-0.05$) ACC to CNTRL and in comparison to the mean IFS-ACC indicate a better forecast skill to the end of the calculation period.

### 5.1.4 Details of the decay and reintensification scenarios

The two groups of scenarios - decay and reintensification - have completely different impact on the evolution of the midlatitude flow in downstream regions. This is shown here in more detail with the decay of Jangmi represented by the CNTRL run, which is close to the actual evolution. This scenario is contrasted against IW21 (5° east- and northward shift of Jangmi at initialisation time), representing the strong reintensification of Jangmi and subsequent Rossby wave triggering. We then briefly discuss scenarios IW25 and IW28 that exhibit the contrasting evolutions while the initial longitudinal position of Jangmi differs by only 1.25°.

#### 5.1.4.1 Scenario with decay of Jangmi

The period following the ET of Jangmi in CNTRL is characterised by the development of a frontal wave-like weak extratropical cyclone. This cyclone propagates quickly across the Pacific along a baroclinic zone. The upper-level flow is dominated by a very broad ridge. At 12 UTC 30 September 2008 Jangmi is located at about 125°E, 29°N south of the main trough with an axis from about 128°E, 40°N to 138°E, 50°N (Figure 5.14a). The central pressure of Jangmi has increased to 1001 hPa. The tropopause lifting and weak ridgebuilding due to the outflow of Jangmi and the interaction with the baroclinic zone led to an acceleration of the jet core wind speed to more than 70 $\mathrm{m\,s^{-1}}$ over northern Japan. The midlatitude wave guide over the Central Pacific is rather zonally oriented. A very weak downstream trough is evident at about 160°E. One day later a new extratropical cyclone develops at about 148°E, 36°N (1003 hPa) while Jangmi weakens and tracks southeastward (Figure 5.14b). The cyclogenesis is located ahead of the main trough with an axis from about 140°E, 40°N to 150°E, 50°N and in the southwestern entrance region of the jet core that moved to about 160°E during the last 24 hours. The downstream ridge extends from 135°E to 170°E and the weak downstream trough is hardly detectable at about 175°E. At 12 UTC 02 October 2008 the new extratropical cyclone at about 180°E, 40°N (Figure 5.14c) becomes located south of the jet streak. It exhibits diabatic Rossby wave characteristics as it propagated quickly along the baroclinic zone. At this time Jangmi has no closed surface isobar. The midlatitude flow is dominated by a broad ridge that expands over the Central Pacific. The downstream trough can no longer be detected. At 12 UTC 03 October 2008 the frontal-wave like cyclone continued its eastward propagation. With a minimum pressure of 1001 hPa at 161°W, 42°N it becomes embedded in a central low south of Alaska. During the entire period the upper-level flow in the Eastern Pacific close to the North American west coast is characterised by a broad slowly propagating trough associated with a northward moving extratropical cyclone (minimum pmsl 990 hPa).

#### 5.1.4.2 Scenario with reintensification of Jangmi

The upper-level flow shows a very different evolution, when Jangmi reintensifies. Downstream of the ET system a pronounced Rossby wave train is triggered and a secondary downstream low develops. At the end of the calculation period Jangmi becomes the major low in the Northeast Pacific.

In scenario IW21 at 12 UTC 30 September 2008 Jangmi at 148°E, 39°N is reintensifying ahead of the midlatitude trough (Figure 5.15a). The strong reintensification began 12 hours earlier and the central pressure decreased by 7 hPa to 978 hPa. Downstream and north of Jangmi a very intense ridgebuilding occurs due to the strongly enhanced divergent outflow of the reintensifying cyclone. The ridgebuilding is reflected in an extended region of very low PV air and an strongly enhanced midlatitude PV gradient over the Central Pacific (from 145°E to 150°W). The midlatitude jet streak is zonally oriented to the north of this low PV air and accelerated to more than 90 m s$^{-1}$ near 175°E. At around 170°E low PV air is also advected southward into the circulation of the subtropical anticyclone. A very weak initial deflection of the jet is seen at around 180°E.

The reintensification of Jangmi ended at 06 UTC 1 October 2008 (964 hPa, Figure 5.4b). At 12 UTC 1 October 2008 Jangmi is a very strong extratropical cyclone (965 hPa) at 162°E, 49°N (Figure 5.15b). Downstream of Jangmi a pronounced ridge is present with an axis at around 175°E. The ridgebuild-

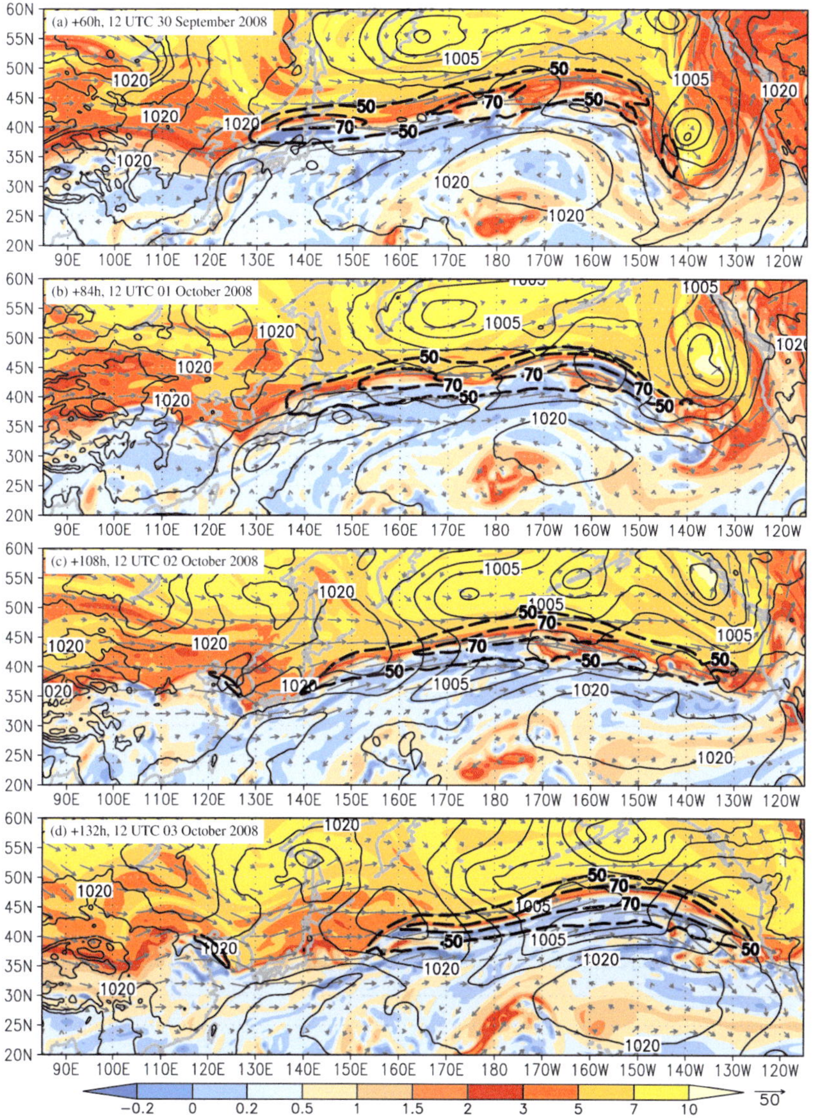


Figure 5.14: PV (in PVU, shaded), wind (grey vectors and black dashed with a $20\,\mathrm{m\,s^{-1}}$ contour interval for $|\vec{v}| > 50\,\mathrm{m\,s^{-1}}$) at the isentropic level of $\Theta$=340K, and pmsl (black contours every 5 hPa) for the CNTRL scenario. Every 24 hours starting at 12 UTC 30 September 2008.

ing amplified a broad downstream trough, that exhibits a secondary axis at about 162°W. The jet core wind speed decreased but still exceeds 70 m s$^{-1}$ in a wide area between the first downstream ridge and the first downstream trough. Ahead of the secondary trough axis a moderate extratropical cyclone develops (<995 hPa). Further to the east an extratropical cyclone at 141°W, 47°N intensified to a central pressure below 985 hPa which is 10 hPa less than on the previous day. A second downstream ridge develops south of the cyclone leading to a thinning of the trough at 125°W, which becomes a PV streamer. Thus a Rossby wave train developed with a broader first downstream ridge-trough couplet and a narrower second downstream ridge-trough couplet.

On the next day, the first downstream trough amplifies at around 145°W and transforms into a PV streamer between the first (axis centred around 160°W) and second (axis centred around 125°W) downstream ridge (Figure 5.15c). An almost meridionally oriented northerly jet with wind speed exceeding 70 m s$^{-1}$ is evident at 153°W and flanks the western side of the PV streamer. The moderate cyclone intensified weakly (<995 hPa) in the "poleward" jet exit region which is also at the southern tip of the PV streamer. The downstream cyclone at about 142°W, 50°N became a major low in the Northeast Pacific and further deepened to below 980 hPa. Jangmi is a prominent extratropical cyclone at 179°W, 52°N.

By 12 UTC 3 October 2008 the downstream cyclone has been absorbed by Jangmi (at 158°W, 52°N, 978 hPa; Figure 5.15d). Jangmi became the major low in the Northeast Pacific. With ongoing upstream ridgebuilding and advection of low PV air, the PV streamer narrows and weakens at about 132°W. The moderate cyclone ahead of this PV streamer decays and tracks northward (<1000 hPa). In the CNTRL scenario the moderate cyclone is not present and an extratropical cyclone in the Eastern Pacific is about 10 hPa less intense than the cyclone developing ahead of the second downstream trough.

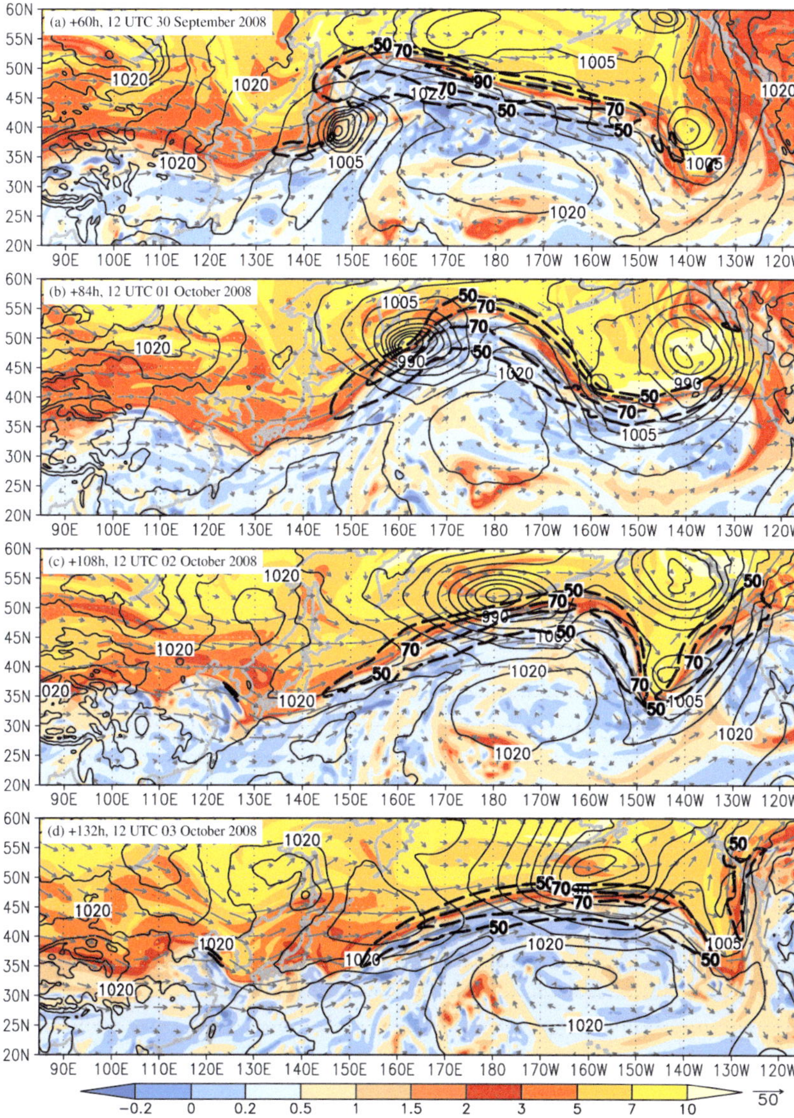


Figure 5.15: As Figure 5.14 but for the IW21 scenario (Jangmi relocated 5°E and 5°N).

### 5.1.4.3 Two neighbouring scenarios

The neighbouring scenarios IW25 and IW28 in which Jangmi's initial meridional position differs by only 1.25° develop into the two contrasting scenarios (Figure 5.16). In scenario IW25 Jangmi decayed and a new extratropical cyclone developed. In scenario IW28 Jangmi was initially located 1.25° (about 128 km) further eastward than in IW25. However, in this scenario Jangmi re-intensified and crossed the Pacific. In scenario IW28 a Rossby wave train is triggered at the end of the calculation period (Figure 5.16b). A downstream cyclone at 137°W, 43°N with a central pressure of less than 985 hPa is associated with the first downstream trough. Contrary, in scenario IW25 a broad ridge with a smaller amplitude is present over the Central Pacific (Figure 5.16a). A strong cyclogenesis downstream is not present. Thus relatively small variations in the initial position of a tropical cyclone prior to ET may have a significant impact on the evolution of the downstream midlatitude flow.

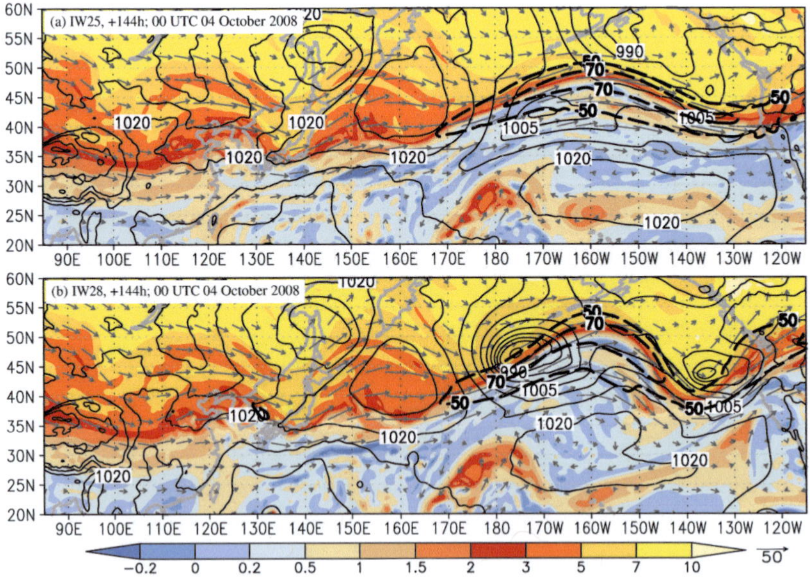


Figure 5.16: As Figure 5.14 but showing scenarios IW25 (a) and IW28 (b) at 00 UTC 04 October 2008 (+144).

In summary the discussion of the downstream development in the various Jangmi scenarios showed that the relative position of the tropical cyclone and the midlatitude wave guide shortly before ET is crucial for the subsequent evolution of the downstream flow. When Jangmi moved into the midlatitudes rather in phase with the approaching midlatitude trough, only a weak frontal wave-like cyclone developed that crossed the Pacific along with a broad small amplitude ridge. However, when Jangmi was located ahead of the midlatitude trough it strongly reintensified and the associated ridgebuilding eventually triggered a RWT or even intense PV streamers downstream. In turn, these upper-level midlatitude features triggered extratropical cyclogenesis, that may have impact on North America as a remote response to the ET in the North Pacific.

### 5.1.5 A possible steering mechanism for a TC during ET

The role of the ambient flow in the different tracks and different downstream impact of the relocated storms is investigated. During the tropical phase the primary mechanism for tropical cyclone motion is the steering trough the environmental flow represented by the deep layer mean wind (Chan and Gray 1982; Dong and Neumann 1986). Various authors investigated the optimal layer average and found that this depends on the intensity of the storm, the ocean basin and the direction of motion (e.g. prior or after recurvature). All these studies showed that a layer mean wind is better suited for the TC track prediction than a single level wind. Velden and Leslie (1991) indicate the 300-850 hPa vertical averaged wind to have the optimum predictive skill for the TC track. For a comparison with previous studies a background flow is required. The NOTC simulation constitutes a set of meteorological fields without the perturbation caused by Jangmi. This scenario can be considered as a representation of the environmental flow during the entire simulation period. We justify this by the fact that the anomaly correlation coefficient at all levels (Figure 5.13) initially decreases rapidly for NOTC during the ET of Jangmi and the interaction with the midlatitudes but afterwards remains constant. This suggests that the skill reduction is primarily due to the lack of the impact of Jangmi in this scenario. If other model errors occur in NOTC during the simulation these should be similar

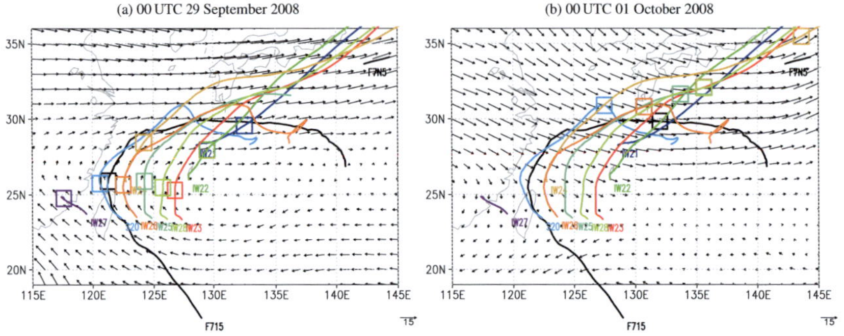


Figure 5.17: 300-850 hPa layer mean wind (length proportional to the wind speed in m $s^{-1}$ as indicated by the scale below panels). The position of Jangmi at the time shown is indicated by the open square on the track. Tracks smoothed with a 10 point running mean, therefore miss the 5 first and last time steps.

to the errors in the Jangmi scenario simulations for which we need an estimation of the background flow.

For the Jangmi scenarios the 350-800 hPa deep layer mean wind direction south of 27°N correlates well with the cyclone motion at 00 UTC 29 September 2008 (Figure 5.17a) for cyclones shortly before recurvature (all except IW21, IW22, IW24). However, two days later when all cyclones have recurved and undergone extratropical transition this is no longer evident (Figure 5.17b). This is similarly evident when the mean layer wind is averaged over other layers, e.g. 500-700 hPa (not shown). A set of 144 hour forward trajectory calculations based on the NOTC data started at 00 UTC 28 September 2008 from boxes along 23°N (the initial latitude of Jangmi) gives more insight into the tropical cyclone motion (Figure 5.18). The boxes are centred around the different initial positions of Jangmi in the scenarios (see caption of Figure 5.18). In the concept of steering by ambient flow the trajectories in the NOTC simulation represent possible tropical cyclone tracks. The westernmost trajectories (black and violet in Figure 5.18b) reflect the track of scenario IW27/I27a with Jangmi making landfall in China at about 119°E. Amongst a possible northeastern track the trajectories emerging west of 126°E (violet, blue, and orange in Figure 5.18b) also indicate air parcels that move southward back to the tropics. Thus these trajectories are able to represent the tracks of the storms, that decay and show a south-

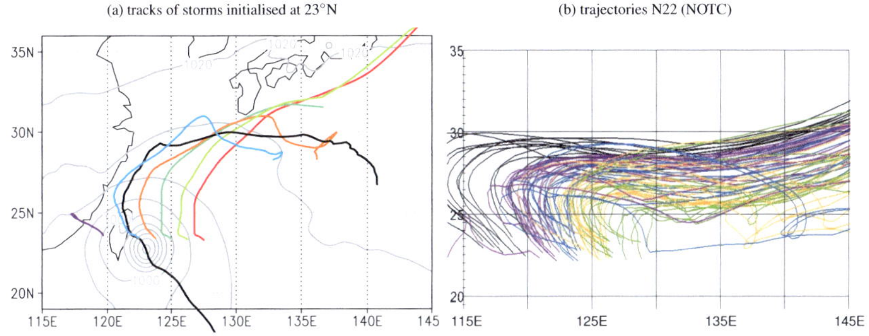


Figure 5.18: (a) as Figure 5.4a but showing only the tracks of the storms relocated in west-east direction (IW23,IW25-IW28, CNTRL (S20), analysis (F715)) and without the position/time markers. The tracks miss the 5 first and last forecast hours due to a 10 point running mean. (b) Every 50th trajectory from different calculations based on data of the NOTC scenario. The same domain as in (a) is shown. Trajectories start in box 22-24°N, 0-10000 m amsl. Zonal extent of boxes: black 115-120°E; violet 120-122°E; blue 122-124°E, orange 124-126°E, green 126-130°E. Only data above 3000 m amsl shown. The colours correspond to the scenarios. The black trajectories show air parcels emerging west of the westernmost scenario (IW27).

eastward track at the end of their lifecycle (CNTRL, NOTW, IW26, IW25). The trajectories east of 126°E (green in Figure 5.18b) do not indicate a south-eastward motion. All air parcels emerging here head northeastward, as seen for the tropical cyclone scenarios IW28, IW23. However, the trajectories remain much further south than the actual tracks (compare Figure 5.18(a) and (b)). It can be concluded that the 350-800 hPa deep layer mean background flow can describe the tropical cyclone motion during the tropical stage until recurvature and early stages of ET (until around 30 September 2008). However, all of the trajectories are well south of the actual track of the reintensifying storms. Thus another explanation for the track at later stages of ET is needed.

Emanuel (2005) illustrates the track of a TC in a stationary ambient flow field, that contains a saddle point (Figure 5.19). TCs that differ slightly in the initial position tend to have dramatically different tracks and motion speeds.

A more detailed investigation of this behaviour is given by Scheck et al. (2011b) in their idealised study of the interaction of a tropical cyclone with a frontal wave. They identified a bifurcation point on the trough axis south of the wave guide. At this bifurcation point the zonal velocity of the background flow matches the phase speed of the wave. Very small deviations of the cyclone loc-

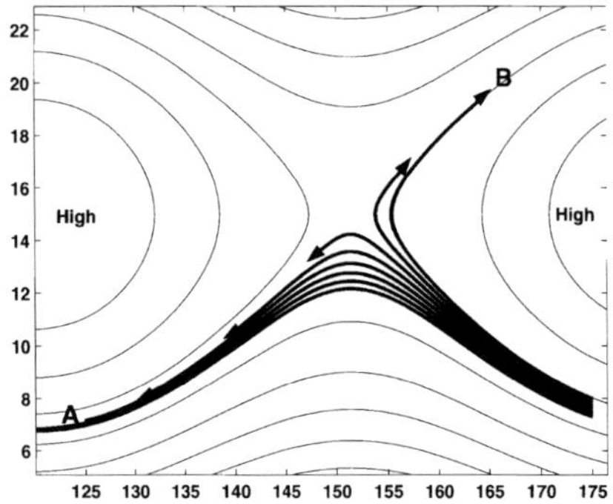


Figure 5.19: Example for hurricane track uncertainty from Emanuel (2005) Figure 29.4: Eight simulations of the movement of a hurricane, using the same model but starting from eight slightly different (20 miles) estimates of its initial location at lower right. The contours represent a stationary streamfunction of the environmental flow.

ation from this bifurcation point determine whether the cyclone moves towards the next trough-ridge couplet downstream and is attracted poleward or towards the ridge-trough couplet upstream, where it is steered equatorward. This is illustrated in Figure 5.20. The reasons for this behaviour become obvious from the definition of the bifurcation point. The trough-relative streamfunction is defined as the streamfunction $\Psi_r$ relative to the zonal phase speed $c$ of the midlatitude trough: $\Psi_r = \Psi - cy, \Psi$ : stream function of total (background) flow. In the fol-

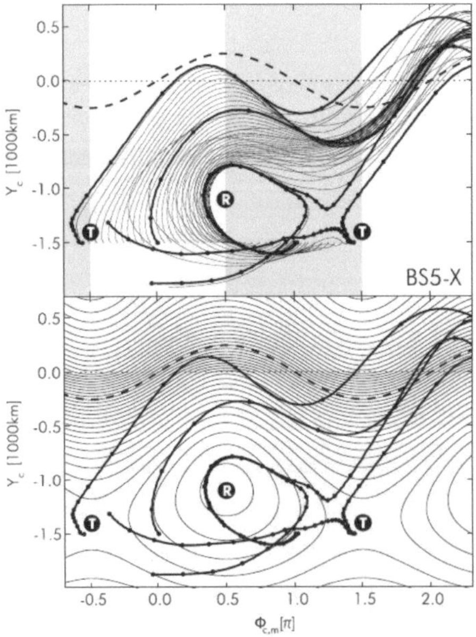


Figure 5.20: Example for tracks of cyclones with different initial longitudinal position from Scheck et al. (2011b, Figure 5): (top) cyclone tracks (solid lines) in the frame moving with the phase speed of the wave. Thick lines are used to highlight the cases of an initial phase shift (in multiples of $\Pi$) $\Phi_{c,m}$ =1.0, 1.02, 1.43, and 1.44 and the start of each day is marked with a dot. The initial latitudinal position of the cyclones is $Y_c = -1500$ km south of the mean jet (dotted). The dashed line is the preexisting frontal wave. The locations of the saddle points in the trough-relative stream function are marked by T. (bottom) The thick lines are as at top; the thin lines are contour lines of the wave streamfunction which is constant in the frame moving with the wave.

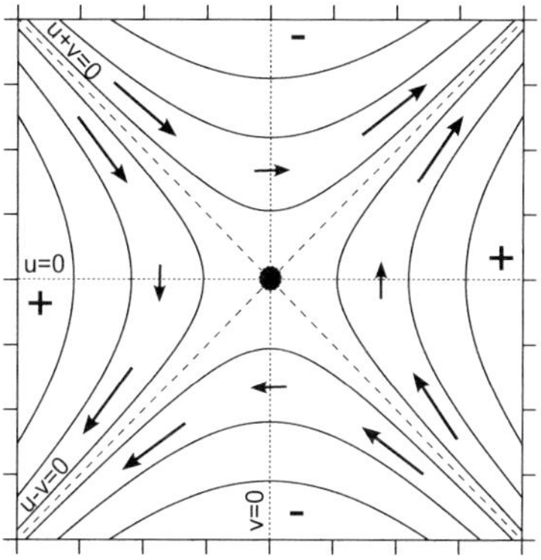


Figure 5.21: Schematic sketch for the trough-relative streamfunction $\Psi_r$ (solid contours) and flow (arrows, dashed lines) in the vicinity of a bifurcation point (marked by the black dot). The + and - signs mark maxima (in the west and east) and minima (in the north and south) of $\Psi_r$. Contours of trough-relative wind components $u_r + v_r = 0$ and $u_r - v_r = 0$ shown as long dashed lines, and of $u = 0$ and $v = 0$ as short dashed lines. The arrow length is a qualitative measure for the wind speed.

lowing we assume that a TC tracks like a Lagrangian air parcel that moves in the flow field associated with $\Psi_r$. For the thought experiment we further assume $\Psi_r$ to be stationary so that streamlines are trajectories. The bifurcation point is defined as a saddle point in the trough-relative stream function (Figure 5.21). It is located between a maximum of $\Psi_r$ to the west and east and a minimum of $\Psi_r$ to the north and south.

The trough-relative wind components in zonal and meridional direction vanish at the bifurcation point ($u_r = -\partial\Psi_r/\partial y = u_\Psi - c = 0, v_r = \partial\Psi_r/\partial x = 0$). In a simple thought experiment we consider possible tracks of air parcels that approach the bifurcation point and are advected solely by the flow field associated with $\Psi_r$. In the quadrants to the northeast and southwest of the bifurcation point the flow is directed away from the bifurcation point (Figure 5.21). In the northwestern and southeastern quadrants the flow is directed towards the bifurcation point. At first sight it remains unclear whether an air parcel near the diagonals will track to southern or northern latitudes. The contour of $u_r + v_r = 0$ separates the streamlines in these quadrants. Air parcels to the northeast of this contour ($u_r + v_r > 0$) move into the northeastern quadrant and thus have a final northward track. This emerges from the fact that, for example, the eastward wind component of the flow northeast of $u_r + v_r = 0$ in the northwestern quadrant is larger than the southern wind component and thus air parcels will be advected faster to the northeastern quadrant than to the southwestern quadrant. This is analogously valid in the southeastern quadrant and southeast of the $u_r + v_r = 0$ contour. Thus the contour $u_r + v_r = 0$ constitutes a critical line for the track of air parcels being advected in the flow field associated with the trough-relative streamfunction. To its northeast ($u_r + v_r > 0$) air parcels will reach the northeastern quadrant and subsequently be advected northward. To its southeast ($u_r + v_r < 0$) air parcels will reach the southwestern quadrant and subsequently be advected southward. If we assume that a TC is advected like an air parcel this behaviour also determines the TC tracks.

In the previous discussion the theoretical existence of a bifurcation point for the track of a TC was explained. As in the idealised barotropic simulations of Scheck et al. (2011b) the tracks in the Jangmi scenarios were sensitive to the TC initial position and diverged when the storm was shifted slightly (cf. Figures 5.20 and 5.4a). Thus it is intriguing to investigate if such a bifurcation point is present in the more realistic Jangmi scenarios. Here the midlatitude wave is not analytically prescribed and the motion speed of the trough needs to be estimated from the unperturbed environmental flow constituted by the NOTC scenario. However, the structure of the midlatitude trough that passes Japan from 29 September 2008 to 2 October 2008 in this realistic scenario is much more complicated than in idealised studies. In fact, it develops where a northern and southern branch of the midlatitude jet merge (see the two branches of moderate intense jet streams west of 120°E and north of 30°N in Figure 4.34). During this period the trough exhibits two consecutive main trough axes in its northern part and several secondary trough axes in the south. The trough axes emerge from waves propagating along the northern and southern jet branches.

The trough propagation speed is estimated using Hovmöller diagrams of relative vorticity and meridional wind component (Figure 5.22). The broader main trough stands out when the 100-300 hPa layer average and a zonal average from 40°N to 48°N are taken (Figure 5.22a). The trough appears as a band of high relative vorticity starting in average at about 105°E 00 UTC 28 September 2008 and being traceable to 170°E at 12 UTC 03 October 2008. The zonal translation speed is approximately 0.5° $h^{-1}$. However, the dominant relative vorticity signal is modified by two trough axes passing through the northern part of the trough during Jangmi's recurvature (29 September 2008 to 01 October 2008).

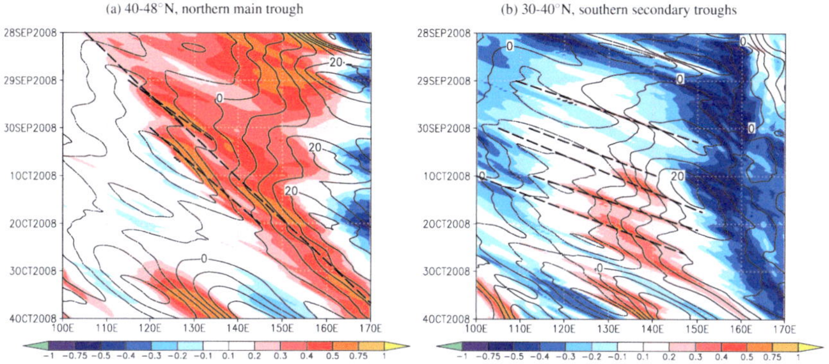


Figure 5.22: Hovmöller diagrams of relative vorticity (shaded in $10^{-5}\,s^{-1}$) and meridional (v-) wind component (black with a $5\,m\,s^{-1}$ contour interval). Vertically averaged from 100-300 hPa (a) and 200-300 hPa (b) (with data every 25 hPa), zonally averaged from 40-48°N (a) and 30-40°N (b). Trough axes appear as maxima of relative vorticity and the mean location is indicated with long dashed lines. Data from NOTC simulation (0-144 hours).

Thus the broad trough exhibits two subsequent main trough axis. The first is reflected in a maximum of relative vorticity starting at around 116°E at 00 UTC 29 September 2008. The clear signal in relative vorticity decays at around 134°E at 06 UTC 30 September 2008. A second trough axis appears reflected in a strong maximum of relative vorticity starting at around 120°E at 00 UTC 30 September 2008. It propagates somewhat more slowly than the first axis and merges with the signal of the main trough at around 145°E, 00 UTC 02 October 2008. For both trough axes the average zonal propagation speed is higher than $0.5°\,h^{-1}$ (about $0.6°\,h^{-1}$ and $0.52°\,h^{-1}$, respectively). On the 340 K isentropic level (200-300 hPa where PV>2 PVU, see Figure 4.4) the first main trough axis travelling from the northern jet branch into the midlatitude trough appears at 12 UTC 29 September from around 121°E, 45°N to 132°E, 51°N (black dashed line in Figure 5.23a). It is hardly detectable 18 hours later when the second main axis arrived (long black line in Figure 5.23b). At 00 UTC 01 October 2008 the second main axis is identifiable from 131°E, 39°N to 142°E, 50°N (Figure 5.23c). It is remarkable that both main axis arriving from the northern jet branch untilt when they become the axis of the broad midlatitude trough.

The secondary axes in the southern half of the main trough are clearer in the Hovmöller diagram at slightly lower levels (200-300 hPa) and averaged further

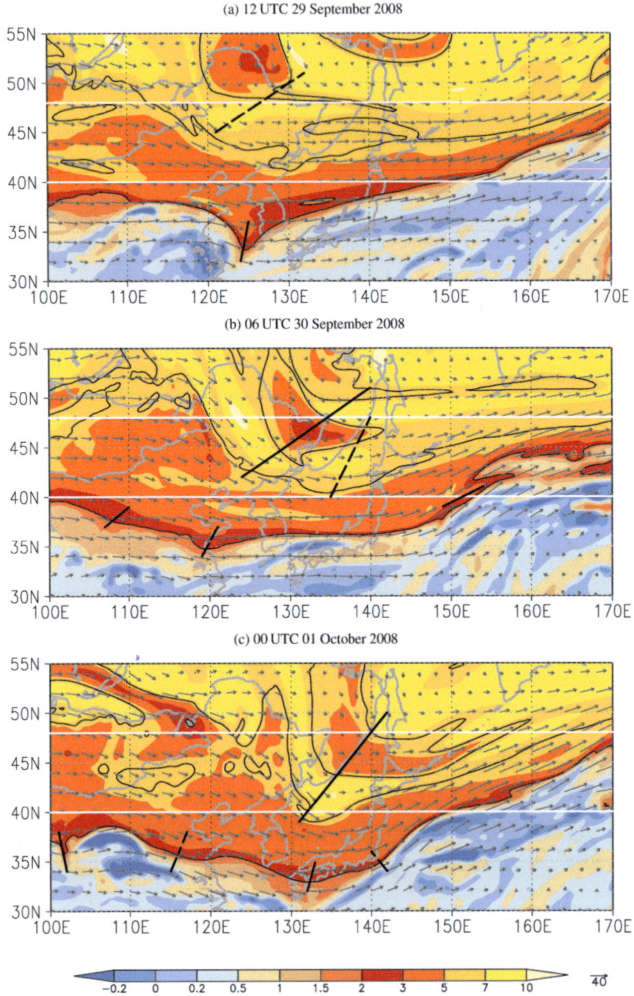


Figure 5.23: 340 K isentropic PV (shaded in PVU, 2 and 6 PVU contours (black)) from the NOTC simulation at times indicated above the panels. The white horizontal lines mark the regions for the zonal average in Figure 5.22. The black tilted lines around the 6 PVU contour mark the first (dashed) and second (solid) main trough axes. The black tilted lines around the 2 PVU contour mark several secondary troughs.

south from 30°N to 40°N (Figure 5.22b). In the period 00 UTC 29 September 2008 to 00 UTC 01 October 2008 five consecutive trough axes appear reflected in maxima of relative vorticity emerging between 105°E-115°E. The first secondary trough is prominent in the 2 PVU contour at around 124°E, 35°N at

12 UTC 29 September 2008 (Figure 5.23a) and associated with the first main trough axis. This is the secondary trough ahead of which Jangmi reintensified and recurved in scenarios IW21, IW22. Four other secondary troughs travel along the southern edge of the main trough until 00 UTC 01 October 2008. The zonal propagation speed of the secondary troughs is two to three times faster than the zonal propagation of the main trough.

The complicated midlatitude trough structure with two main and five secondary trough axes in the period of Jangmi's ET makes it hard to determine a unique zonal propagation speed of the broad midlatitude trough and the exact trough-relative streamfunction. The zonal propagation speed of the main trough is estimated by $c = 0.5^\circ\,h^{-1}$. which is the average zonal propagation speed determined for the main trough in the layer 100-300 hPa from Figure 5.22a. The associated trough-relative streamfunction $\Psi_r$ has been calculated every six hours based on the NOTC data.

$\Psi_r$ vertically averaged in the layer 100-300 hPa depicts two subsequent bifurcation points during the period 29 September 2008 to 02 October 2008 (Figure 5.24). In geographical coordinates (Figure 5.24a) a first bifurcation point approaches south of the first main trough axis at 12 UTC 28 September 2008 and is located in about the area of the pronounced secondary trough (northern black track in Figure 5.24a labelled JB1). Similar to the propagation of the main

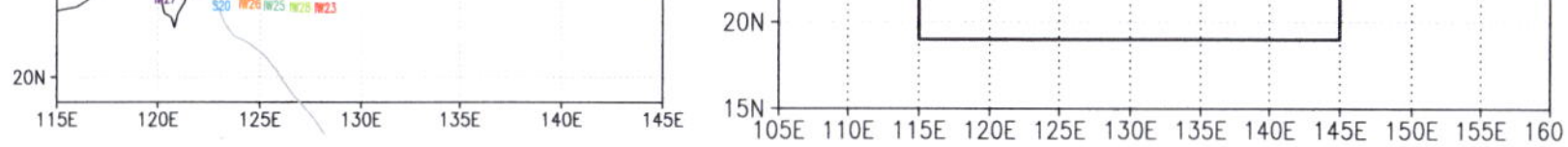


Figure 5.24: (a) as Figure 5.4(a) but analysis in grey and with the tracks of the bifurcation points associated with the first and second main trough axis (black, labelled JB1, JB2). Position of Jangmi and of the bifurcation point is indicated at 12 UTC 29 September 2008 (open square). (b) Tracks of Jangmi and of the bifurcation points in the frame moving with constant zonal propagation speed of $0.5^\circ\,h^{-1}$. The tracks are shown relative to their geographical location at 06 UTC 01 October 2008 (marked with open square). The black outlined box indicates the subregion shown in (a).

trough axis it moves to about 137°E at 00 UTC 30 September 2008 where it disappears. In this period only the early reintensifying TCs in scenarios IW21, IW22 track (north)east of the first bifurcation point. This is better evident in the frame moving with constant zonal propagation speed (Figure 5.24b). The first bifurcation point becomes almost stationary (rightmost black track in Figure 5.24b) and the storms in IW21 and IW22 (blue and green) recurved well to its northeast. A second bifurcation point follows at about 122°E at 06 UTC 30 September 2008 (southern black track in Figure 5.24a) south of the approaching second main trough axis. From 06 UTC 30 September 2008 to 18 UTC 01 October 2008 the second main trough axis passes over Japan and with it the second bifurcation point (Figure 5.24a). However, afterwards its position jumps between 130°E and 141°E when the secondary troughs pass. In the frame moving with constant zonal propagation speed (Figure 5.24b) the reintensifying storms in IW24, IW23 and IW28 (yellow, red, light green) recurve well to the northeast of the second bifurcation point, while the decayers remain or decay in the vicinity of it.

The position of the bifurcation point has been detected objectively in the trough-relative streamfunction (relative to the estimated constant propagation speed $0.5° \, h^{-1}$) at locations where $u_r = v_r = 0 \, m \, s^{-1}$ south of a local minimum in $\Psi_r$. This is illustrated at 12 UTC 29 September 2008 for the bifurcation point associated with the first main trough axis (Figure 5.25a). The bifurcation point is located at 125.5°E, 31.0°N. In the reference frame moving with constant zonal speed $0.5° \, h^{-1}$ the tracks of Jangmi in IW21 (blue) and IW22 (green) are in the northeastern quadrant. Jangmi in IW24 (dark yellow) passes close to the bifurcation point. In all other scenarios the storms remain well south of it. At a later time (06 UTC 01 October 2008, Figure 5.25b) the relative tracks of the early recurving Jangmis (IW21, IW22 and IW24) are well in the northeastern quadrant of the second bifurcation point (134.75°E, 29.75°N) associated with the second main trough axis. In the decaying scenarios CNTRL (light blue), analysis (black), and IW26 (orange) the tracks remain west of or close to the bifurcation point. Although they are located northeast of the $u_r + v_r = 0 \, m \, s^{-1}$ contour at this time, they track westward later. It is harder to discriminate the

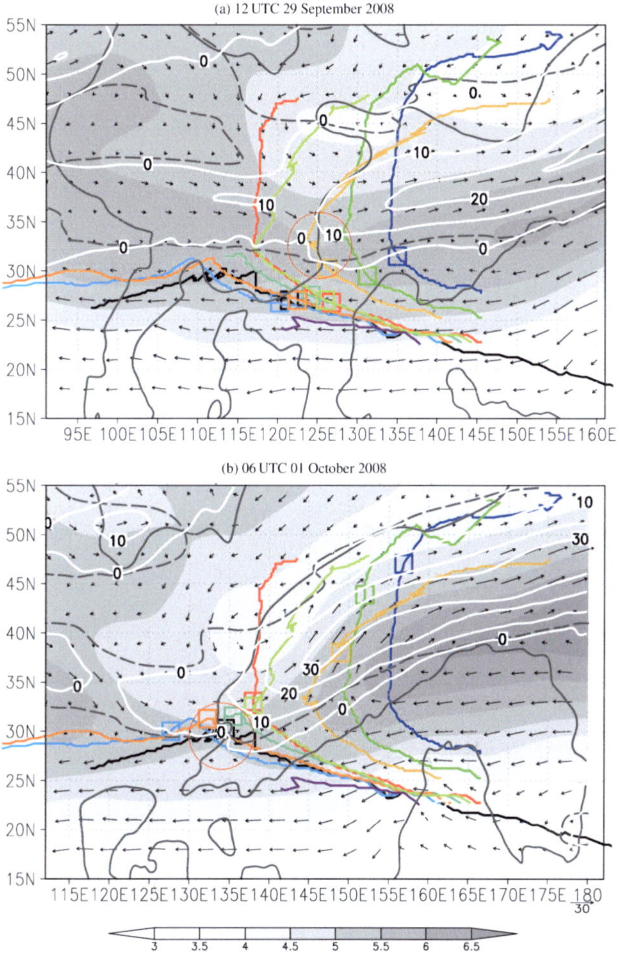


Figure 5.25: Trough-relative stream function $\Psi_r$ (shaded every $0.5 \cdot 10^6\,\mathrm{m^2 s^{-1}}$) vertically averaged from 100-300 hPa and associated trough-relative non-divergent wind vectors (scale for wind speed in $\mathrm{m\,s^{-1}}$ below the panels). Contours: $u_r = 0\,\mathrm{m\,s^{-1}}$ (dark grey dashed), $v_r = 0\,\mathrm{m\,s^{-1}}$ (dark grey solid), $u_r + v_r = 0, 10, 20, 30, 40, 50$ $\mathrm{m\,s^{-1}}$ (white solid). Tracks are shown relative to the estimated zonal trough propagation speed of $0.5°\,\mathrm{h^{-1}}$ and centred on the TC position (open square) at the time shown. The red circles highlight the bifurcation points.

tracks of scenarios IW23 (red), IW28 (light green) and IW25 (turquoise). These move close to northeast of the bifurcation point. A better discriminator than $u_r + v_r > 0$ at the location of the storm seems to be some higher threshold, e.g. Jangmi in IW25 (turquoise) is southwest of the contour $u_r + v_r = 10\mathrm{m\,s^{-1}}$ and in IW23 (red), IW28 (light green) it is northeast at this particular time. Jangmi

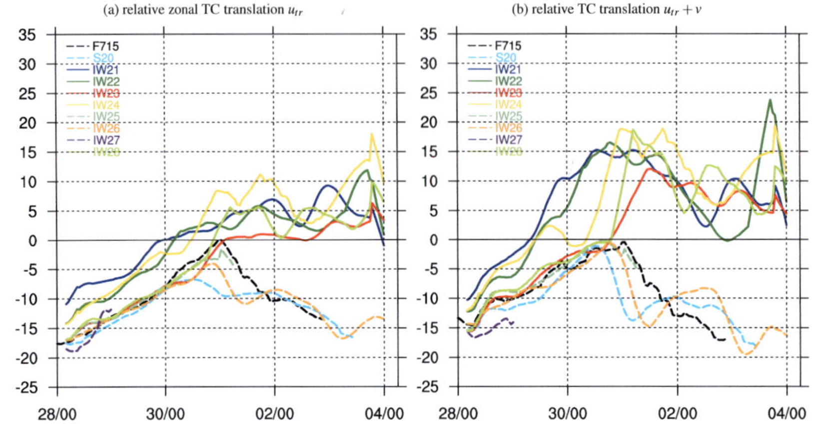


Figure 5.26: Translation speed (in $\mathrm{m\,s^{-1}}$) of Jangmi in the frame relative to the trough propagating with constant zonal speed of $c = 0.5^{\circ}\,\mathrm{h}^{-1}$ calculated from one-hourly track data. The different scenarios are coloured, reintensifiers with solid contours, decayers dashed. (a) zonal relative translation speed $u_{tr}$, (b) sum of zonal and meridional component $u_{tr}+v$ ($v_{tr}=v$).

in IW25 (turquoise) exhibits a westward track and soon decays. In IW23 (red) and IW28 (light green) the relative zonal translation speed of Jangmi becomes positive and the storm subsequently recurves and reintensifies. The translation velocity components in the frame moving with constant zonal propagation speed (Figure 5.26) allows the scenarios to be separated into decayers and reintensifiers. A trough-relative zonal translation component $u_{tr} \geq 0\mathrm{m\,s^{-1}}$ indicates recurvature in the frame moving with constant zonal propagation speed and the storms subsequently reintensify (Figure 5.26a). The sum of relative zonal and meridional translation component $u_{tr}+v > 0\mathrm{m\,s^{-1}}$ constitutes an even better discriminator for the different scenarios (Figure 5.26b cf. discussion of Figure 5.21).

The unstationary position of the bifurcation point (also in the frame moving with constant zonal phase speed) makes an investigation of the track of Jangmi in the reference system relative to the position of the bifurcation point difficult. Nevertheless, a subset of Jangmi's track in the different scenarios relative to the bifurcation point is discussed for the period of the first and second main trough axis (Figure 5.27). The bifurcation point in the mean streamfunction for the first main trough axis (Figure 5.27a) clearly separates the tracks. The final part (about one third) of the tracks in the early reintensifying scenarios IW21 (blue) and IW22 (green) are to the northeast of the $u_r + v_r = 0\,\mathrm{m\,s^{-1}}$ contour. IW24 (dark yellow) moves through the bifurcation point and all other tracks remain

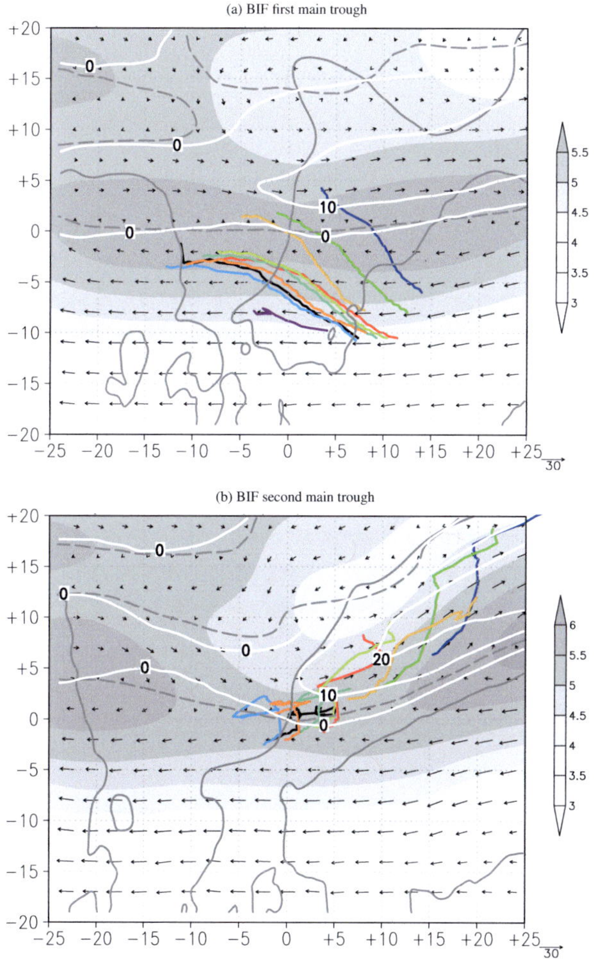


Figure 5.27: As Figure 5.25 but temporal average of $\Psi_r, u_r, v_r$ centred on the position of the bifurcation point. Vertical average from 100-300 hPa. (a) temporal average 12 UTC 28 Sep 2008 - 00 UTC 30 Sep 2008, (b) 06 UTC 30 Sep 2008 - 18 UTC 01 Oct 2008. The tracks of Jangmi in the diverse scenarios are relative to the position of the bifurcation point.

well south of it. The tracks relative to the bifurcation point in the mean streamfunction for the second main trough axis (Figure 5.27b) fit in the conceptual picture also. The tracks of all reintensifying storms (IW21-IW24, IW28) are well in the northeastern quadrant and well beyond the $u_r + v_r = 10\mathrm{m\,s}^{-1}$ contour. The tracks of the decaying storm (CNTRL (light blue), analysis, (black), IW26 (orange), IW25 (turquoise)), although located to the north of the bifurcation point and partly beyond the $u_r + v_r = 0\mathrm{m\,s}^{-1}$ contour, rather circle around before decaying in the vicinity of the bifurcation point and remain in the region $u_r + v_r < 10\mathrm{m\,s}^{-1}$.

Finally we highlight the mean trough-relative streamfunction during the entire period 12 UTC 28 Sep 2008 - 00 UTC 03 Oct 2008 when a bifurcation point was identifiable (Figure 5.28). The temporal mean bifurcation point in the trough-relative streamfunction composite relative to the bifurcation point (Figure 5.28a) clearly separates the decaying from the reintensifying scenarios, although the tracks in the frame relative to the bifurcation point appear ragged. The last part of the tracks in the reintensifying scenarios are all located well in the northeastern quadrant and within the $u_r + v_r = 10\mathrm{m\,s}^{-1}$ contour. The tracks of IW23 and IW28 (red and light green) initially are grouped with the decaying scenarios but translate quickly to the northeast of the bifurcation point after recurvature. The temporal average of $\Psi_r$ in the same period (without relocation to the position of the bifurcation point Figure 5.28b), reveals a mean bifurcation point at 133°E, 30°N. In the temporal mean it is unclear at which position the tracks in the frame moving with the estimated constant zonal propagation speed should be fixed. In contrast to Figure 5.25, we therefore discuss the actual tracks in geographic coordinates (Figure 5.28b). Jangmi passes close to the mean bifurcation point in all these scenarios. The recurving scenarios (IW21-IW24, IW28) move in the northeastern quadrant and their tracks become located in the region where $u_r + v_r > 10\mathrm{m\,s}^{-1}$. The decaying scenarios IW27 (violet), CNTRL (light green), and analysis (black) remain southwest of the mean $u_r + v_r = 0\mathrm{m\,s}^{-1}$ contour, although some of them come close to the bifurcation point. Decaying scenarios IW26 and IW25 (orange and aquamarine) also enter the quadrant northeast of the bifurcation point. However they remain well outside the mean

$u_r + v_r = 10\,\mathrm{m\,s^{-1}}$ contour and decay in the vicinity of the bifurcation point. It is remarkable that the tracks of the reintensifying storms at later stages are comparable to the streamlines of the mean trough-relative streamfunction in particular for IW21 and IW22 (blue and green). This suggests that $\Psi_r$ at 100-300 hPa constitutes some kind of steering for the extratropical cyclones. We note also,

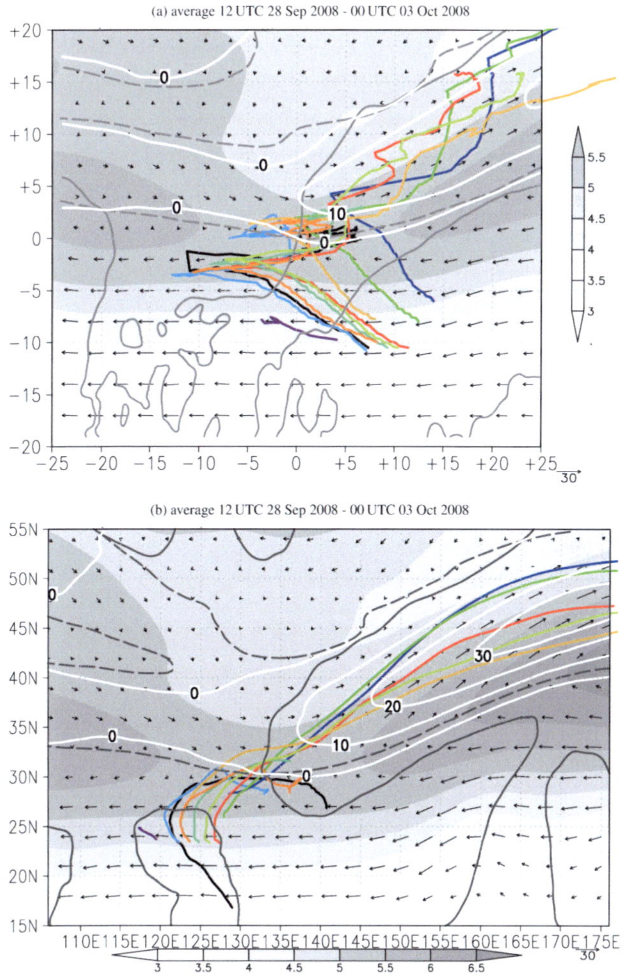


Figure 5.28: (a) as Figure 5.27 but the temporal average is from 12 UTC 28 Sep 2008 to 00 UTC 03 Oct 2008. (b) as Figure 5.25 but a temporal average of $\Psi_r, u_r, v_r$ is taken from 12 UTC 28 Sep 2008 to 00 UTC 03 Oct 2008 and the tracks are real tracks in geographic coordinates.

that in all decaying scenarios the extratropical cyclogenesis immediately downstream of Jangmi occurred in the northeastern quadrant of the bifurcation point associated with the second main trough axis (not shown).

The investigation of the 100-300 hPa streamfunction relative to an estimated constant zonal trough propagation of $0.5^{\circ}\,h^{-1}$ revealed the existence of two saddle points south of two northern main trough axes. It is remarkable that despite no continuous saddle point for the main trough, this bifurcation point allows the Jangmi scenarios to be discriminated into decayers and reintensifiers. The obvious investigation of the tracks in the frame relative to the exact, position of the bifurcation point is hindered because of the unclear exact trough propagation. A cleaner relative track is achieved in the frame moving with the assumed constant zonal propagation speed. In both frames of reference we found a necessary condition for Jangmi to recurve is that it becomes located to the northeast of the mean bifurcation point, in accordance with our thought experiment and the findings of Scheck et al. (2011b). The reintensifying storms seem to track beyond a critical line that is determined by a threshold of $u_r + v_r > 0$. Once being located well to the northeast of the bifurcation point and this critical line, storms are able to move northeastwards towards the midlatitude front and reintensify. Otherwise storms are repelled and eventually move back to the tropics. Also, the zonal translation speed for all reintensifying storms equalled or exceeded the assumed constant zonal propagation speed of the trough at later stages.

This section is concluded with a discussion of the parameters discriminating the different scenarios. As outlined earlier, during the tropical stage Jangmi's track is steered by the environmental deep layer mean wind (Figure 5.17). The foregoing discussion gave evidence that during ET the trough-relative streamfunction at upper-levels constitutes a steering for the transforming cyclone and enables us to separate the scenarios into intensifiers and discriminators. The temporal evolution of the sum of the wind components of these steering flows centred on the location of Jangmi is shown together with the sum of the translation components of Jangmi (Figure 5.29). Results are similar if the layer wind is taken from a different level (e.g. 650 hPa or 500-700 hPa, not shown) rather than 300-850 hPa. For the trough-relative flow it emerged from Figures like

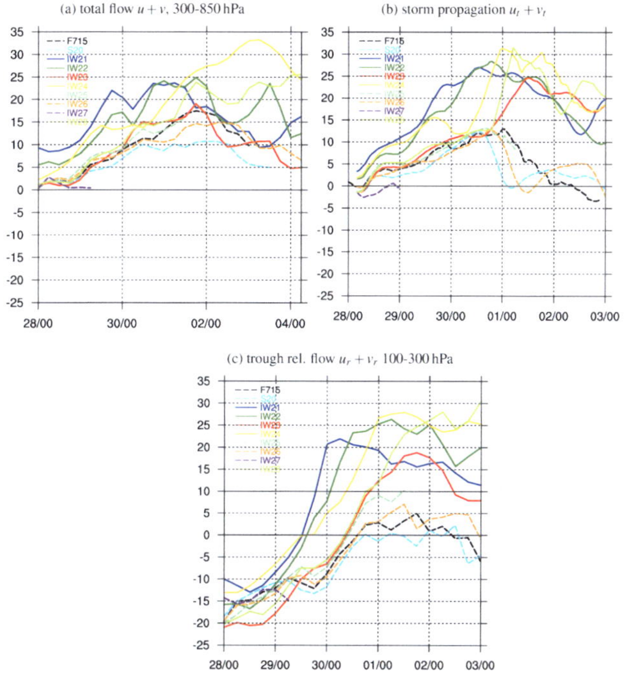


Figure 5.29: Sum of the wind components (in m s$^{-1}$) along the tracks of Jangmi in the scenarios. (a) total background flow $u+v$ vertically averaged from 300-850 hPa. (c) non-divergent flow $u_r+v_r$ in the frame relative to the trough moving with constant zonal propagation speed of $c$ =0.5° h$^{-1}$ vertically averaged from 300-850 hPa. The data in (a) and (c) is six-hourly from NOTC and the average of a 2°x2° box centred on the storm location. (b) sum of the translation vector of Jangmi ($u_t+v_t$, in m s$^{-1}$) calculated from one-hourly track data, smoothed with a 10 point running mean. Reintensifiers are shown with solid contours, decayers dashed. X-axis indicates DD/HH day (DD) and hour (HH) in UTC of September/October 2008.

5.29c but for different layers that the 100-300 hPa vertical average is best suited for describing the steering (not shown). The same illustration for the individual vector components are given in the Appendix Figure A.1.

The combination of both zonal and meridional components provides a clearer discrimination of the cyclone motion. As outlined in the thought experiment (Figure 5.21), cyclones that track in the region northeast of the $u_r+v_r=0$ contour ($u_r+v_r>0$, at cyclone position) should move into the northeastern quadrant of the bifurcation point and subsequently be attracted into the midlatitudes. Looking at the position of the storms in relation to the bifurcation point we speculated that $u_r + v_r = 10\,\mathrm{m\,s^{-1}}$ at the position of the storm is a threshold for reintensifying storms. The temporal evolution of $u_r + v_r$ at the cyclone position (Figure 5.29c) confirms this behaviour. The quantity $u_r + v_r$ exceeds a threshold of $10\,\mathrm{m\,s^{-1}}$ only for reintensifying storms. Comparing $u_r + v_r$ with the actual storm translation $u_t + v_t$ also reveals that following the time of recurvature in trough-relative coordinates ($u_{tr} = 0$, at 30/00Z IW21, 30/06Z IW22, 30/12Z IW24, 01/00Z IW23 and IW28; Figure 5.26a) or when the decaying storms are close to the bifurcation point ($u_{tr} \approx 0$, at around 01/00Z analysis, CNTRL and IW25 and IW26 Figure 5.26a) $u_r + v_r$ (Figure 5.29c) is a very good representation of the actual storm translation $u_t + v_t$ (Figure 5.29b) for at least 24 h. Quantitatively this is less distinct for the decayers. We conclude that when an ET system approaches a bifurcation point in the trough-relative streamfunction its track is predominantly governed by the trough-relative non-divergent upper-level flow. The deep layer mean wind $u + v$ (Figure 5.29a) hardly gives any indication of the actual cyclone propagation.

In summary, despite various limitations regarding the determination of the actual midlatitude trough propagation, a simple criterion in the environmental background flow was distilled to distinguish whether the environment enables an ET system to recurve and reintensify or not. In accordance to previous studies on TC motion we found the deep layer mean wind to be the predominant steering flow during the tropical and early stages of ET until around recurvature in geographical coordinates. However, during later stages of ET the 100-300 hPa upper-level flow relative to the (estimated constant) propagation of the midlatitude trough governs the steering and reintensification of the ET system. When the ET system moves into an environment where the sum of the trough-relative non-divergent wind components $u_r + v_r$ exceeds a certain threshold, $10\,\mathrm{m\,s^{-1}}$ in the case of Jangmi, the storms recurved and reintensified, others decayed. The threshold is higher than expected from a simple thought experiment which is most likely due to the constraints of the determination of the trough propagation. It also has to be kept in mind that in the individual scenarios ridgebuilding downstream of Jangmi modifies the amplitude of the approaching trough and eventually triggers the formation of secondary troughs.

Consequently the environmental flow as represented by the NOTC scenarios constitutes only a first approximation of the actual environmental flow. Our criterion is much simpler than the discriminating parameters proposed by Ritchie and Elsberry (2007) based on Petterson-Smebye type-B cyclogenesis. With the bifurcation point in the trough-relative streamfunction we also give a simple explanation of the physical mechanism that determines the systematic separation of the scenarios. Our study further shows the existence of this mechanism proposed by Scheck et al. (2011b) in realistic three-dimensional baroclinic cases.

We should remember also that the Jangmi scenarios exhibit rather small differences of their initial position compared to previous studies (Ritchie and Elsberry 2007; Riemer and Jones 2010). The maximum distance between a decaying scenario and reintensification scenario is about 610 km (IW25, IW21) and the minimum distance only about 125 km (IW25,IW28). Such track "errors" are frequent in current numerical weather prediction models (see discussion of Figure 5.3). The sensitivity study shown here highlights the link between the uncertainty in track forecast and in the forecast for the midlatitude flow downstream of an ET event. Due to the large sensitivity to the TC initial position, the correct track of an ET system is crucial for the correct prediction of the midlatitude flow evolution. As a consequence of this sensitivity and the difficulties in the track forecast, the correct forecast of the downstream impact during ET becomes a very challenging task for numerical weather prediction. Given a forecast for background flow the trough-relative stream function highlight regions that are favourable for a northeastward track of an ET system, and regions where a TC would decay. TCs that move to the northeast of the bifurcation point in trough-relative streamfunction would be expected to reintensify. Knowing the forecast track of an ET system relative to the bifurcation point gives an indication of the probability of this track and thus guidance for operational weather forecast. If the storm moves close to the bifurcation point in the forecast will be more sensitive to the correct track predication, than when the storm passes well away of the bifurcation point.

## 5.2 The ET of Hurricane Hanna (2008)

As a next case study we investigate Hurricane Hanna in the North Atlantic in September 2008. As in the case of Jangmi, Hanna recurves into a rather zonal midlatitude flow to the north between the subtropical ridge and a primary low to its northeast. In contrast to Jangmi, a more pronounced trough approaches upstream and Hanna reintensifies. During and following the ET of Hanna a chain of weather events occurred leading to the formation of a PV streamer and the triggering of a Mediterranean cyclogenesis event. The crucial impact that repeated diabatic processes associated with the evolution of interacting cyclonic systems have on the modification of the midlatitude Rossby wave guide have been investigated by the research group PANDOWAE in a joint case study which was managed by the first and second author of Grams et al. (2011). Parts of this study are repeated here but a more detailed focus is on the ET sensitivity experiment using PV surgery. In the following, we discuss four cyclonic systems: (Ex-)Hanna undergoing ET, a short-lived cyclone developing downstream of Hanna (downstream cyclone), an extratropical cyclone forming upstream of Hanna near Newfoundland (upstream cyclone), and the Mediterranean cyclone "Olivia".

### 5.2.1 Synoptic overview

#### 5.2.1.1 Tropical stage

At 12 UTC, 28 August 2008 Hanna formed from a tropical wave at about 57°W, 19°N (analysis track labelled HHJ in Figure 5.30). The tropical storm headed westnorthwestward to the north of the Leeward Islands. At 00 UTC, 2 September 2008 Hanna reached maximum intensity with a central pressure of 977 hPa and maximum sustained winds of 39 m s$^{-1}$ (Brown and Kimberlain (2008)). It made a loop north of Haiti and on 4 September 2008 was back on its northwestward track passing the Bahamas and Florida off the coast. Hanna made landfall close to the border of North and South Carolina at 07 UTC, 6 September 2008 and recurved along the US East Coast. It was officially declared extratropical at

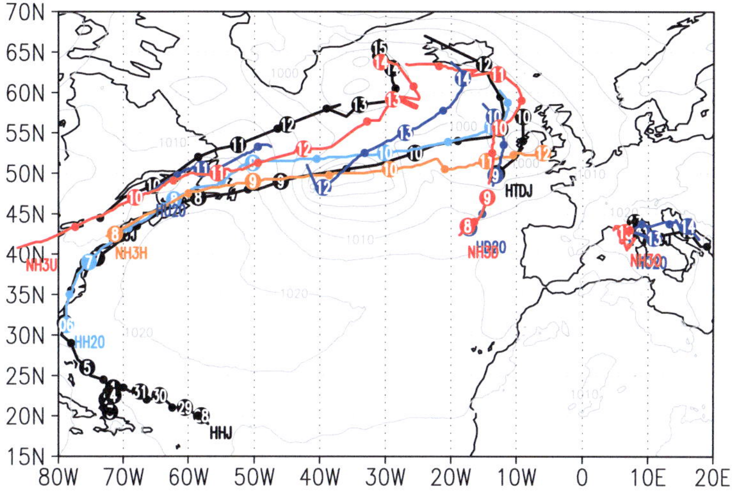


Figure 5.30: Tracks of the cyclones during the ET of Hurricane Hanna. The cyclones are (Ex-)Hanna (H), the upstream cyclone (U), the downstream cyclone (D) and Mediterranean cyclone Olivia (O). Tracks from analysis data are in black and labelled HHJ, HTUJ, HTDJ, HTOJ. Tracks from the CNTRL run are in blue colours and labelled HH20, HU20, HD20, HO20. Tracks from the NOTC experiment are in red colours and labelled NH3H, NH3U, NH3D, NH3O. The thick points, labelled with the day of August/September 2008 mark 00 UTC times, the thin points mark 12 UTC times. Pressure at mean sea level (pmsl) from the analysis data is shown at 12 UTC 09 September 2008. (b) Minimum pmsl (in hPa) in the centre of the discussed cyclones during the period of the COSMO simulations. The same labels as in (a) are used.

06 UTC, 7 September 2008. At this time structural changes indicate the development of a warm sector over New England.

#### 5.2.1.2 Extratropical stage

The synoptic evolution during and following the ET of Hanna is shown with the help of 320 K isentropic PV in the ECMWF analysis data complemented by WCB intersection points within $\pm$ 2.5 K of the 320 K isentropic level (Figure 5.32, cf. Figure 4 from Grams et al. 2011). This isentropic layer typically slopes from about 550 hPa on the southern side of the midlatitude PV gradient to about 300 hPa on the northern side and thus represents the evolution at mid to upper levels. The WCB intersection points are based on 48 hour forward trajectory calculations using six-hourly ECMWF analyses. The air parcels ascend more than 600 hPa in 48 hours. The trajectory calculations

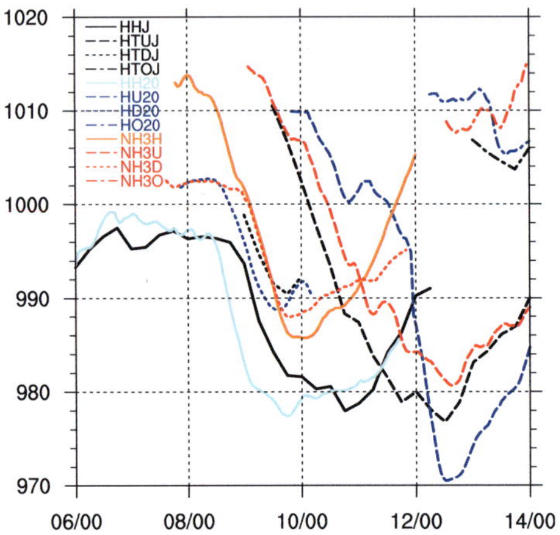


Figure 5.31: Minimum pmsl in the centre of the discussed cyclones during the period of the COSMO simulations. The same labels as in (a) are used.

are started every six hours and intersection points of all these calculations valid at a particular time are shown.

As Hanna approaches the midlatitudes the central pressure (from ECMWF analysis) increased to about 998 hPa on 7 September 2008 (Figure 5.31). Hanna (labelled $L_{Hanna}$, Figure 5.32a) moves into a broad ridge and its sustained outflow broadens the ridge at the western flank. During the next days a trough approaches from the west and interacts with Hanna, initiating an extratropical reintensification on 9 September 2008 (Figure 5.32c and black solid line (HHJ) in Figure 5.31). The ridgebuilding immediately downstream of Hanna is enhanced by the advection of low PV air in WCBs triggered ahead of the reintensifying cyclone. The air parcels within this WCB (as marked by the black crosses in Figure 5.32c) align along the region of lowest PV at the ridge and to the northwest of Ex-Hanna's surface centre. As a consequence of the ridgebuilding, a downstream trough amplifies and ahead of it a short-lived cyclone develops (labelled $L_{ds}$, Figure 5.32c).

This downstream cyclone caused moderate rainfall in parts of Ireland and the British Isles. In the following two days the northern part of Ex-Hanna's trough wraps up cyclonically supported by the WCB that partly advects low PV air cyclonically around Ex-Hanna and partly extends the ridge downstream of Ex-Hanna far to the north (Figure 5.32e). The southern portion of Ex-Hanna's trough elongates southeastward and becomes a PV streamer. At the same time a new extratropical cyclone develops upstream of Ex-Hanna near Newfoundland (labelled $L_{upstream}$, Figure 5.32e, HTUJ in Figure 5.30 and Figure 5.31 ). It intensifies ahead of a midlatitude trough and builds a ridge upstream of (Ex)-

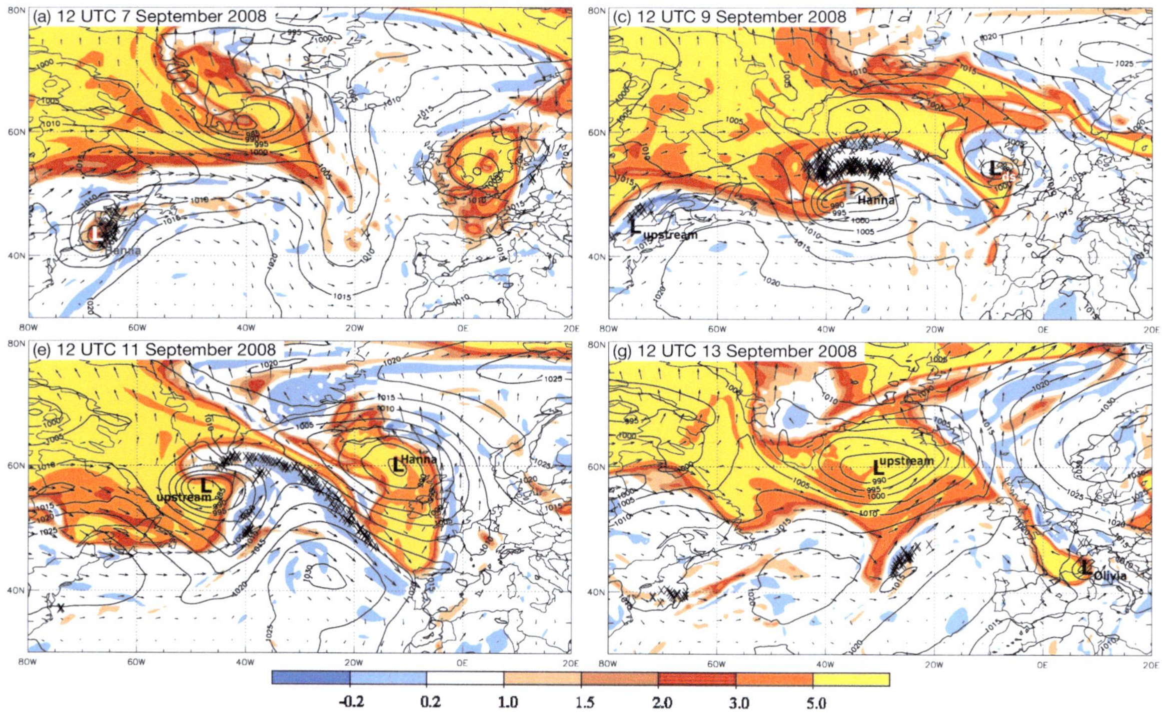


Figure 5.32: Adapted from Figure 4a,c,e,g of Grams et al. (2011): PV (in PVU, shaded) and wind vectors (black) on the 320-K isentropic level, pmsl (black contours every 5 hPa). PV and pmsl are from the ECMWF operational analyses. The crosses mark the intersection of WCB trajectories with the isentropic layer between 317.5 K and 322.5 K. Specifically every 30th point is displayed in (a) and (c), every point in (e), and every 2nd point in (g). The "L" labels mark the surface position of the cyclones. The plots were originally produced by Maxi Böttcher and Heini Wernli (ETHZ).

Hanna. The ridge building is induced by the advection of low PV air in a WCB triggered ahead of the upstream cyclone (see the alignment of intersection points along the midlatitude PV gradient). Along with the upstream ridge building this WCB also elongates Ex-Hanna's trough southeastward at its western flank (Figure 5.32e). While Ex-Hanna had only a relatively weak direct impact on Europe, convection triggered ahead of the PV streamer lead to severe thunderstorms with heavy precipitation in western Europe. Finally, another WCB is triggered ahead of the cold front of the upstream cyclone (intersection points at 25°W, 45°N in Figure 5.32g). The WCB narrows the PV streamer so that it cuts off. The cut-off moves into the western Mediterranean and triggers the cyclogenesis of Olivia (labelled $L_{Olivia}$, Figure 5.32g). The cut-off also causes large-scale heavy precipitation north of the Alps. Olivia itself caused severe mesoscale convective systems over the western Mediterranean region and the Adriatic Sea. The impacts in terms of precipitation are discussed in more detail in Section 5.2.5

In summary, the interplay of WCBs linked to Hanna and the upstream cyclone crucially modified the midlatitude wave guide such that a PV streamer forms over the eastern North Atlantic and evolves into an upper-level cut-off low triggering a Mediterranean cyclogenesis. Grams et al. (2011) further showed that the advection of tropical warm air masses against a baroclinic zone by Hanna triggered the WCBs. Along the WCB trajectories diabatic PV production occurred at lower and mid levels and diabatic PV reduction at upper levels. This low PV air had a significant impact on the ridgebuilding downstream of Hanna. A simple sensitivity experiment that interrupts this process is the removal of Hanna prior to the interaction with the baroclinic zone. The direct impact of Hanna on this chain of events will be investigated in the following by comparing a COSMO run with and without Hanna in the initial conditions. We start with a brief description of the technical details of this PV surgery application.

## 5.2.2 Technical details of the PV surgery

Two COSMO simulations were computed using the standard COSMO configuration (Section 3.1.2). The simulations differ only in the initial conditions. Six-hourly ECWMF analyses (with 0.25° horizontal resolution) serve as boundary conditions. The model runs start at the end of the tropical stage at 00 UTC 6 September 2008 when Hanna reached the US East Coast. Similar to the determination of the initialisation date for Jangmi at this time Hanna is well south of the midlatitude wave guide and an interaction of Hanna and the midlatitude flow was not seen in the isentropic PV charts. The horizontal domain covers the entire North Atlantic and major parts of Europe. Information on the COSMO runs are summarised in Table A.4.

The control simulation (CNTRL=SH20, SH20 is the label for the experiment) is initialised from the unmodified ECMWF analysis. The positive PV anomaly linked with Hurricane Hanna and the associated anomalies in the meteorological fields have been removed at 00 UTC 6 September 2008 in the initial conditions for the no-TC simulation (NOTC=NH30). Details of the PV inversion and the anomaly definition are summarised in the Appendix, Table A.5.

### 5.2.3 Verification of the control experiment

The track of Hanna in the control run (HH20, light blue in Figure 5.30) shows weak deviations from the analysed track (HHJ, black in Figure 5.30). Major differences occur between 8 and 10 September 2008, when the track is up to 4° further north. During that period Hanna also reintensifies about 12 hours earlier than in the analysis (Figure 5.31). However, the magnitude of the intensity change is simulated very well. The downstream cyclone (HD20, blue short dashed control track; HTDJ, black short dashed analysis) passes Ireland off the coast and further westward. It is detectable 24 hours earlier than in the analysis data, but the intensification in the control run after 00 UTC 09 September 2008 is almost identical (Figure 5.31). Although the upstream cyclone (HU20, blue control track; HUJ, black analysis track) develops at about the same location (65°W, 48°N) its subsequent development in the control run differs significantly from the analysis. In the control scenario an explosive intensification period occurs on 12 September 2008 linked to a DRW-like structure that appears to the southeast of the original cyclogenesis. The cyclone centre transitions

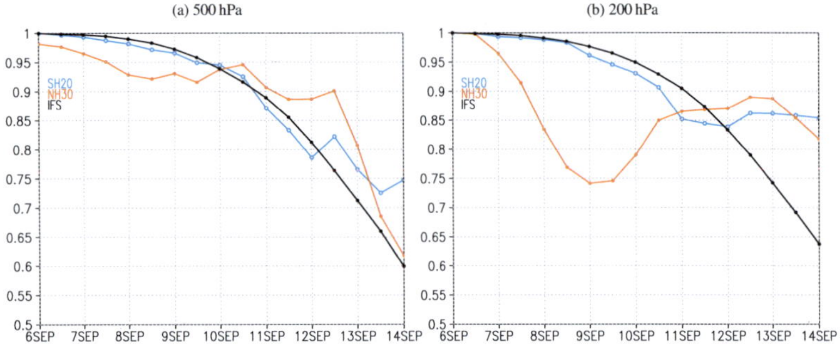


Figure 5.33: Anomaly correlation coefficient (ACC) of geopotential at 500 hPa (a) and 200 hPa (b). Data is from the CNTRL (SH20, light blue) and NOTC (NH30, orange) simulation. The ACC is calculated on the subdomain 80°W-20°E, 25°N-55°N) every 12 hours. The ECMWF analyses interpolated on the COSMO grid are used as the analysis data and the daily mean values of ERA-INTERIM interpolated on a 0.25°x0.25° horizontal grid is used as the climatology. Additionally the mean ACC for all deterministic ECMWF IFS forecasts during 1 August 2008 to 1 November 200 calculated for the same domain is shown in black.

to the DRW structure and its subsequent track is more south- and westward. The Mediterranean cyclone Olivia (HO20, blue) develops in the Ligurian Sea at about the same location as in the analysis on 12 September 2008 (HOJ, black) and propagates quickly eastward reaching about the same intensity.

The different evolution of the upstream cyclone is the main constraint on the control experiment. For the Jangmi scenarios differences between the analysis and the COSMO control simulation were increasing after the 4 day (96h) forecast range. This is similarly evident for the Hanna control simulation. The anomaly correlation coefficient for CNTRL at mid (500 hPa, Figure 5.33a) and upper (200 hPa, Figure 5.33b) levels is above 0.93 until 10 September 2008 (96 hour forecast). After the genesis of the upstream cyclone and the interaction with the unrealistic DRW the ACC for 500 hPa geopotential decreases to 0.79 at 00 UTC 12 September 2008. For the rest of the simulation period the ACC reduces less with time and at upper levels the ACC even increases slightly, indicating the rather useful representation of the cut-off process on 13 September 2008. Thus the control run can be considered as a reference simulation for the quantification of Hanna's interaction with the midlatitudes during the early stages of ET. However, Hanna's impact on the downstream evolution can only be estimated using CNTRL.

### 5.2.4 The impact of Hanna on the the midlatitude flow evolution: PV surgery

From 7 September 2008 WCBs associated with Hanna's outflow transport and advect very low PV air towards the midlatitude wave guide and initiate ridge-building downstream of the tropical storm. As presented by Grams et al. (2011) and similar to Jangmi the low level TC circulation advects tropical air towards the baroclinic zone where it ascends slantwise resulting in a diabatically enhanced transport of low PV air to jet level. At upper-levels the WCBs support the poleward advection of low PV air in the TC outflow.

At 00 UTC 8 September 2008 in the control simulation (CNTRL; Figure 5.34a) very low PV air at jet level expands from Hanna (62°W, 46°N) eastward along the southern edge of the midlatitude wave guide ($\approx$ 2 PVU). The advection and transport of low PV air by Hanna's interaction with the baroclinic zone and upper-level flow lead to a significant amplification of the ridge downstream. The northern edge of the ridge is at about 55°N where a pronounced jet streak with wind speed of more than 80 m s$^{-1}$ extends from 60°W to 30°W. Further downstream, a trough is elongated to 30°W, 26°N. In contrast, in the experiment with Hanna removed from the initial conditions (NOTC; Figure 5.34b) the downstream trough is weaker and its southern tip is located significantly further east at about 22°W, 30°N. The ridge over the Atlantic shows less amplification with the northern edge at around 48°N, about 700 km south south of that in the control experiment. The jet streak is weaker also ($|\vec{v}| \approx 60\,\mathrm{m\,s^{-1}}$). The ridge-building is reflected in a broad region of very low negative difference PV on 345 K (<-7 PVU) in the jet region (Figure 5.35) where the absolute wind speed difference exceeds 40 m s$^{-1}$. The amplified downstream trough appears as a region of positive difference PV (up to 5 PVU) at 25°W. The more westward location ($\approx$ 9°longitude) of the southern edge of the trough in the control scenario is obvious when directly comparing the 345 K 2 PVU contour in both runs (light blue and orange contours in Figure 5.35). The joint diabatic and advective (PV-) frontogenetic enhancement of the tropopause PV gradient $\nabla_h PV$ is evident when comparing $\nabla_h PV$ on 345 K in the scenarios (Figure 5.36). In the

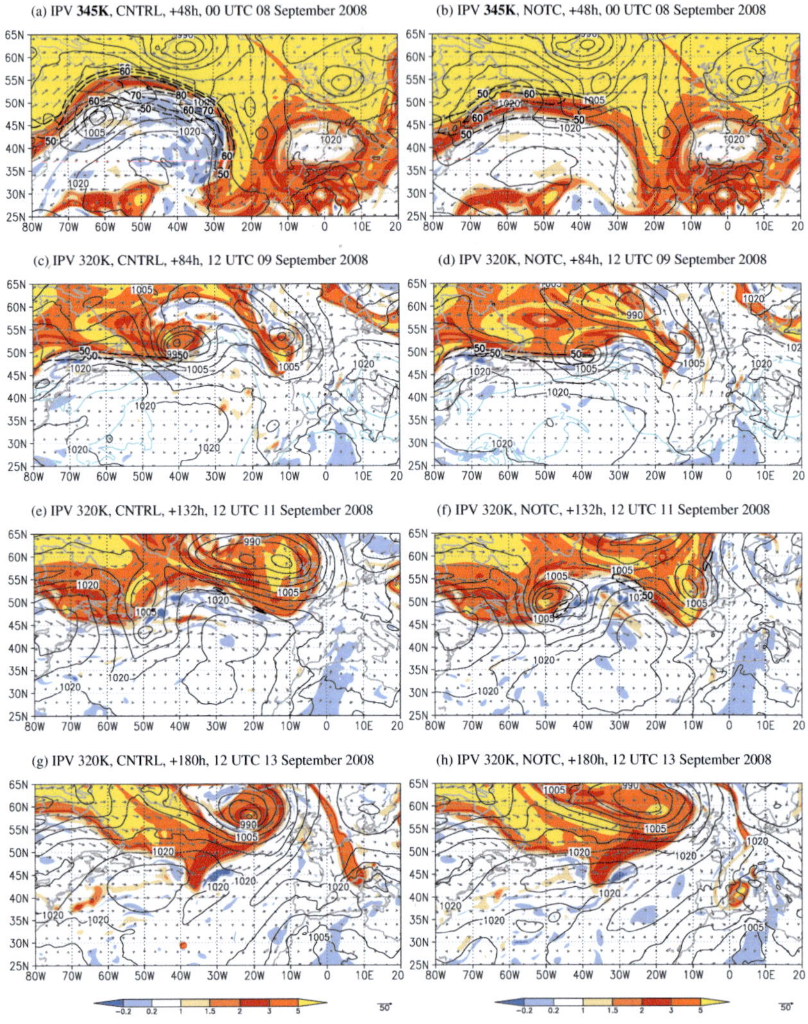


Figure 5.34: PV (in PVU, shaded), wind (grey vectors and black dashed with a $10\,\mathrm{m\,s^{-1}}$ contour interval for $|\vec{v}| > 50\,\mathrm{m\,s^{-1}}$) at the 345 K (a,b) and 320 K (c-h) isentropic level, and pmsl (black contours every 5 hPa). In panels (c),(d) additionally the 2 PVU contour on 345 K is shown. On the left data from the control simulation is plotted and on the right data from the NOTC simulation. The forecast hour and valid times are indicated above the panels.

control simulation an almost uninterrupted band of $\nabla_h PV$ >3 PVU $(100\mathrm{km})^{-1}$ extends from 70°W, 45°N downstream of Hanna to 30°W, 35°N reflecting the ridgebuilding and the amplification of the downstream trough (Figure 5.36a). The maximum PV gradient is 7 PVU $(100\text{km})^{-1}$ at 40°W, 55°N, slightly north of the jet streak. In contrast, the weaker ridge in the NOTC simulation appears as an intermittent band of $\nabla_h PV > 2$ PVU $(100\text{km})^{-1}$ from 70°W, 45°N to 30°W, 50°N (Figure 5.36b). In the NOTC run the banded structure of the PV gradient and consequently the jet do not extend meridionally at around 28°W in the region of the transition from the ridge to the downstream trough. The maximum at 45°W, 50°N is below 5 PVU $(100\text{km})^{-1}$. Thus the tropopause PV gradient at 345 K increases by up to 2 PVU $(100\text{km})^{-1}$ in the presence of Hanna's circulation impinging on the baroclinic zone and this enhancement is attributable to the diabatic PV reduction and advection of low PV air at upper-levels. The resulting difference in the jet streak maxima is about 20 m $s^{-1}$. The crucial impact of Hanna on the upper-level flow is reflected in the 200 hPa anomaly correlation coefficient. Whereas the ACC is high in the control run during the first 96 hours, it decreases dramatically at 200 hPa during the first 72 h when Hanna is removed (Figure 5.33b).

At lower levels in the NOTC experiment a trough in the 1015 hPa contour at 72°W, 42°N indicates an extratropical cyclogenesis at about the location of Hanna (Figure 5.34b and Figures 5.30,5.31). Also, a secondary low at about 35°W 50°N is more pronounced in the NOTC experiment (Figure 5.35a). The downstream cyclone develops at 18°W, 44°N in both simulations (Figure 5.34ab, Figures 5.30, 5.31).

At 12 UTC 09 September 2008 an extratropical cyclone is present in the NOTC experiment slightly south of the location of Hanna in the control run (Figure 5.34cd and Figure 5.30) and its central pressure is 8 hPa higher (Figure 5.31). The mid-level ridgebuilding is weaker (Figure 5.34d). In the control experiment low PV air of Hanna's downstream ridge wraps cyclonically around the reintensified storm (Figure 5.34c). The wrap-up and northward shift of the low PV air are reflected in a very strong negative 320 K difference PV (<-6 PVU) north of Hanna (Figure 5.37a). This cyclonic wrap-up is important for the southeastward extension of Hanna's trough (Grams et al. 2011). At upper-levels a pronounced downstream ridge and trough are evident in the control experiment

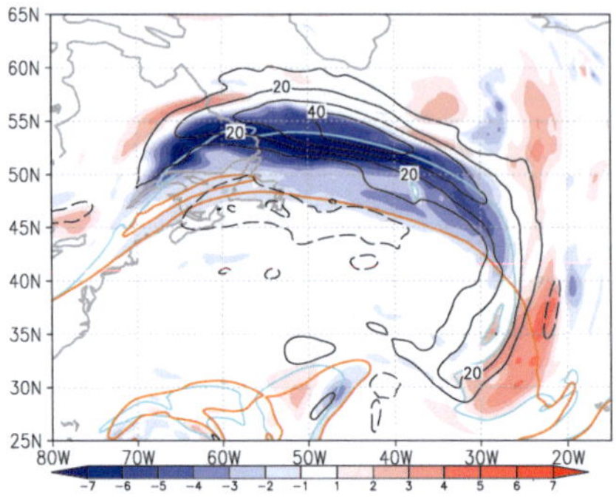


Figure 5.35: Difference CNTRL minus NOTC of PV (shaded in PVU) and of wind speed (black contours with a $10\,\mathrm{m\,s^{-1}}$ contour interval) on the 345 K isentropic level at 00 UTC 08 September 2008 (+48h). The 2 PVU contour from CNTRL (cyan) and NOTC (orange) is also shown. Note that the section zooms on the ridgebuilding.

(a) $\nabla_h PV$ at 345 K CNTRL , +48h, 00 UTC 08 September 2008

(b) $\nabla_h PV$ at 345 K NOTC, +48h, 00 UTC 08 September 2008

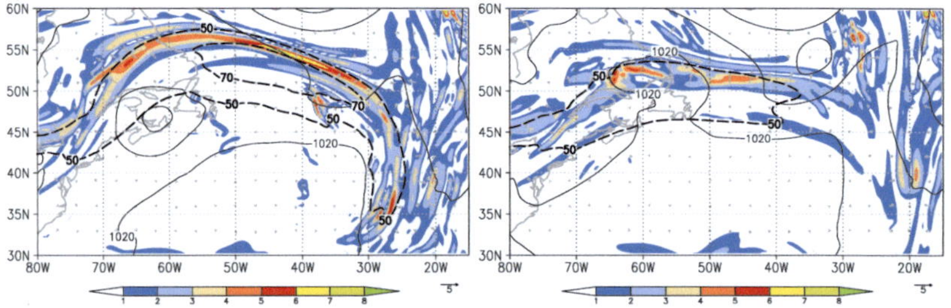


Figure 5.36: Gradient of 345 K isentropic PV $\nabla_h PV$ (shaded, every 1 PVU $(100\,\mathrm{km})^{-1}$ and grey vectors) at 00 UTC 08 September 2008 along with the pressure at mean sea level (black with a 10 hPa contour interval) and the wind speed $|\vec{v}| > 50\,\mathrm{m\,s^{-1}}$ on 345 K (black dashed every $20\,\mathrm{m\,s^{-1}}$) at 00 UTC 08 September 2008. Left 48 hour forecast from CNTRL scenario, right from NOTC. Note that the section shown zooms on the ridgebuilding.

from 50°W to 20°W (345 K isentropic 2 PVU contour, blue in Figure 5.34c).

The ridge and southern part of the trough axis tilted anticyclonically compared to the previous day indicating an initial anticyclonic wave breaking. The downstream trough extends as a PV streamer along the North African coast. This ridge-trough couplet is much broader and less amplified in the NOTC run (345 K isentropic 2 PVU contour in Figure 5.34d). In the subsequent days the downstream trough forms an cut-off low at upper levels off the coast of Morocco and affects Madeira (Section 5.2.5). A similar development is evident in the analysis data (not shown). As a consequence of the anticyclonic tilting in CNTRL, the northern part of the downstream trough at 345 K is located further to the east at 45°N, which is also reflected on the 320 K level (Figure 5.37a). Consequently, the downstream cyclone (located at 13°W, 53°N, Figure 5.34c,d) in the CNTRL

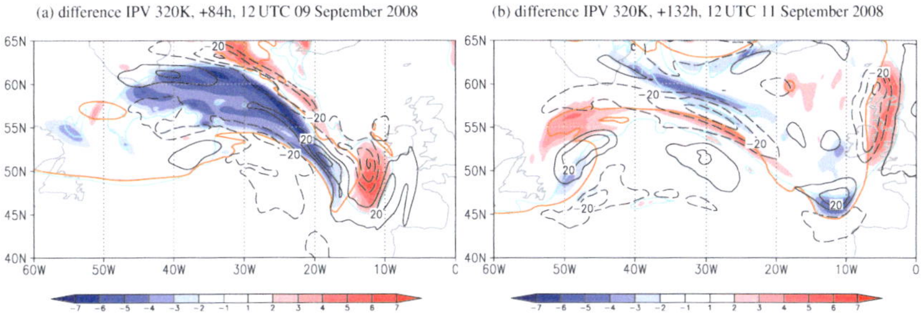


Figure 5.37: As Figure 5.35 but at 320 K and at 12 UTC 09 September 2008 (a) and 12 UTC 11 September 2008 (b). Note that the section shown zooms on the a different region compared to Figure 5.34.

scenario moves slightly more slowly and closer to the western Irish coast, although the intensity is comparable in both simulations (Figures 5.30,5.31). The genesis of the upstream cyclone is indicated in both experiments by a closed 1015 hPa pmsl contour at 78°W, 44°N (Figure 5.34c,d).

Two days later at 12 UTC 11 September, (Figure 5.32e), Hanna moved to about 12°W, 59°N in the control run and the associated trough extends more south- and eastward (cf. Figure 5.34e, Figure 5.30). Lower 320 K isentropic PV in the region 40°W, 60°N to 20°W, 55°N reflects the cyclonic wrap-up of low PV air. In contrast to the analysis, the upstream cyclone in CNTRL has a second centre at 50°W, 44°N attributable to a DRW-like structure approaching to its south. In the NOTC scenario (Figure 5.34f) the extratropical cyclone (10°W, 53°N), that developed instead of Hanna has a more westward track and heads towards Ireland (Figure 5.30). This cyclone remains secondary to a steering low over Iceland and does not exhibit a cyclonic wrap-up of low PV air as seen for Hanna. Also, the associated trough has a more meridional orientation and is located further westward compared to the analysis or CNTRL scenario. The development of the upstream cyclone at about 50°W, 51°N is similar to the analysis, though it is less intense and located slightly further south (Figures 5.30 5.31). The rather good representation of the upstream cyclone in NOTC is reflected in relatively high ACCs after 00 UTC 10 September 2008 exceeding the ACC in CNTRL (Figure 5.33).

For a better understanding of the impact of Hanna on the Mediterranean cyclogenesis, the evolution of the upstream trough that transforms into a PV streamer is highlighted in more detail. In the CNTRL scenario and in the analysis (Figure 5.34e and Figure 5.32e) the cyclonic wrap-up of low PV air leads to a reduction of the stratospheric high PV values. This results in a tropopause fold from around 40°W ,60°N to 20°W ,55°N (Figure 5.34e and Figure 5.37b) north of the region where ridgebuilding associated with the upstream cyclone occurs. Thus the PV gradient is reduced and the associated jet streak is weaker (absolute difference $< -30\mathrm{m\,s^{-1}}$, Figure 5.37b). The joint action of the anticyclonic ridgebuilding by the upstream cyclone to the south and the cyclonic wrap-up by Hanna to the north lead to a more eastward elongation of Hanna's trough in the control run than in NOTC (cf. Figure 5.34e and f). In NOTC the ridgebuilding associated with the upstream cyclone leads to a more amplified ridge trough couplet (Figure 5.34f). Consequently, the trough expands further south than in CNTRL. This is reflected in the negative 320 K difference PV at around 12°W 46°N in Figure 5.37b. Qualitatively the 320 K difference PV at 12 UTC 11 September 2008 (Figure 5.37b) depicts the inverse picture to the 345 K difference PV three days earlier. At 00 UTC 8 September 2008 (Figure 5.35) the advection of low PV air in CNTRL by Hanna's outflow and the associated ridgebuilding result in a more amplified and anticyclonically tilting downstream trough. The cyclonic wrap-up of this low PV air during the following two days (Figure 5.37b) counteracts the ridgebuilding of the upstream cyclone and results in a more eastward elongation of Hanna's trough in the control run. Under the influence of the upstream ridgebuilding alone, the trough in NOTC has a more meridional amplification and is located about 5° less to the east than in CNTRL.

In the subsequent development Hanna's trough cuts off in both scenarios (Figure 5.34g,h cf. Figure 5.32g). However, in the NOTC scenario this cut-off is located at 3°E, 41°N southwest of the cut-off in the analysis and control experiment (8°E, 45°N). The cut-off in the NOTC scenario also only triggers a weaker and rather stationary Mediterranean cyclone (Figures 5.30, 5.31). These differences in the NOTC scenario result in a low ACC at the end of the simulation period (Figure 5.33).

It can be concluded that the ridgebuilding linked to Hanna and the subsequent wrapping-up of low PV air modify the upper-levels so that the PV streamer and cut-off developing from Hanna's upstream trough have a more eastward location. Thus Hanna may have modified the exact location of the Mediterranean cyclone Olivia. Due to the complex orography in the Mediterranean region this has an crucial impact on the triggering of heavy precipitation. During the earlier stages of ET, the PV surgery experiment showed that Hanna modified the upper-levels due to a strong ridgebuilding and amplification of the downstream trough. As in the Jangmi case study, this resulted in an increased tropopause PV gradient and a significant acceleration of the midlatitude jet.

### 5.2.5 Downstream impact

During the period of Hanna's ET Europe was affected by several high impact weather events in terms of heavy precipitation (cf. section 4.2 of Grams et al. 2011). On 9 and 10 September 2008 the downstream cyclone brought moderate rain of around 20 mm in 24 hours to the British Isles and southern Scandinavia, and Ex-Hanna caused some rainfall in Ireland and in the UK (Figure 5.38a). As this rainfall follows a flooding event in the UK at the beginning of September 2008 it had a severe impact on the UK. During the following day thunderstorms were triggered in France ahead of Ex-Hanna's cold front with rainfall amounts of up to 70 mm in 24 hours (Figure 5.38b). On 12 September 2008 large-scale heavy precipitation (up to 88 mm in 24 hours) occurred in central Europe north of the cut-off that developed from Ex-Hanna's PV streamer (Figure 5.38c). During and after 13 September 2008 the Mediterranean cyclone Olivia lead to orographically and convectively enhanced severe precipitation in the Alps, northern Italy and the Balkans exceeding 200 mm in 24 hours at some locations (Figure 5.38d).

Although the upstream cyclone develops differently from the analysis, in the CNTRL simulation the direct impact of the downstream cyclone and Ex-Hanna on the British Isles, of Ex-Hanna's cold front on France and of the cut-off and

Mediterranean cyclogenesis on central Europe and the Mediterranean constitute a realistic scenario (cf. Figures 5.38 and 5.39). The major differences in the observed precipitation and in the CNTRL scenario are: the frontal system associated with the downstream cyclone is located further southward in CNTRL and consequently brings moderate rainfall to central Europe and not to southern Scandinavia (cf. Figures 5.38a and 5.39a). The convection triggered ahead of and at Ex-Hanna's cold front is more confined to southern France in CNTRL and not to eastern France as seen in the observations (cf. Figures 5.38b and 5.39b). The large-scale precipitation north of the cut-off is less pronounced in the CNTRL scenarios. Maximum rainfall amounts occur over the northern

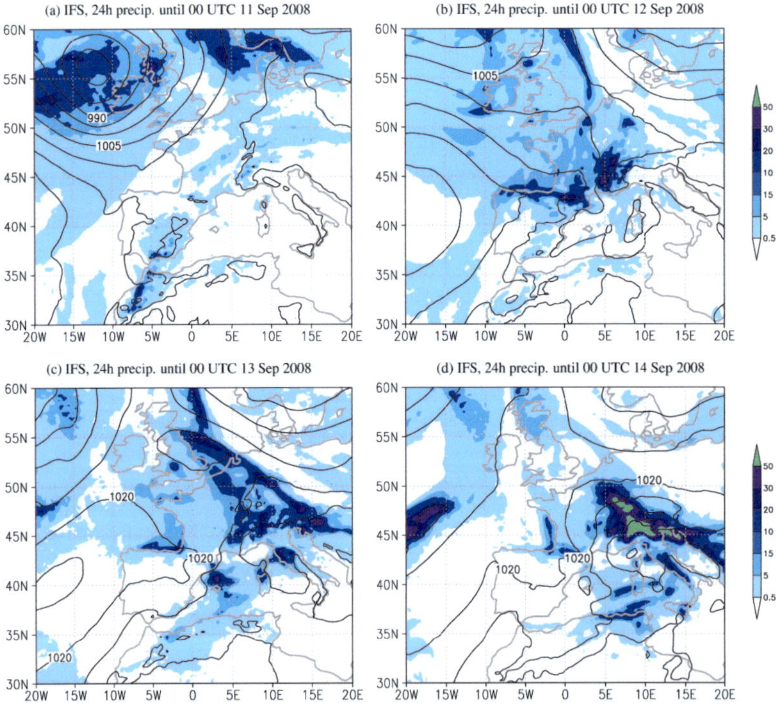


Figure 5.38: 24 hour accumulated precipitation (shaded in kgm$^{-2}$) in the IFS "analysis" and pmsl (black with a 5 hPa contour interval). The 24 hour accumulation period ends at the date indicated above each panel. To be compared with Figure 5.39 and Figure 5 from Grams et al. (2011). The IFS "analysis" precipitation is calculated by adding the total precipitation at forecast hour 12 in the time shifted twelve hourly IFS forecasts beginning at 00 UTC 06 September 2008.

Alps and are rather linked with the cyclogenesis of the Mediterranean cyclone (cf. Figures 5.38c and 5.39c). The precipitation due to Olivia is slightly shifted eastward in the CNTRL simulation (cf. Figures 5.38d and 5.39d).

Comparing the difference in 192 hour accumulated precipitation between the control scenario and the scenario without Hanna (Figure 5.40) reveals crucial differences between the scenarios in three distinct regions. A first region of significantly more precipitation in the presence of Hanna appears over the British Isles and Brittany (up to 50 mm more in Wales). This results from the joint impact of the downstream cyclone and of Ex-Hanna's frontal systems. The downstream cyclone moves slower and closer to the western Irish coast in the control scenario. Ex-Hanna moves northward off the coast of the British Isles

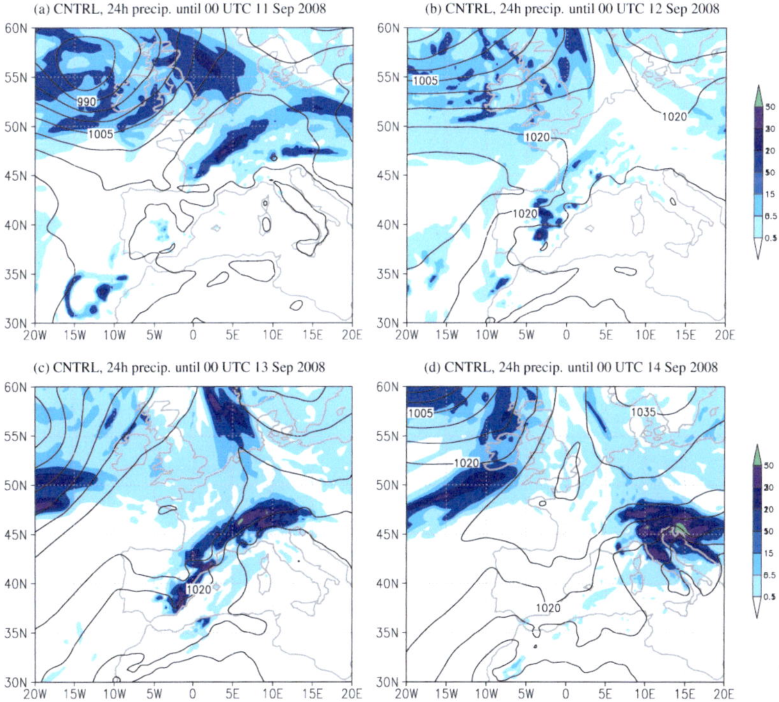


Figure 5.39: 24 hour accumulated precipitation (shaded in $kgm^{-2}$) in the CNTRL experiment and pmsl (black with a 5 hPa contour interval) from the control experiment. The 24 hour accumulation period ends at the date indicated above each panel. To be compared with Figure 5.38. Note the different domains.

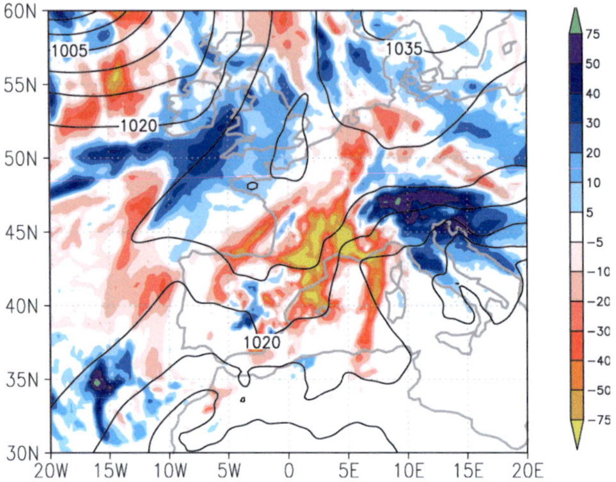


Figure 5.40: Difference of the total accumulated precipitation (shaded in $kgm^{-2}$) in the control minus NOTC experiment until 00 UTC 14 September 2008 and pmsl (black with a 5 hPa contour interval)at 00 UTC 14 September 2008 from the control experiment. Blue colours indicate more precipitation in the control scenario with the impact of (Ex-)Hanna, while red colours indicate more precipitation in the NOTC scenario without Hanna.

in the control scenario whereas the extratropical cyclone, developing instead of Hanna in the NOTC scenario, decays when reaching Ireland. Therefore the frontal systems linked to the downstream cyclone and Ex-Hanna move more distinctly across the British Isles in the control than in the NOTC scenario and consequently bring more precipitation. A second major difference occurs in the Mediterranean region. In the presence of Hanna, the precipitation linked with the Mediterranean cyclone appears similar to the observations over northern Italy, the Alps and the Balkans. In contrast, high rainfall amounts are evident in southern France and northeastern Spain in the NOTC scenario due to a more westward position of the Mediterranean cyclone and of the upper-level cut-off. In the Mediterranean region rainfall amounts depend critically upon the interaction of the circulation and the complex orography. Thus a different position of weather systems exhibits a crucial impact on local rainfall amounts. In the NOTC scenario precipitation in southern France is widely more than 75 mm higher than in the CNTRL scenario especially over the mountain ranges. A third interesting region is at about 15°W, 34°N off the Moroccan coast. In the

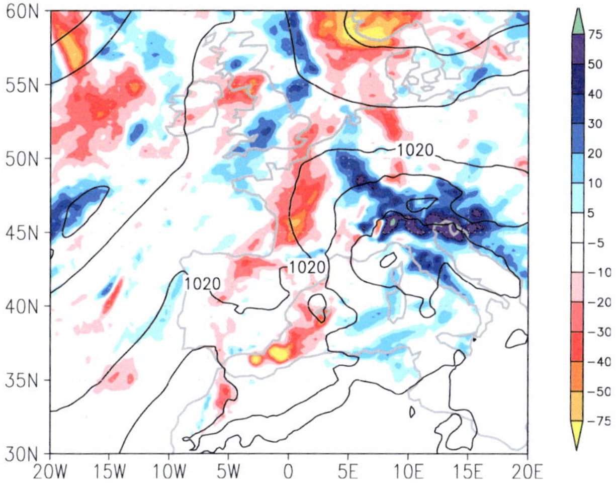


Figure 5.41: As Figure 5.40 but the difference between the accumulated precipitation (shaded in mm) in the IFS "analysis" and the accumulated precipitation after 192 forecast hours in the IFS deterministic forecast initialised at 00 UTC 06 September 2008 and at 00 UTC 14 September 2008 (black with a 5 hPa contour interval). The 192h total precipitation from the IFS "analysis" is calculated by adding the total precipitation at forecast hour 12 in the time shifted twelve hourly IFS forecasts beginning at 00 UTC 06 September 2008.

vicinity of Madeira rainfall amounts are of up to 75 mm higher in the control scenario than in the NOTC. This is due to the cut-off of the downstream trough which is not present in the NOTC scenario.

The precipitation forecast differences in the operational ECMWF IFS forecast (Figure 5.41) were of the same order of magnitude as the differences between CNTRL and NOTC (Figure 5.40). The IFS forecast also underestimates the precipitation in southwestern England, Italy and the Balkans (Figure 5.41). The IFS forecast overpredicts precipitation east of Spain. The correlation of these patterns with the differences between the CNTRL and NOTC scenario may indicate uncertainties on the exact transformation of Ex-Hanna's trough into a PV streamer in the IFS forecast. In the Madeira region neither the IFS analysis nor the IFS forecast indicate strong precipitation, linked to the cut-off of the downstream trough. However, it has to be noted that precipitation of up to 75 mm was observed by the TRMM satellite rainfall radar during the period 7-9 September 2008.

In summary, three crucial impacts on the precipitation patterns over Europe can be attributed to the ET of Hanna based on the CNTRL and NOTC experiment. Widely 10-20 mm more precipitation occurs over the British Isles and up to 50 mm more in Wales associated with the downstream cyclone, that moves slower along and closer to the western Irish coast, and with Ex-Hanna. The ET of Hanna is crucial for a more eastern location of the Mediterranean cyclone Olivia. Rainfall amounts were widely 40-50 mm and locally more than 75 mm higher over northern Italy, the Alps and the Balkans when Hanna was present in the initial conditions. The ET of Hanna helps the downstream trough to cut off which results in high rainfall amounts off the Moroccan coast in the vicinity of Madeira.

## 5.3 The ET of Typhoon Choi-wan (2009)

Typhoon Choi-wan formed in the western North Pacific in September 2009. It reached maximum intensity at around 16 September 2009 with a minimum central pressure of 915 hPa and maximum sustained winds of 54 $\mathrm{m\,s^{-1}}$ (10 min mean, JMA). A NASA reported even states a maximum sustained wind of 82 $\mathrm{m\,s^{-1}}$ (1 min mean)[1]. During ET the midlatitude environment was characterised by Rossby wave activity. A pronounced Rossby wave train occurred downstream of Choi-wan accompanied by an early outbreak of cool air over large areas in central North America.

On 18 September 2009 Choi-wan recurved ahead of a midlatitude trough with an axis at around 135°E (Figure 5.42). In the subsequent days a Rossby wave train developed with a pronounced first downstream trough over the Central Pacific (axis at 160°W, 00 UTC 21 September 2009, Figure 5.42) and a second downstream trough over North America that cuts off on 23 September over the central USA (axis at 100°W, 00 UTC 23 September 2009).

The impact of Choi-wan on the midlatitude flow evolution and the triggering of the polar air outbreak in North America is investigated with the help of a COSMO sensitivity experiment. After some details on the COSMO scenarios,

[1] http://www.nasa.gov/mission_pages/hurricanes/archives/2009/h2009_Choi-Wan.html retrieved at 6 August 2011

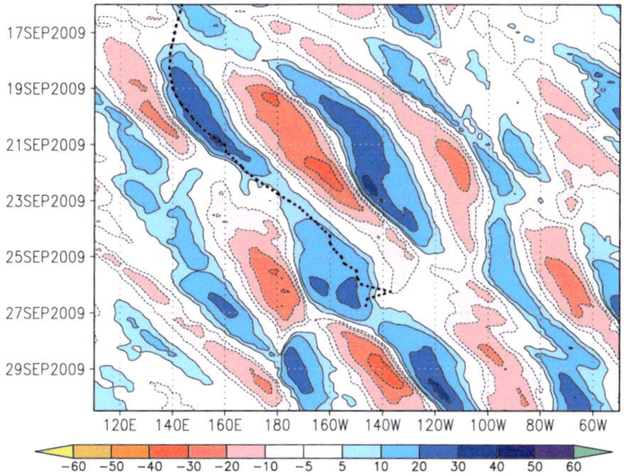


Figure 5.42: Hovmöller diagram of the meridional wind component at 250 hPa (shaded every 10 $\mathrm{m\,s^{-1}}$) in the ECMWF analysis. Data meridionally averaged from 25°N to 55°N. Track of Choi-wan in the analysis (dotted black line).

an overview of Choi-wan's ET is given in Section 5.3.2 followed by a more detailed quantification of Choi-wan's role in the evolution of the midlatitude flow (Section 5.3.3) and a discussion of the downstream impact in North America (Section 5.3.4).

### 5.3.1 Technical details of the PV surgery

The midlatitude impact of Choi-wan is investigated by comparing a COSMO run with (CNTRL) and without (NOTC) Choi-wan in the initial conditions. The COSMO simulations are computed with the same configuration except for the initial conditions. Choi-wan is removed from the initial conditions of the NOTC run using the PV surgery technique (Section 3.2.4). Six-hourly ECWMF analyses (with 0.25° horizontal resolution) serve as boundary conditions. The horizontal model domain covers the entire North Pacific and North America. The northern edge of the model domain is set to 78°N to avoid an impact of the northern boundary condition on the evolution of the Rossby wave train. As a consequence, the horizontal resolution is $\approx 6\,km$ in the northern model domain and a numerical time step of 10 s is required. The model runs are initialised one day prior to recurvature at 00 UTC 17 September 2009. Information on the COSMO model runs are summarised in the Appendix, Table A.6.

The control simulation (CNTRL) is initialised from the unmodified ECMWF analysis and the no-TC simulation (NOTC) from the ECMWF analysis from which Choi-wan has been removed. Choi-wan was still a strong tropical cyclone and well south of the midlatitude wave guide at the time of the TC removal (00 UTC 17 September 2009). An interaction of Choi-wan and the midlatitude flow was not seen in the ECMWF analysis isentropic PV prior to the initialisation time. Details of the TC removal of Choi-wan are given in the Appendix in Table A.7

### 5.3.2 Overview of ET Choi-wan

The CNTRL scenario gives a very good representation of the the analysis during the first 6 days. This is reflected in a high anomaly correlation coefficient

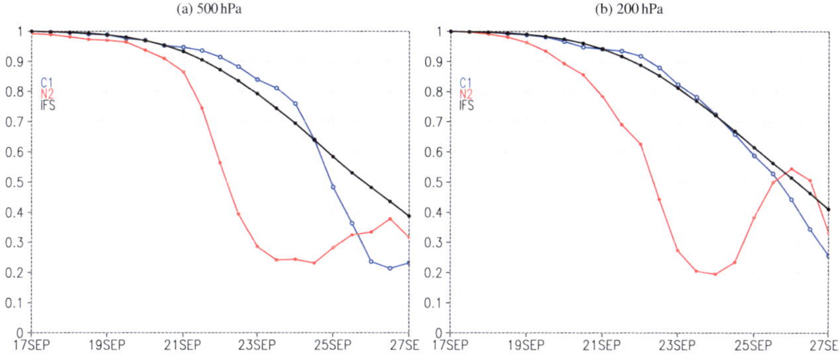


Figure 5.43: Anomaly correlation coefficient (ACC) of geopotential at 500 hPa (a) and 200 hPa (b). Data is from the CNTRL (blue) and NOTC (red) scenarios introduced in Section 5.3.1. The ACC is calculated on the subdomain shown in Figures 5.62 (120-300°E, 20-60°N) every 12 hours. The ECMWF analyses are used as the analysis data and the daily mean values of ERA-INTERIM interpolated on a 0.25°x0.25° horizontal grid is used as the climatology. Additionally the mean ACC for all deterministic ECMWF IFS forecasts during 1 August 2009 to 1 November 2009 calculated for the domain 110°E-300°E, 20°N-60°N is shown in black.

$ACC > 0.8$ until 00 UTC 23 September 2009 (+144 h) for geopotential at mid and upper levels (Figure 5.43). Afterwards the ACC decreases significantly and indicates no useful forecast skill ($ACC < 0.6$) after 00 UTC 25 September 2009. In terms of the ACC the control simulation exhibits a better forecast skill until 24 September 2008 compared to the average deterministic forecast skill.

The analysed evolution of the low-level flow at the beginning of ET is dominated by the extratropical reintensification of Choi-wan during which Choi-wan interacted with a preexistent extratropical cyclone and has a second period of rapid extratropical reintensification to 964 hPa until 12 UTC 21 September 2009 (Figures 5.44 and 5.45). At 00 UTC 20 September 2009 (Figure 5.46a) the extratropical cyclone developed at the baroclinic zone to the northeast of Choi-wan. During 20 September 2009 (Figure 5.46b) Choi-wan, which still exhibits a warm core ($\Theta_e > 355\,K$), approaches this extratropical cyclone. At 00 UTC 21 September 2009 both cyclonic systems merged and a period of re-intensification started (Figure 5.46c). The warm core dissolved. At 12 UTC 21 September 2009 (Figure 5.46d) Choi-wan is a mature extratropical cyclone and subsequently crosses the Pacific as a steadily decaying but strong extratropical cyclone until 26 September 2009, when it reaches Alaska at 148°W, 59°N.

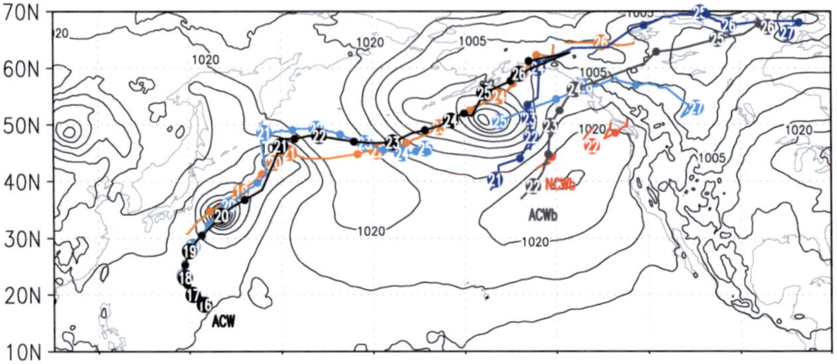


Figure 5.44: Tracks of Typhoon Choi-wan during ET and of the downstream cyclone CWb. Tracks from analysis data in black colours, labelled ACW, ACWb. Tracks from control scenario in blue colours. Tracks from no-TC scenario in red colours. Thick points 00 UTC times, labelled with the day of September 2009, thin points 12 UTC times. Pressure at mean sea level (black every 5 hPa) from the analysis data shown at 00 UTC 20 September 2009.

This evolution is a compound ET in the classical description of Matano and Sekioka (1971), in which a TC interacts with the frontal zone of an existent extratropical cyclone. As Agustí-Panareda (2008) pointed out, compound ET must be differentiated into an interaction with an extratropical system upstream or downstream of the recurving TC. The interaction with the cold front of an pre-existent extratropical cyclone downstream, will result in an intensification of the merger system as the TC forces ascending motion over the extratropical system. This is what we see in the case of Choi-wan (Figure 5.46). In contrast, when interacting with the warm frontal system as in the case of Hurricane Gert (1999) studied by Agustí-Panareda (2008),the TC will force descending motion over the extratropical system forming upstream, which would result in a weakening of the extratropical development.

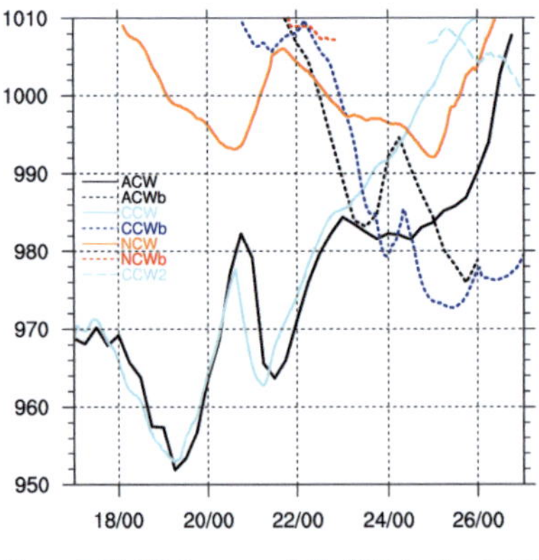


Figure 5.45: Minimum pmsl (in hPa) in the centre of Choi-wan and cyclone CWb. The same labels and colours as in Figure 5.44 are used.

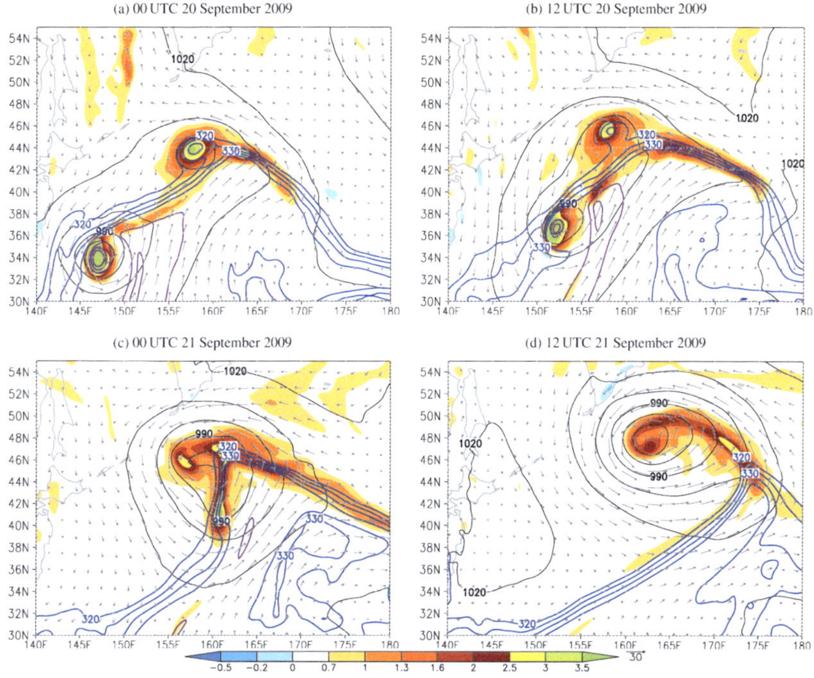


Figure 5.46: Low-level PV (in PVU shaded), equivalent potential temperature (blue contours 320-335 K every 5 K, and purple, 350-360 K every 5 K), and wind vectors (grey) vertically averaged from 975-800 hPa every 25 hPa. Pmsl (black with a 10 hPa contour interval). Data from the ECMWF analysis shown in a 12 hourly sequence from 00 UTC 20 September 2009 (a) to 12 UTC 21 September 2009 (d).

In the control scenario the evolution of Choi-wan is similar until 23 September 2009 (light blue in Figures 5.44 and 5.45 see also Figure 5.43). Afterwards Choi-wan at around 170°W 45°N merges with a cyclone that develops at 150°W, 50°N on 25 September 2009. In the NOTC scenario the extratropical cyclone, with which Choi-wan merges on 20 September 2009, forms also and exhibits a similar track to extratropical Choi-wan in the analysis (orange in Figures 5.44 and 5.45). However, it is weaker and experiences a period of weakening during 21 September, when Choi-wan merged with the extratropical cyclone and reintensified. The track differs from that of Choi-wan after 23 September 2009.

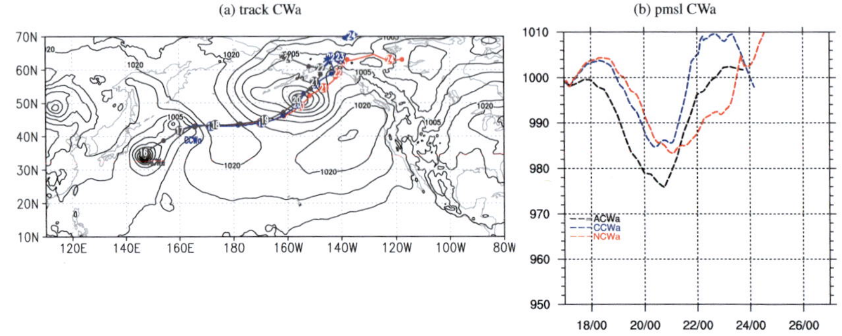


Figure 5.47: (a) as Figure 5.44 but track of cyclone CWa and pmsl shown at 00 UTC 17 September 2009. (b) as Figure 5.45 but pmsl for cyclone CWa.

An extratropical cyclone (labelled CWa) forms on 16 September 2009 ahead of the trough that becomes the first downstream trough (grey in Figure 5.47a). When the first downstream trough amplifies (cf. Figure 5.48a,b) it intensifies from 1000 hPa at 00 UTC 18 September 2009 to 976 hPa at 18 UTC 20 September 2009 (black in 5.47b). On 23 September 2009 cyclone CWa reaches Alaska at about the same location as Choi-wan three days later. The track of CWa in CNTRL is very close to the analysis until 23 September 2009, though the intensification is delayed about 24 hours and not as intense (blue in Figures 5.47a and 5.47b). In the NOTC scenario the intensification is also delayed 24 hours and the track is more to the east after 20 September 2009.

An extratropical cyclone (CWb) that develops later ahead of the amplifying first downstream trough is important in terms of the downstream impact. It forms at around 150°W 39°N on 21 September 2009 (1010 hPa; grey and black dashed in Figures 5.44 and 5.45). On a northward track CWb intensifies to a minimum pressure of 984 hPa at 12 UTC 23 September 2009 reaching the North American coast on 24 September 2009. It subsequently transitions into an extratropical cyclone, that develops over northern Canada. The evolution and intensification of CWb is similar in the CNTRL scenario but the track is about 10° to the right of the track in the analysis (blue in Figures 5.44 and 5.45). In contrast, in the NOTC scenario this cyclone is much weaker and develops as a secondary cyclone to CWa (red dashed in Figure 5.45). The track is more westward and hits the North American coast near Vancouver Island (red intermittent in Figure 5.44).

The upper-level flow during and following the ET of Choi-wan is dominated by a pronounced Rossby wave train over the northern Pacific (Figure 5.48). At 00 UTC 19 September 2009 Choi-wan, located at 140°W, 28°N, moved ahead of a midlatitude trough (upstream trough) with an axis at 135°E. This is evident consistently in the analysis and CNTRL data. In the NOTC scenario the trough is located 5° further eastward, is broader and the ridge is less amplified. Further east a first downstream trough is located at about 180°E followed by a second downstream ridge-trough couplet with an axis at about 155°W and 135°W.

Two and a half days later a poleward, cyclonic wave breaking of the upstream trough occurred and, as for Hanna, the first downstream ridge wraps cyclonically around Choi-wan (at 163°E 47.5°N) in the analysis and CNTRL scenario (Figure 5.48b). The wave breaking is not present in NOTC and the upstream trough and first downstream ridge are much further east and less amplified. The first downstream trough amplified at 160°W. The amplification is stronger in the CNTRL scenario than in the analysis. To the east a very broad and strongly amplified second downstream ridge over the eastern Pacific and western North America follows and a second downstream trough is located over central North America. The amplitude of this trough-ridge-trough pattern is much weaker and shifted about 5°-15° eastward in the NOTC scenario. This suggests that the phase speed of the Rossby wave (propagating westward relative to the mean flow) in the analysis/CNTRL scenario may have been accelerated due to the amplification by Choi-wan's ET. The begin of stronger differences between the NOTC scenario and the actual development is reflected in a decrease of the ACC starting on 21 September 2009 (Figure 5.43). Interestingly, the ACC for the NOTC scenario is significantly below the mean IFS-ACC through the entire calculation period. This indicates a major forecast skill reduction when the tropical cyclone is not present in the analysis. We note that the interaction of Choi-wan with the upper-level midlatitude flow slows down the eastward propagation of the Rossby wave train, consistent with the findings of Riemer et al. (2008) in an idealised setup with a straight jet. Furthermore, the ET of Choi-wan leads to

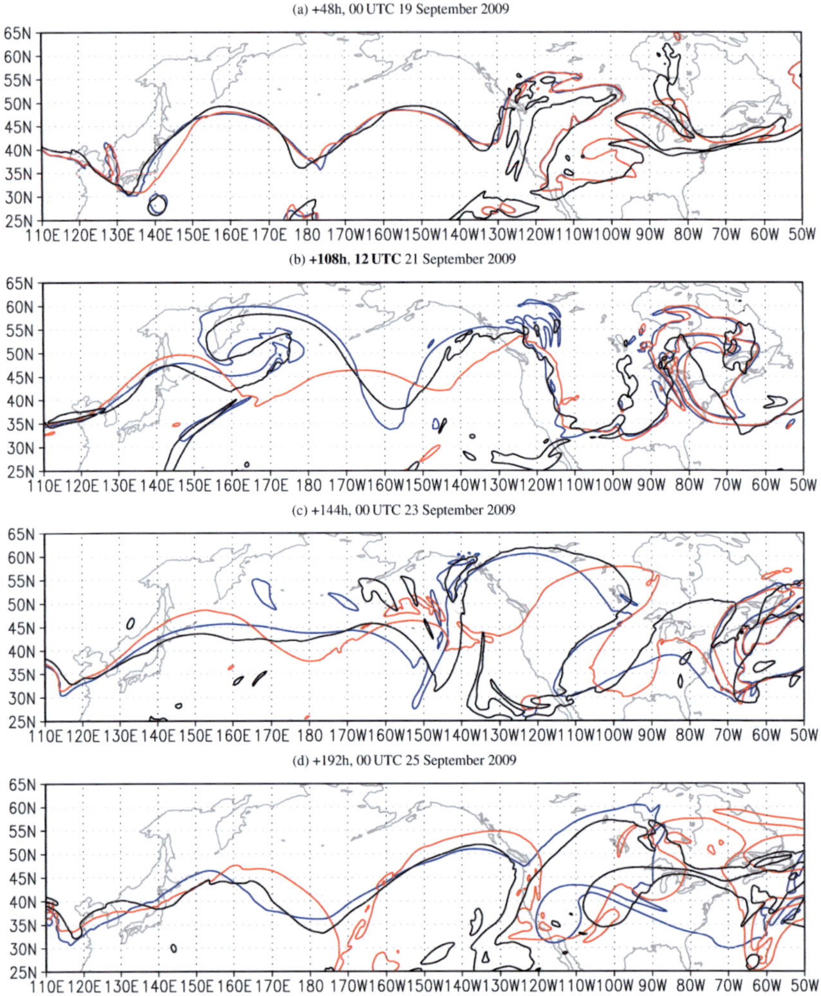


Figure 5.48: 340 K isentropic PV (2PVU contour) from the analysis (black), control (blue) and NOTC (red) scenario. The times are indicated above each panel. Note the different latitudinal extent of (a) and (b)-(d)

an amplification of the Rossby wave train, which was also reported by Riemer and Jones (2010) for an idealised simulation of the interaction of a TC with a prescribed baroclinic wave.

At 00 UTC 23 September 2009 the second downstream trough has tilted significantly and starts to cut off over the central western USA (Figure 5.48c). This is similarly evident in the analysis and CNTRL scenario. The second downstream ridge is strongly amplified and expands over the western USA, southwestern Canada, and parts of the East Pacific. Further upstream the first downstream trough, narrows into a PV streamer located at around 150°W in CNTRL and 145°W in the analysis. In the NOTC scenario the first downstream trough is broader and the trough-ridge pattern shifted about 10° eastward. The second downstream trough is less tilted, is located more than 10° eastward compared to the analysis/CNTRL scenario and has already cut off. The ACC for NOTC fell below the level of meaningful forecast skill (Figure 5.43).

Two days later major differences between the analysis and the COSMO scenarios occur (cf. Figure 5.43). In the analysis, the cut-off of the southern portion of the second downstream trough is located at around 103°W (Figure 5.48d). In the CNTRL scenario the cut-off has moved westward to around 115°W. Whereas in the NOTC scenario the cut-off is located further east at around 90°W and merged with the first downstream trough, which is located over the western USA with an axis at around 112°W. Due to the major differences in the CNTRL simulation the ET of Choi-wan based on the COSMO simulations is only discussed until 00 UTC 23 September 2009. In the analysis data the northern portion of the second downstream trough becomes a PV streamer over the western North Atlantic with an axis at 60°W at 00 UTC 26 September 2009 (Figure 5.42). During the subsequent days this PV streamer cuts off and triggers a subtropical cyclogenesis (not shown).

### 5.3.3 Quantification of Choi-wan's downstream impact

At initialisation time (Figure 5.49a) Choi-wan is well south of a broad trough with an axis at 140°E. This trough becomes the first downstream trough in the subsequent days. Cyclone CWa forms ahead of it at 160°E, 42°N. This is similarly evident in the NOTC scenario (5.51a) in which Choi-wan could be removed without altering the midlatitude flow.

At 00 UTC 19 September 2009 Choi-wan reintensified ahead of the southern tip of a pronounced midlatitude trough that approached upstream (Figure 5.49b).

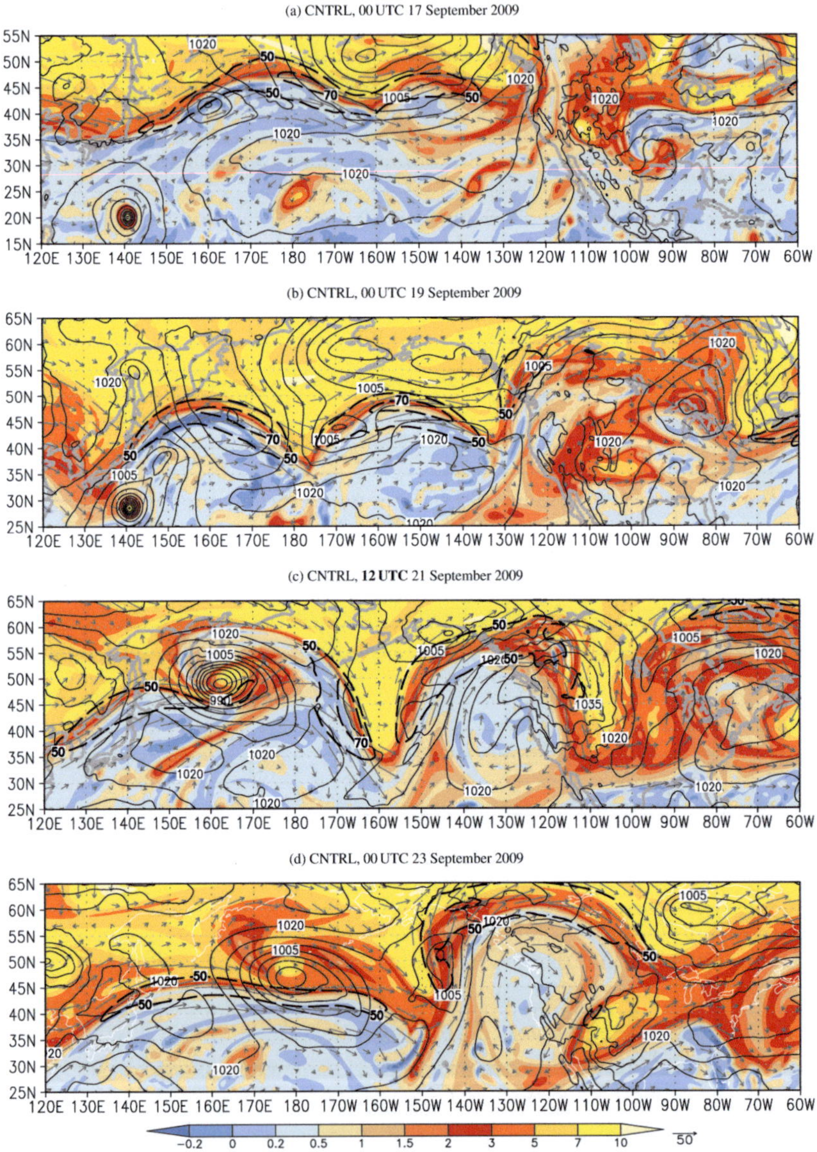


Figure 5.49: PV (in PVU, shaded), wind (grey vectors and black dashed with a $20\,\mathrm{m\,s^{-1}}$ contour interval for $|\vec{v}| > 50\,\mathrm{m\,s^{-1}}$) at the isentropic level of $\Theta$=340K, and pmsl (black contours every 5 hPa) shown for the control scenario. The time is indicated above each panel. Note the different latitudinal extent of (a) and (b)-(d).

A downstream ridge formed with an axis at 160°E, and the first downstream trough at 180°W became narrower. These upper-level features are also present in the NOTC scenario but the upstream trough is broader and has an axis slightly further east (Figure 5.51b). As the first downstream trough is at about the same location the downstream ridge is smaller. Thus the outflow of Choi-wan acts to broaden the downstream ridge at its western flank and to narrow the upstream trough. This is reflected in the 340 K difference PV of $< -4$ PVU to the north of Choi-wan (Figure 5.50a). In this region the midlatitude jet streak extends towards Choi-wan reflecting an enhancement by Choi-wan's outflow. This behaviour is similar to that seen for Jangmi and Hanna, but less distinct as the midlatitude jet streak is also present in the NOTC scenario. An extratropical cyclone forms ahead of the upstream trough in both scenarios. In CNTRL it is located

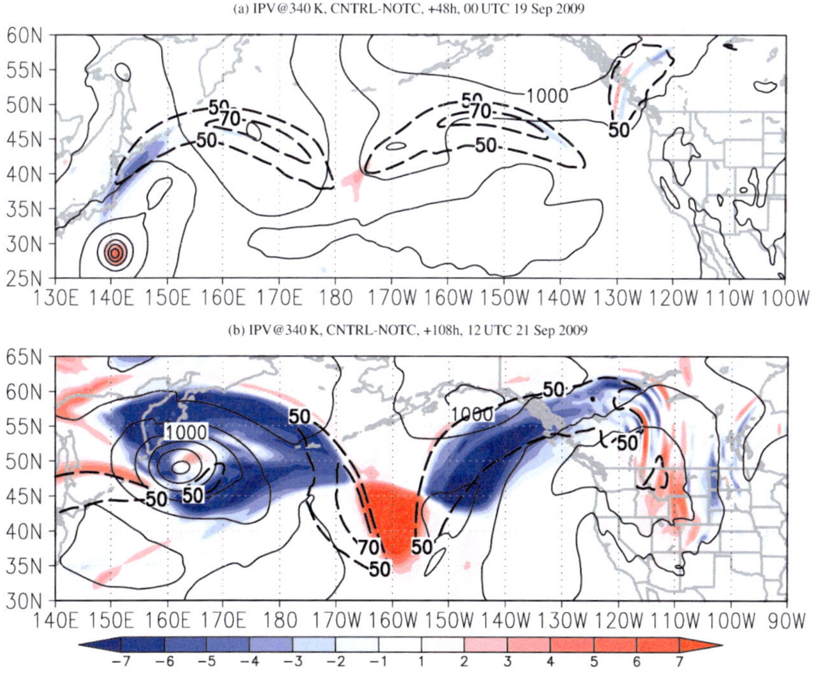


Figure 5.50: Difference CNTRL minus NOTC of PV (shaded) and wind vectors (grey) on the 340 K isentropic level, along with wind speed $|\vec{v}| > 50\,\mathrm{m\,s^{-1}}$ on 340 K (thick black dashed every $20\,\mathrm{m\,s^{-1}}$) and pmsl (black every 10 hPa) from the CNTRL simulation. Time indicated above the panels.

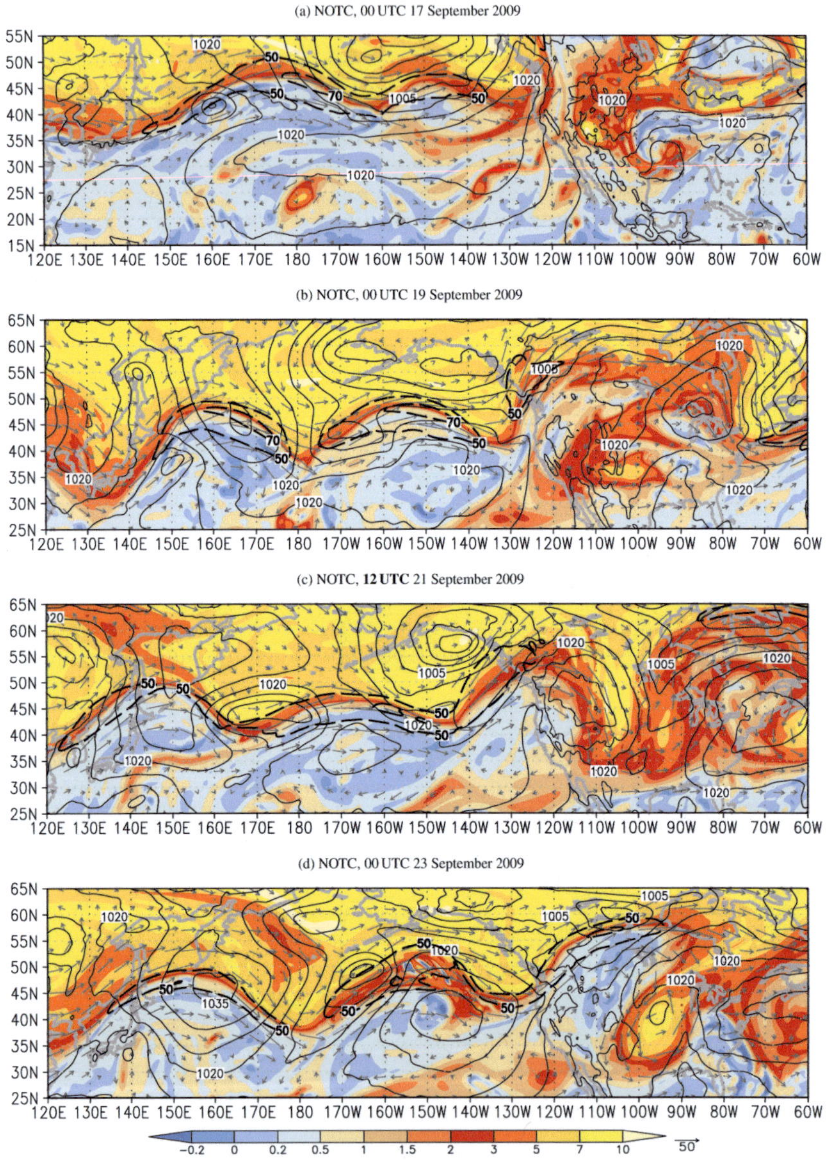


Figure 5.51: As Figure 5.49 but data from NOTC scenario.

at 148°E 38°N, whereas in NOTC at 150°N, 38°N (Figure 5.49b, Figure 5.51b). Cyclone CWa is at the same location at 170°W, 44°N in both scenarios ahead of the first downstream trough. Further to the east the second downstream ridge-trough couplet formed in both scenarios.

At 12 UTC 21 September 2009 dramatic differences occur between CNTRL and NOTC. In CNTRL (Figure 5.49c), Choi-wan experienced a reintensification of 20 hPa in 18 hours to 963 hPa (06 UTC) when merging with the extratropical cyclone to its northeast whereas the cyclone in NOTC (Figure 5.51c) reached a minimum pmsl of 990 hPa twelve hours earlier (Figure 5.45). Along with this reintensification a cyclonic poleward wave breaking occurred with low PV air of the downstream ridge wrapping cyclonically around the upstream trough (Figure 5.49c). The cyclonic wrap-up is reflected in 340 k difference IPV of $< -5$ PVU south of Choi-wan's centre (Figure 5.50b). The northward extension of the downstream ridge is seen as difference IPV below $-5$ PVU to the north of Choi-wan. The wrap-up is similar to that seen for the ET of Hanna, although the implications for the downstream evolution are different. The first downstream trough (160°W) and the second downstream ridge-trough couplet (130°W,105°W) amplified significantly in CNTRL. This indicates that ridge-building similar to that downstream of Choi-wan occurs downstream of the intensifying extratropical cyclone CWb at 150°W, 43°N ahead of the first downstream trough. In NOTC (Figure 5.51c) this trough is broader and located further east (145°W). The cyclogenesis of CWb is not yet evident. The second downstream ridge-trough couplet is much less amplified in NOTC and located 5°-10° further east with the axes at around 120°W and 100°W. The different amplification and shift of the downstream trough-ridge-trough pattern is also reflected in the double pattern of negative and positive 340 k difference IPV ($< \pm 7$ PVU, Figure 5.50b) over the Pacific. This suggests a higher westward group velocity of the Rossby wave train in the reference simulation (westward propagating, relative to the mean flow). Interestingly, the intensity of the jet streaks is similar in both scenarios. The fact, that the areal extent of the negative PV anomalies in Figure 5.50b is larger than the positive PV anomalies indicates that diabatic processes play a major role in the diabatic PV reduction at upper-levels. Thereby the basic process of low PV air transport to upper-level, as described in the case of Jangmi, is much weaker in the NOTC simulation.

At 00 UTC 23 September 2009, blocking by the broadened second downstream ridge is evident over the Eastern Pacific and western North America (CNTRL, Figure 5.49d). The first downstream trough to the west has narrowed (axis at 150°W) and ahead of it the cyclone CWb at 146°W, 51°N intensifies as it moves northward. To the east, the southern part of the second downstream trough tilted dramatically and starts to form a cut-off over the western USA at 107°W 37°N. In the NOTC scenario the first downstream trough and second downstream ridge are shifted 15°-20° westward. The first downstream trough is broader with an axis at 130°W. Ahead of it cyclone CWb is reflected in a weak surface trough in the 1010 hPa contour near Vancouver Island at 125°W, 50°N. The southern part of the second downstream trough hat cut off at 95°W. Also note the different upper-level flow over the West and Central Pacific.

Despite the more amplified first downstream trough in CNTRL after 20 September 2009, the Hovmöller diagrams of 250 hPa meridional wind indicate no major difference between the CNTRL run and the analysis before 00 UTC 23 September 2008 (not shown). The enhancement of the first downstream ridge-trough

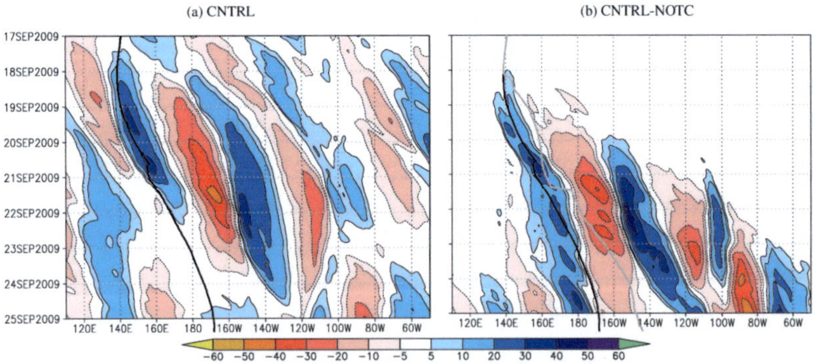


Figure 5.52: Hovmöller diagrams of the meridional wind component at 250 hPa (with a 5 m s$^{-1}$ contour interval) in the control simulation CNTRL (a), and the wind component difference at 250 hPa (with a 5 m s$^{-1}$ contour interval) between CNTRL and NOTC (b). The data is meridionally averaged from 25°N to 55°N. Forecast from 00 UTC 17 September 2009 (0 h) to 00 UTC 25 September 2009 (192 h) shown. The black solid line in marks the track of Choi-wan in CNTRL, in (d) in grey of cyclone NCW in NOTC.

couplet following the ET of Choi-wan in the CNTRL run (Figure 5.52a) is reflected in a 250 hPa meridional wind of up to $\pm 40\,\mathrm{m\,s^{-1}}$ centred around 160°W at 12 UTC 21 September 2009. The 250 hPa meridional difference wind (Figure 5.52b) reveals that the ET of Choi-wan triggers a dramatic amplification and acceleration of the downstream Rossby wave train. Downstream of Choi-wan the group velocity of the Rossby wave train increases on 19 September 2009, when the first downstream ridge builds. The group velocity of the wave packet in difference wind is about $30\,\mathrm{m\,s^{-1}}$ (80° in 2.5 days).

In summary, this more detailed view of the CNTRL and NOTC scenarios showed that Choi-wan acts to significantly modify the upper-level Rossby wave train. In the presence of Choi-wan the Rossby wave train is more amplified, exhibits a higher group velocity and consequently propagates slower westward. Due to the stronger amplification a much more intense downstream cyclone (CWb) develops ahead of the first downstream trough. A cut-off developing from the second downstream trough is located further west when Choi-wan is present.

### 5.3.4 Implications for high impact weather in North America

The westward shift in the location of the second downstream ridge-trough couplet and the subsequent upper-level cut-off low during 20 to 26 September 2009 has some implications for high impact weather in the the central and western USA.

The northerly upper-level flow at 110°W from 20 to 22 September 2009 lead to a very intense temperature drop during 20-21 September 2009 in the western Great Plains, USA and over the Rocky Mountains. For example, at Denver Airport (104.67°E, 39.86°N) the daily maximum temperature was above 25°C until 20 September 2009 and reached 27.8°C at 20 UTC 20 September 2009 (14 local time (LT), Figure 5.53). During the night the temperature initially dropped to 9.4°C at 8 UTC (2 LT) 21 September 2009, reflecting the usual nocturnal cooling. However, a slight increase to 11.1°C during the next two hours indicates a deviation from the diurnal cycle in that night. At this time the low-level wind began to turn to Northerlies (Figure 5.54) and remains from the north until 22

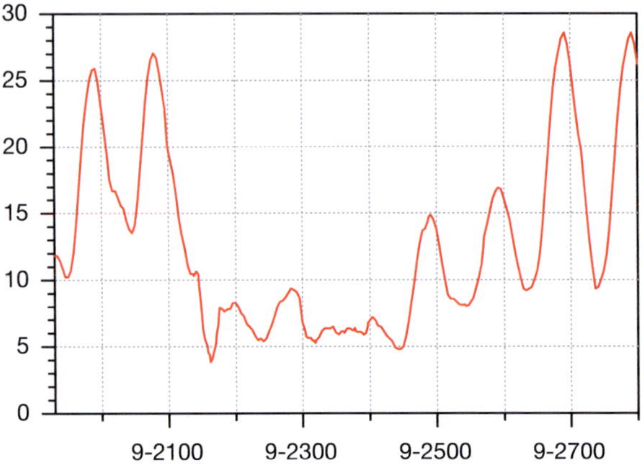


Figure 5.53: Hourly observed 2 m temperature (in°C) at Denver Airport, Colorado, USA from 07 UTC 19 September 2009 to 00 UTC 28 September 2009. Data smoothed with a 5 point running mean. Local time is UTC-6 h. Thanks to Julian Quinting for plotting the data.

September 2009. Associated with this, the temperature drops sharply to 3.3°C in the late morning of 21 September 2009 (16 UTC (10 LT), Figure 5.53). The maximum at this day is as low as 9.4°C at 23 UTC (17LT). This is more than 18 K lower than the previous day. The temperature remains below 10°C for the next three days. The maximum temperature on 23 September 2009 only reaches 7.2°C (0 UTC+1d (18 LT)). After 26 September 2009 the temperature curve is back on the usual late September diurnal cycle with daily maxima above 25°C and nocturnal minima around 10°C.

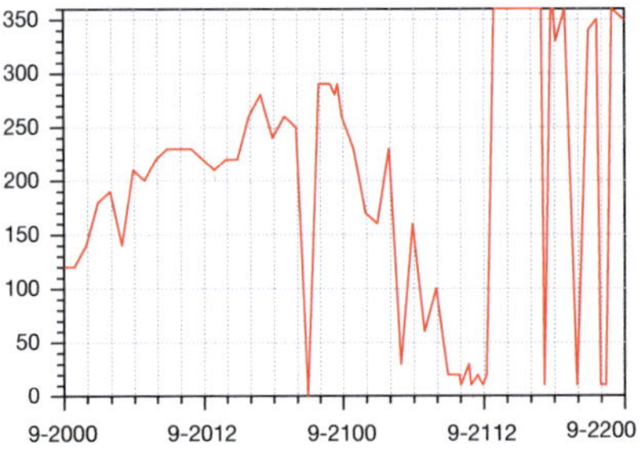


Figure 5.54: Hourly observed wind direction (in °) at Denver Airport, Colorado, USA from 00 UTC 20 September 2009 to 00 UTC 22 September 2009. Local time is UTC-6 h. Thanks to Julian Quinting for plotting the data.

In the following, the surface weather situation in North America is discussed with focus on the period until 23 September 2009, as the COSMO reference simulation (CNTRL) constitutes a very good representation of the actual analysis until that day (cf. Figure 5.43). This allows for a robust quantification of the surface impact of the westward shifted Rossby wave train during the ET of Choi-wan.

At 23 UTC 22 September 2009 (the approximate time of the diurnal temperature maximum in the western USA), the upper-level cut-off low in CNTRL appears as a region of very low 500 hPa temperature (<-20°C) centred at 107°W, 37°N (Figure 5.55a cf. Figure 5.49d). Cool air (<-10°C) at 500 hPa extends southward to around 30°N. The 500 hPa wind vectors reflect the cyclonic circulation around the cool air mass. The centre of the ridge is located at about 123°W, 47°N. To its southwest a warmer air mass with temperatures above -5°C at 500 hPa is advected northward along the US Northwest Coast (Figure 5.55a). In contrast, the upper-level cut-off low is smaller and located further northeast at 95°W, 39°N in the NOTC simulation (Figure 5.55b cf. Figure 5.51d). Also, the ridge has a smaller amplitude and is much more tilted. The associated warm air anomaly is concentrated in the centre of the ridge at about 115°W, 40°N. The shift of the upper-level cold air in the cut-off low has major consequences for the surface temperature. In CNTRL the northerly flow over an extended time period (21-23 September) results in a 2 m temperature maximum below 9°C over southern Wyoming, western Nebraska, Colorado, western Kansas, New Mexico, northwestern Texas and western Oklahoma (110°W-100°W, 30°N-42°N, Figure 5.55c). The lowest 2 m temperatures (<-3°C) occur over elevated terrain in the Rocky Mountains. Interestingly, the cool air is accompanied by a surface high pressure system, leading to a southward advection of the cold air at its eastern side. The cool air reaches the Mexican border at about 105°W.

Surface observations south of Denver reflect the southward spread of the cold air from 21 to 23 September 2009 (Table 5.2). One day after the temperature maximum in Denver dropped to 12°C on 21 September 2009, the cool air reached New Mexico. For example, in Albuquerque the maximum 2 m temperature was 28°C on 21 September and dropped to 17°C on 22 September. A similar temperature fall is evident further south in El Paso, Texas, from 33°C on 21 September to 23°C on 22 September. The higher maximum compared to Denver shows, that the air mass is mixed when advected southward, and the coldest air is rather close to the centre of the cut-off low further north (cf. Figure 5.55c).

In contrast, southeasterly flow in the northwestern USA (northern California, Oregon, Washington) advects the warm continental air mass northwestward. Surface observations in that region indicate an unusual "heat wave" spreading northward from 21 to 23 September 2009 (Table 5.2). The southernmost of three representative stations, Portland, Oregon, USA, has a maximum temperature of 27°C on 21 September. In Seattle, Washington, USA, the maximum temperature rose from 22°C on 21 September to 31°C on 22 September. One day later the warm air reached Vancouver, British Columbia, Canada, where maximum temperature peaked at 27°C on 23 September 2009 which constitutes a record for that day. Note that the "heat wave" begins earlier and is therefore longer at the more southern stations (Table 5.2). To the east of the cut-off southerly flow advects a subtropical air mass northward, leading to rather warm late summer temperatures of 26°C in Columbia, Missouri, on 21 September 2009 (Table 5.2).

In the NOTC scenario (Figure 5.55d) the 2 m temperature anomaly linked to the upper-level cut-off west of the Great Lakes is not as pronounced as over the western USA in CNTRL. The 2 m temperature is below 15°C in the region from 103°W-90°W, 34°N-48°N. Here the terrain is lower, so that the upper-level cold air has not such a dramatic impact as over the elevated terrain in the west. Also, the the cool surface air spreads over a smaller area. As the upper-level ridge is located over the western USA, 2 m temperature widely exceeds 20°C there.

Table 5.2: Maximum (Tx) and minimum (Tm) 2 m temperature in (°C) from 19 to 26 September 2009 at airport stations (IATA code, elevation): From south to north in the northwestern USA: Portland, Oregon, USA (PDX, 15 m amsl), Seattle, Washington, USA (SEA, 132 m amsl), Vancouver, British Columbia, Canada (YVR, 4 m amsl); from north to south in the southwestern USA: Denver, Colorado, USA (DEN, 1655 m amsl), Albuquerque, New Mexico, USA (ABQ, 1630 m amsl), El Paso, Texas, USA (ELP, 1206 m amsl); in the Midwest Columbia, Missouri, USA (COU, 215 m amsl).

| | PDX | | SEA | | YVR | | DEN | | ABQ | | ELP | | COU | |
|---|---|---|---|---|---|---|---|---|---|---|---|---|---|---|
| day | Tx | Tm | Tx | Tm | Tx | Tm | Tx | Tm | Tx | Tm | Tx | Tm | Tx | Tm |
| 19 | 21 | 11 | 18 | 14 | 19 | 14 | 27 | 8 | 23 | 12 | 28 | 16 | 26 | 13 |
| 20 | 22 | 8 | 20 | 11 | 17 | 11 | 29 | 9 | 26 | 13 | 30 | 17 | 23 | 16 |
| 21 | **27** | 9 | 22 | 12 | 18 | 9 | **12** | **3** | 28 | 11 | 33 | 18 | 26 | 16 |
| 22 | **31** | 20 | **31** | 12 | 22 | 11 | **11** | **4** | **17** | **7** | **23** | **13** | 21 | 15 |
| 23 | **31** | 14 | **29** | 13 | **27** | 10 | **8** | **4** | **20** | **7** | **23** | **13** | 21 | 14 |
| 24 | 21 | 11 | 20 | 13 | 17 | 11 | 17 | 4 | 20 | 7 | 25 | 14 | 21 | 17 |
| 25 | 24 | 9 | 22 | 10 | 18 | 9 | 18 | 8 | 26 | 9 | 29 | 16 | 24 | 13 |
| 26 | 23 | 11 | 21 | 10 | 17 | 9 | 29 | 8 | 28 | 11 | 32 | 16 | 21 | 13 |

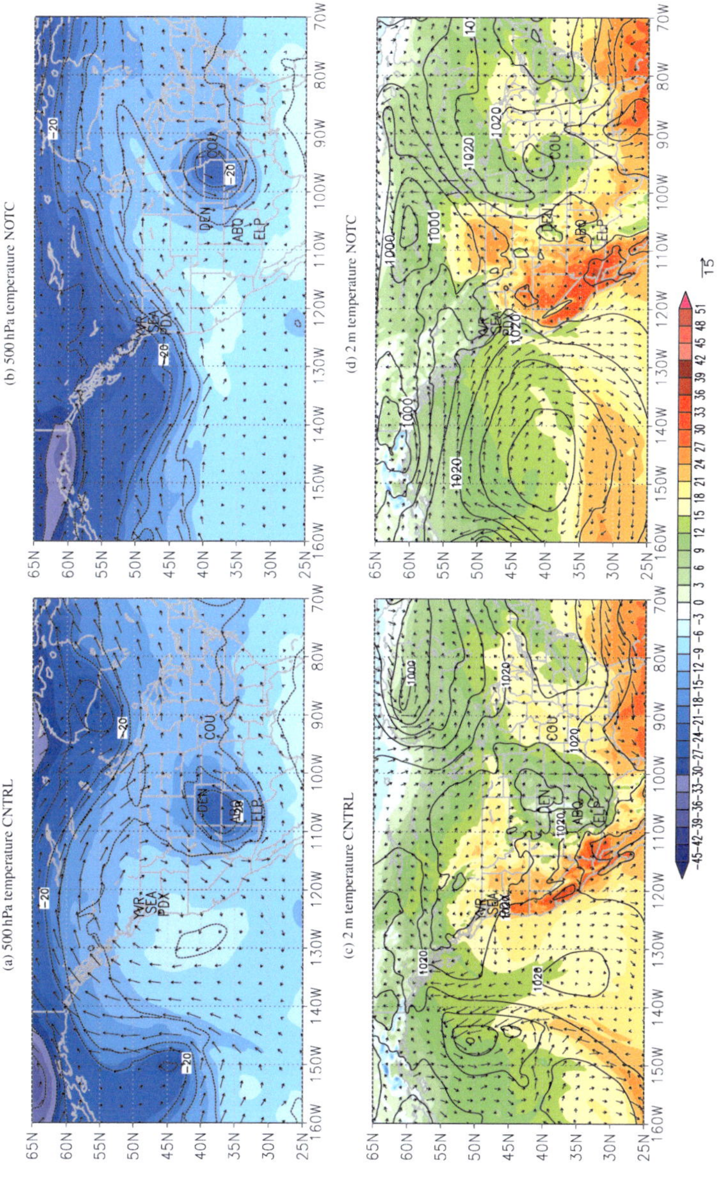


Figure 5.55: (a),(b) 500 hPa temperature (shaded every 3°C and black contours every 5°C) and 500 hPa wind vectors at 23 UTC 22 September 2009. (b),(c) 2 m temperature (shaded every 3°C), 10 m wind vectors, and pmsl (black every 5 hPa) at 23 UTC 22 September 2009. In (a,c) data from the CNTRL run is shown, in (b,d) data is from the NOTC run. The locations of the airport stations given in Table 5.2 are indicated.

The difference temperature CNTRL minus NOTC reflects the foregoing findings (Figure 5.56). In the region of northerly flow between 110°W-100°W a very strong low-level cooling of below -10 K over a broad area and locally below -18 K occurs (Figure 5.56). Almost the entire western USA experiences a negative 2 m temperature anomaly. An exception is the the northwestern corner of the USA and southeast Canada. Here the positive anomaly partly exceeds +15 K and locally reaches +20 K. The region of cooler 2 m temperature southwest of the Great Lakes in the NOTC simulation is reflected in the positive anomaly of >6 K centred at 95°W 37°N. It is remarkable that the spatial extent and the magnitude of this temperature anomaly is much smaller than the cold anomaly linked to the cut-off in CNTRL.

The COSMO 2m temperature at representative stations from the three regions shows the dramatic local effect of the shift in the Rossby wave pattern (Figure 5.57 cf. Figure 5.56). Due to the rather coarse resolution of the COSMO runs and different heights of the representative grid points, the 2m temperature in CNTRL differs from the measurement. Despite this, the COSMO 2m temperature allows for a quantification of the local temperature effect of Choi-wan in the COSMO scenarios.

Denver is initially located at the western edge of the cut-off in the NOTC run and thus experiences a temperature drop on 21 September 2009 in both COSMO

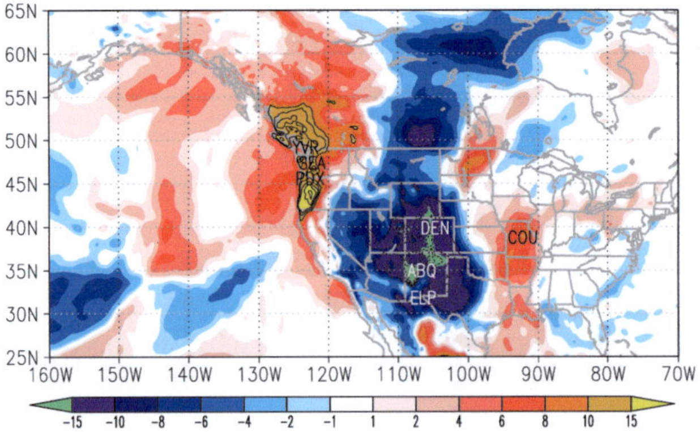


Figure 5.56: 2 m temperature difference CNTRL-NOTC (shaded in K, and contours every 2 K for values $\gtrless \pm 10$ K) at 23 UTC 22 September 2009. The locations of the airport stations given in Table 5.2 are indicated.

simulations (Figure 5.57a cf. Figure 5.53). However, in NOTC the late summer diurnal cycle reestablishes after one day, while in CNTRL the temperature remains below 13°C until 24 September 2009. While there are differences of about 5 K in CNTRL compared to the measurements, the CNTRL run captures the temperature evolution qualitatively correctly. The cool air reaches Albuquerque further south one day later (Figure 5.57b). Here only minor cooling would have occurred without Choi-wan as the cut-off would not have reached as far south. In Seattle (Figure 5.57c) to the west of the cut-off temperatures increase above 20°C from 22 to 24 September 2009, with a maximum of 27°C on 23 September 2009. This reflects the autumn heat wave at the US Northw-

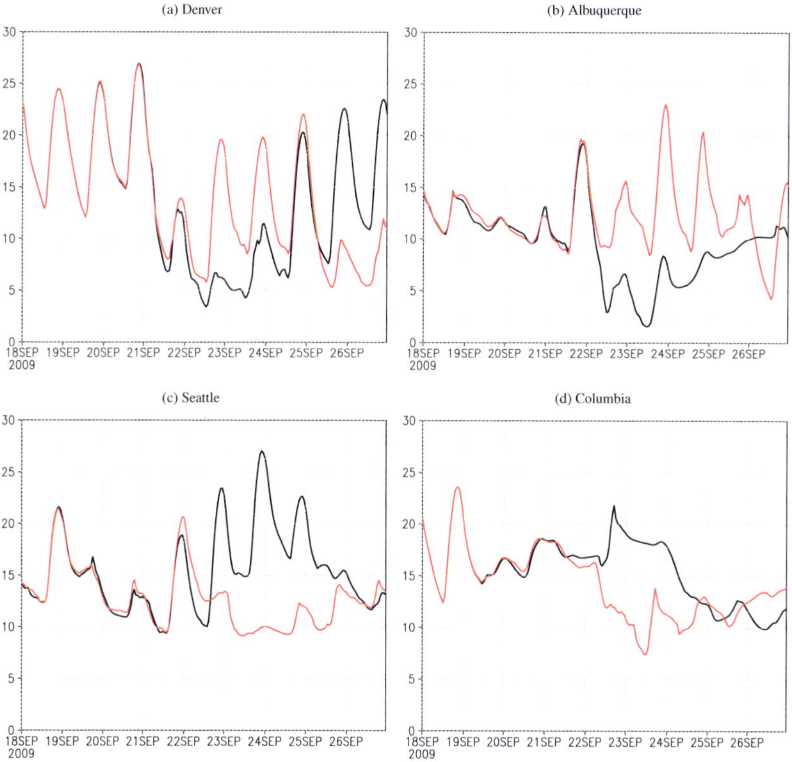


Figure 5.57: COSMO 2m temperature (in °C) from 00 UTC 18 September 2009 to 00 UTC 27 September 2009 at selected airport stations (cf. Table 5.2). Data is from the CNTRL run (black) and the NOTC run (red).

est Coast. Without Choi-wan cool weather with temperatures around 10-15°C would have occurred. In Columbia (Figure 5.57d) to the east of the cut-off temperatures remained at moderate levels during 22-24 September 2009, while without Choi-wan temperatures would have dropped around 5-10 K as the cut-off would have been located further east.

Finally, the impact of the different location of the upper-level cut-off low on the accumulated precipitation in CNTRL and NOTC during the period from 00 UTC 17 September to 00 UTC 24 September 2009 is investigated (Figure 5.58). In CNTRL more precipitation occurred in the region 110°W-95°W, 30-50°N where the northerly flow at the eastern part of the cut-off was evident. The maximum differences of up to 50 mm occur at the southern edge of the northerly flow, reflecting a (cold) frontal structure. A second maximum occurs at 102°W 43°N where the northerly flow of the ridge and the easterly flow around the cut-off converge (cf. Figure 5.55a). In contrast, in the NOTC scenario a much smaller region of enhanced precipitation centred at 95°W, 42°N is associated with the cut-off. However, the cut-off in NOTC dramatically enhances a frontal precipitation region around 87°W. Over the East Pacific, the shift of the western edge of the ridge results in a different steering of the cyclone CWb. In CNTRL it has a northward track (see pmsl in Figure 5.58) and maximum differ-

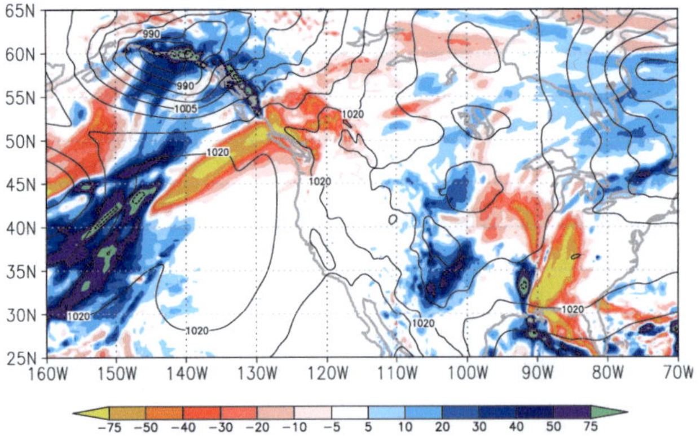


Figure 5.58: 168h accumulated precipitation difference CNTRL-NOTC (shaded in $kgm^{-2}$) until 00 UTC 24 September 2009 and pmsl from CNTRL (black every 5 hPa at 00 UTC 24 September 2009. More precipitation in CNTRL appears in blue colours, more precipitation in NOTC in red colours.

ences in precipitation over land are concentrated at 140°W, 60°N (Figure 5.58). In NOTC CWb is secondary to CWa, has an eastward track and causes high rainfall amounts over land near 125°W, 50°N. Thus without Choi-wan very cool and wet weather in the Vancouver (Island) region would have occurred. Actually the rain has been shifted northwestward and temperatures rose in an autumn "heat wave".

In summary the COSMO sensitivity experiment revealed that the ET of Choi-wan modified the midlatitude upper-level Rossby wave train. The strong outflow of Choi-wan during ET lead to a broadening of the first downstream ridge at the western flank. The modification in the Rossby wave propagated downstream so that the second downstream ridge-trough couplet exhibited a westward shift in the presence of Choi-wan of around 10°. As the second downstream trough cut off a more intense upper-level cyclone developed significantly further southwest over the western USA. At the surface, the cut-off caused an outbreak of cool air east and over the Rocky Mountains, that spread over the entire western USA. However, a small strip in the northwestern corner of the USA became located in the centre of the ridge and experienced an autumn "heat wave". In the vicinity of the cut-off, precipitation occurred over the central USA. At the western flank of the ridge, cyclone CWb and associated precipitation over the Pacific were forced to a more northwestern track leading to heavy precipitation rather in Alaska than near Vancouver Island. This case study highlights the crucial impact that an ET event can have on the weather in downstream regions.

## 5.4 The ET of Typhoon Lupit (2009)

Typhoon Lupit was a strong and long-lived tropical cyclone in the western North Pacific in October 2009. In contrast to Jangmi the primary midlatitude feature during ET was a pronounced trough to the northwest of Lupit (Northwest pattern of ET in the classification of Harr et al. 2000). After recurvature it transitioned into an intense extratropical cyclone ahead of the midlatitude trough. Downstream of the ET system over the Pacific and North America a marked amplified Rossby wave train occurred with a second trough causing an early cold air outbreak in the western USA and the genesis of a strong extratropical cyclone in the eastern USA. The Rossby wave train appears with enhanced southerly flow at upper levels directly downstream of Lupit at around 140°E on 24 October 2009 (Figure 5.59). A pronounced first downstream ridge and trough forms with the trough axis at 180°E at 00 UTC 27 October 2009. The first trough transforms into a PV streamer and a second downstream ridge and trough develop over the eastern North Pacific and North America. On 29 October 2009 a broad trough is evident over the western USA with an axis at around 110°W (Figure 5.59). During the next two days this trough transforms into a narrow PV streamer also. A third downstream ridge-trough couplet forms over the western North Atlantic with a trough axis at 50°W on 31 October 2008. It arises the question as to what extent Lupit does contribute to this downstream development. In the following, a synoptic overview based on the discussion of two-daily ECMWF analyses is given (Section 5.4.1). Then the impact of Lupit on the evolution of the downstream flow is quantified with the help of PV surgery experiments (Sections 5.4.3 and 5.4.4).

### 5.4.1 Synoptic overview

#### 5.4.1.1 Tropical stage

Lupit formed west of the Philippines at around 150°E, 10°N on 14 October 2009. It headed westward as it intensified and initially recurved on 17 October 2009 whilst continuing to intensify. On 19 October 2009 it reached maximum

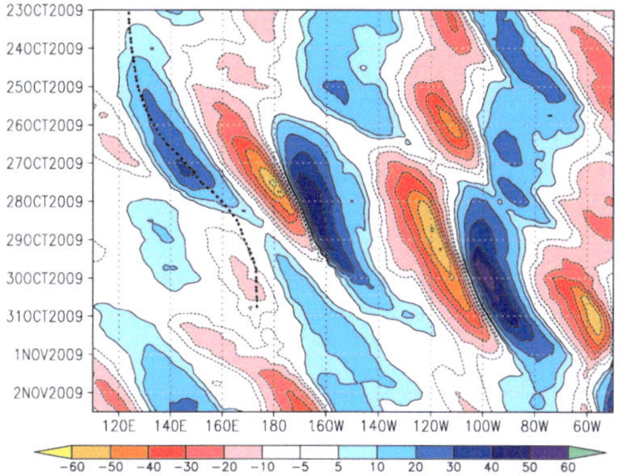


Figure 5.59: Hovmöller diagram of the meridional wind component at 250 hPa (shaded every 10 $m\,s^{-1}$) in the ECMWF analysis. Data meridionally averaged from 25°N to 55°N. Track of Lupit in the analysis (dotted black line).

intensity with a central pressure of 930 hPa and maximum sustained surface wind of more than 45 $m\,s^{-1}$ (JMA). At this stage it curved back westward so that its track depicts a complex S-shape. On 23 October 2009 Lupit finally recurved just before reaching northern Luzon. Under slow decay tropical characteristics remained until around 24 October 2009 when structural changes indicate that Lupit is undergoing ET.

### 5.4.1.2 Extratropical stage

During the extratropical stage Lupit moves steadily northeastwards over the western Pacific (see track in ECMWF analysis (black, labelled AL) in Fig-

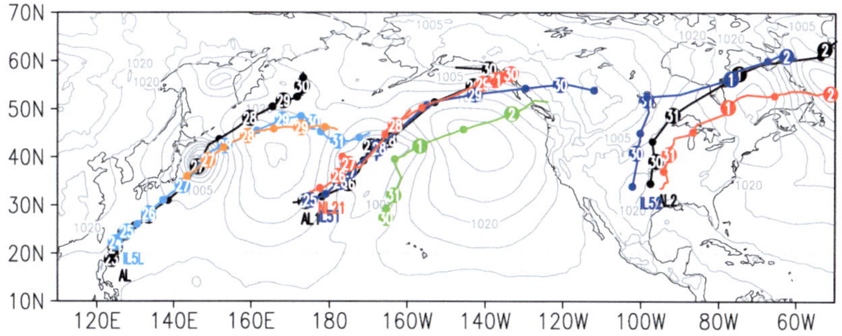


Figure 5.60: Tracks of the cyclones during the ET of Typhoon Lupit. The cyclones are Lupit, the 1st downstream cyclone, and the 2nd downstream cyclone. Tracks from analysis data in black, labelled AL, AL1, AL2. Tracks from control run IL25 in blue, labelled IL5L, IL51, IL52. Tracks from NOTC scenario NL20, labelled NL2L (orange), NL21 (red), NL22 (red), NL2T (green). Thick points 00 UTC times, labelled with the day of October/November 2009, thin points 12 UTC times. Pressure at mean sea level (grey every 5 hPa) from the analysis data shown at 00 UTC 27 October 2009.

ure 5.60). The extratropical transition of Lupit starts with a 48 hour period of decay in central pressure from 970 hPa at 06 UTC 24 October 2009 to 988 hPa at 00 UTC 26 October 2009 (Figure 5.61).

As in Jangmi, the upper-level midlatitudes downstream of Lupit are characterised by a rather zonal flow at the beginning of ET (Figure 5.62a). In contrast to Jangmi, a pronounced midlatitude trough approaches upstream and constitutes the primary midlatitude feature. Following Harr et al. (2000) Lupit will undergo the Northwest pattern of ET and a strong reintensification must be expected.

At 00 UTC 25 October 2009 Lupit is located at 127°E, 24°N ahead of this trough with an axis at about 120°E (Figure 5.62a). Anticyclonic outflow is evident to the Northeast of Lupit and low PV air (<0.2 PVU) extending northeast and eastwards indicates downstream ridgebuilding. At around 165°W, 24°N a weak tropical cyclone to the Northwest of Hawaii is undergoing ET. PV from this TC propagates materially into the midlatitudes during the next 4 days and will have an impact on the development of the first downstream cyclone, which exhibits a closed 1010 hPa contour at 173°E, 31°N (Figure 5.62a and track AL1 (black) in Figure 5.60).

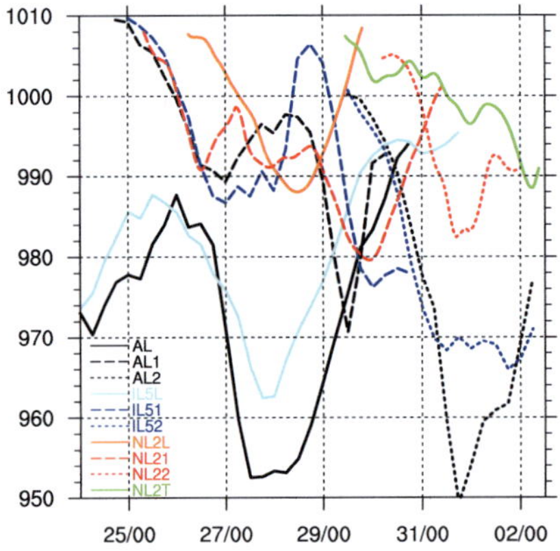


Figure 5.61: Minimum pmsl (in hPa) in the centre of the cyclones presented in Figure 5.60. The same labels and colours are used. One-hourly data from the NOTC scenario smoothed with a 5 point running mean.

From 00 UTC 26 October 2009 to 12 UTC 27 October 2009 Lupit experiences a very strong extratropical reintensification ahead of the midlatitude trough. During the 36 hours period it deepens by 35 hPa to 953 hPa at 12 UTC 27 October 2009 (Figure 5.61). The 340 K isentropic PV indicates a strongly amplified ridge downstream of Lupit (Figure 5.62b) resulting from the advection of very low PV air by Lupit's outflow. A jet streak of more than 80 m $s^{-1}$ extends from 145°E to 160°E at 50°N north of the region of the low PV air. Further east, a first downstream trough amplified with an axis at about 180°W. Ahead of it, the first downstream cyclone intensified. It has multiple centres with the minimum pmsl of 989 hPa at 170°W, 42°N (Figure 5.62b, black dashed in Figure 5.61). The weak ET system in the central Pacific is located at 166°W, 28°N (Figure 5.62b). In the next 24 hours it exhibits a DRW-like northward propagation and moves into the centre of the first downstream cyclone (not shown).

On 29 October 2009 Lupit is a mature and decaying extratropical cyclone east of Kamchatka (169°E, 52°N; Figures 5.60 and 5.61). The first downstream trough narrowed into a thin PV streamer at 165°W (Figure 5.62c). After an initial period of decay, the first downstream cyclone experienced a strong intensification from 998 hPa at 06 UTC 28 September 2009 to 971 hPa at 12 UTC 29 September 2009 (Figure 5.61). This intensification was inaugurated when remnants of the Hawaiian ET system transitioned into the cyclone's centre. Along with this, the first downstream cyclone propagated rapidly across the Pacific and is located at 143°W, 55°N at 00 UTC 29 September 2009 (Figure 5.62c, Figure 5.60). A pronounced second downstream ridge extends over the eastern North Pacific. Further downstream a second trough is located over western North America. Between the second downstream ridge and this trough a broad region of northerly upper-level flow over the western USA is evident (Figure 5.62c). This leads to an early outbreak of cold polar air masses in the western and central USA during the next days. Deflections in the pmsl isobars ahead of the second trough at 100°W, 35°N indicate the cyclogenesis of a second downstream cyclone.

During the next two days Lupit and the first downstream cyclone decayed (Figure 5.62d, Figures 5.60 and 5.61). The first downstream trough/PV streamer becomes embedded in a trough extending from 150°W-130°W. The second downstream ridge moved over the western USA leading to a thinning of the second downstream trough which becomes a PV streamer (Figure 5.62d). On its western side northerly flow reaches the Caribbean advecting polar air into the central USA. Ahead of the PV streamer southerly flow advects subtropical moist air masses from the Caribbean into the eastern USA. The second downstream cyclone formed at the frontal zone between these contrasting air masses (black track labelled AL2 in Figure 5.60) and is undergoing a strong intensification reaching a minimum pressure of 950 hPa at 18 UTC 31 October 2009 (black dotted in Figure 5.61). East of this cyclogenesis event a pronounced ridge forms accompanied by a far southward extending PV streamer over the Central Atlantic at around 50°W.

The second downstream cyclone moved from the Great Lakes to southern Greenland by 00 UTC 2 October 2009, where it became a central low in the Northwest Atlantic region (Figure 5.60, Figure 5.62e). The PV streamer over the Atlantic cut off and triggered a (sub-)tropical transition event near 58°W, 32°N transforming into an extratropical system during the subsequent days.

The foregoing discussion of the synoptic evolution during and after the ET of Typhoon Lupit highlights the impressive sequence of weather events that occurred due to the amplification of a pronounced Rossby wave train. During the investigation period the Rossby wave train propagation could be traced by the subsequent evolution of three downstream ridge-trough/PV streamer couplets. It is remarkably that the first ridge develops when Lupit's outflow imposes a southerly component on the upper-level flow (see track of Lupit in Figure 5.59). In the following, PV surgery experiments are used to quantify the impact of Lupit on the triggering or amplification of this Rossby wave train.

### 5.4.2 Technical details of the TC removal and scenarios

For the quantification of Lupit's midlatitude impact the typhoon is removed from the initial conditions of a COSMO simulation using the PV surgery technique (Section 3.2.4). This simulation (NOTC) is compared against a control simulation. The COSMO simulations are computed with the same configuration except for the initial conditions. Six-hourly ECWMF analysis (with 0.25° horizontal resolution) serve as boundary conditions. The horizontal model domain covers the entire North Pacific and North America. We aim to represent the evolution of the three ridge-trough couplets of the Rossby wave train during the period 23 October 2009 to 01 November 2009 (Figure 5.59). However, in the previous simulations of Jangmi and Hanna it turned out that the COSMO sim-

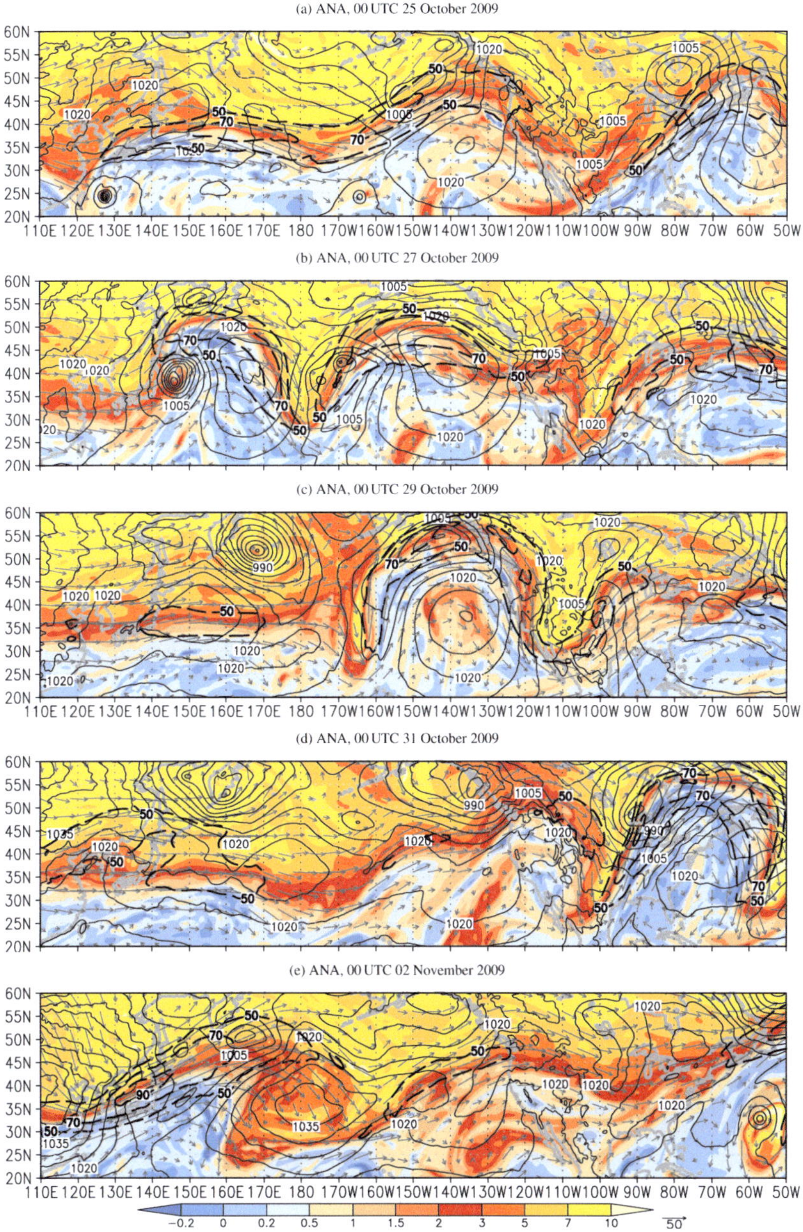


Figure 5.62: PV (in PVU, shaded), wind (grey vectors and black dashed with a $20\,\mathrm{m\,s^{-1}}$ contour interval for $|\vec{v}| > 50\,\mathrm{m\,s^{-1}}$) at the isentropic level of $\Theta$=340K, and pmsl (black contours every 5 hPa) shown for the ECMWF analysis. The time is indicated above each panel.

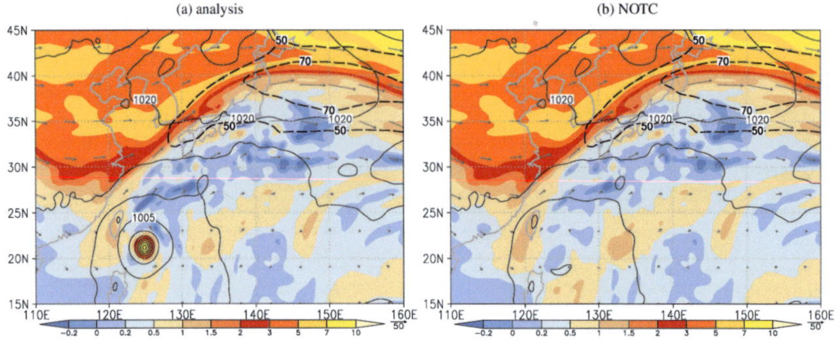


Figure 5.63: PV (in PVU shaded) and wind (grey vectors and black dashed with a $20\,\mathrm{m\,s^{-1}}$ contour interval for $|\vec{v}| > 50\,\mathrm{m\,s^{-1}}$) at the isentropic level of $\Theta$=340K, and pmsl (black contours every 5 hPa) shown for the ECMWF analysis (a) and modified analysis after TC removal (b) at 00 UTC 24 October 2009.

ulations are only of limited value after the four day forecast range. Major ridgebuilding was seen in the analysis after 00 UTC 25 October 2009 (Figure 5.62a,b). Therefore the model runs are initialised at 00 UTC 24 October 2009. Information on the COSMO model runs are summarised in the Appendix, Table A.8.

Although some initial interaction of Lupit's outflow and the downstream flow occurred before initialisation time (see southerly flow near track of Lupit in Figure 5.59), its positive PV tower is still distinct from the approaching midlatitude trough (Figure 5.63a). Thus a clean removal of the positive PV anomaly associated with Lupit in the initial conditions of the NOTC scenario can be carried out (Figure 5.63b). Due to the possible midlatitude impact of Lupit prior to the ini-

Table 5.3: Scenarios with relocated initial position of Lupit: label as used in the text, longitude (in °E) and latitude (in °N) of Lupit's centre, zonal (in °E) and meridional (in °N) shift of the storm centre, zonal/meridional shift in number of grid points to the East/North, brief notes and colour used in Figure 5.64 and 5.65.

| **label** | **lon.** | **lat.** | $\Delta\lambda$ | $\Delta\varphi$ | $\Delta x$ | $\Delta y$ | **remarks** | **colour** |
|---|---|---|---|---|---|---|---|---|
| AL | 124.75 | 21.25 | 0 | 0 | 0 | 0 | ECMWF analysis | black |
| CL20 | 124.75 | 21.25 | 0 | 0 | 0 | 0 | original reference simulation (not used) | light blue |
| NL20 | - | - | - | - | - | - | Lupit removed, no-TC scenario NOTC | - |
| IL21 | 124.25 | 21.75 | -0.5 | +0.5 | -2 | 2 | | yellow |
| IL22 | 124.75 | 21.75 | 0 | +0.5 | 0 | 2 | | aquamarine |
| IL23 | 125.25 | 21.75 | +0.5 | +0.5 | +2 | 2 | | violet |
| IL24 | 124.25 | 22.25 | -0.5 | +1.0 | -2 | 4 | | orange |
| IL25 | 124.75 | 22.25 | 0 | +1.0 | 0 | 4 | used as reference simulation | blue |
| IL26 | 125.25 | 22.25 | +0.5 | +1.0 | +2 | 4 | | green |

tialisation time, the quantification presented here may only represent part of the actual downstream impact of Lupit. Details of the TC removal are summarised in the Appendix, Table A.9.

A control simulation (CL20) is initialised from the unmodified ECMWF analysis and a no-TC simulation (NOTC=NL20) from the modified analysis. Furthermore sensitivity experiments (IL21-IL26) with slightly shifted initial position of Lupit ($\pm$ 0.5°(2 grid points) in zonal direction and + 0.5°, 1.0° (2, 4 grid points) in meridional direction) have been computed. The anomaly definition for the relocation of Lupit is summarised in the Appendix in Table A.10 and an overview about the scenarios is given in Table 5.3.

### 5.4.3 Overview of Lupit scenarios

The anomaly correlation coefficient reveals a poor forecast skill (*ACC* < 0.6) for all scenarios in the forecast range beyond 144 hours (00 UTC 30 October 2009; Figure 5.64). Compared to the mean IFS-ACC this is significantly below the forecast skill reduction, that we would typically expect. A first forecast degradation - higher than in average for the IFS forecasts - occurs on 27 October 2009 when the strong ridgebuilding and amplification of the first downstream

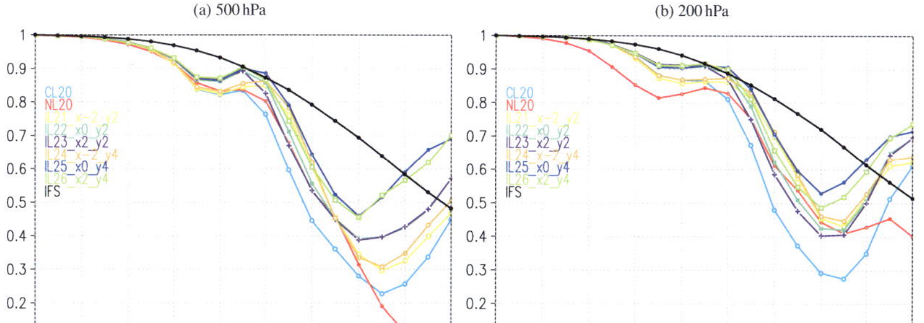


Figure 5.64: Anomaly correlation coefficient (ACC) of geopotential at 500 hPa (a) and 200 hPa (b). Data is from the IL25 and NL20 scenarios introduced in Section 5.4.2. The ACC is calculated on the subdomain shown in Figures 5.62 (110-310°E, 20-60°N) every 12 hours. Additionally, the mean ACC for all deterministic ECMWF IFS forecasts during 1 August 2009 to 1 November 2009 calculated for the domain 110°E-300°E, 20°N-60°N is shown in black.

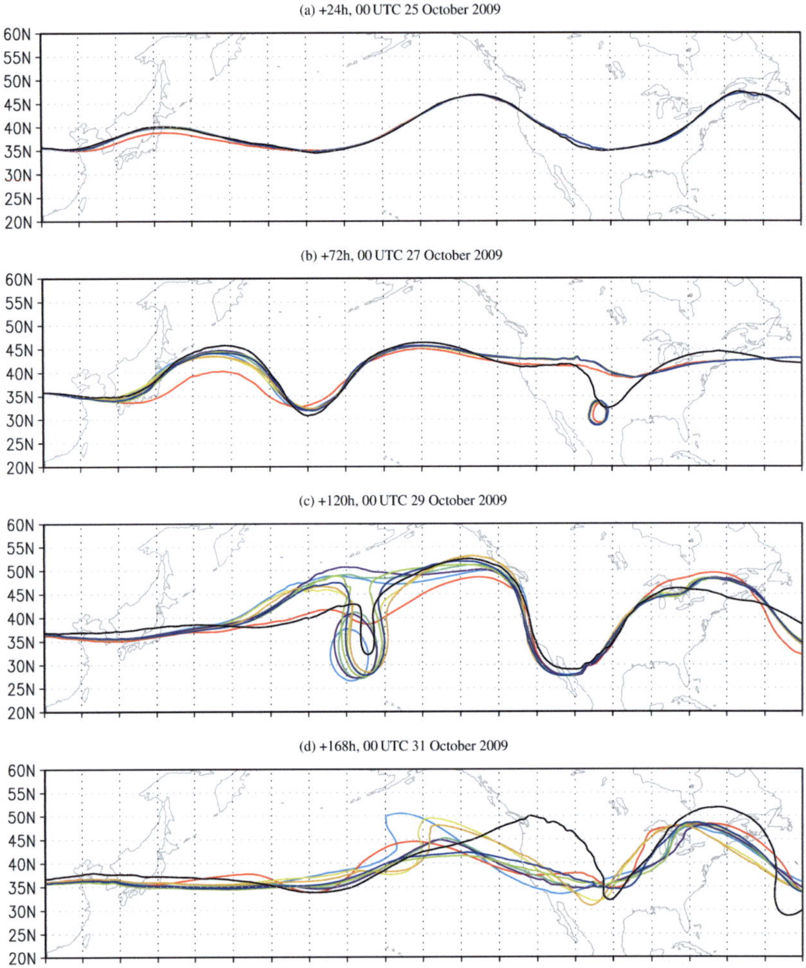


Figure 5.65: 200 hPa geopotential (1200 gpdm contour) from the different scenarios introduced in Section 5.4.2. The scenarios are coloured as in 5.64. The times are indicated above each panel.

ridge-trough couplet occurred. The ACC at upper levels in the NOTC scenario (red in 5.64b) is initially lower than in the original reference simulation CL20 (light blue). However, after 00 UTC 29 October 2009 the original reference simulation exhibits the poorest forecast skill. Lupit in CL20 moves significantly too slowly after 26 October 2009 and too far east after 28 October 2009 (not shown), leading to the reduced forecast skill. Also, the position of Lupit in the ECMWF analysis at 00 UTC 24 October 2009 (124.75°E, 21.25°N, initialisation time) is slightly too far southwest of the position in best track data (124.9°E, 21.4°N). This may explain the track discrepancies and the poor forecast skill. The sensitivity experiments with slightly shifted initial positions aim to address two points: firstly, to improve the track forecast of Lupit and of the first and second downstream cyclones, and secondly to test if the evolution of the downstream flow is sensitive on the exact initial position of Lupit, in particular, to the evolution of the second downstream ridge-trough couplet causing high impact weather in the USA. It turned out that a relocation of Lupit about 1° to the North at initialisation time (scenario IL25) improves the track forecast for the cyclones (cf. blue and black tracks in Figures 5.60 and 5.61) and yields the overall best forecast skill (IL25 blue in 5.64). Although all simulations exhibit an unusually strong skill reduction after 120 hours (00 UTC 29 October 2009), the skill of the other scenarios is between IL25 and NOTC and always above the original reference simulation CL20. Therefore IL25 is used in Section 5.4.4 as the reference simulation for the quantification of Lupit's downstream impact.

At 00 UTC 25 October 2009 (+24 hours), the upper-level flow in the scenarios with Lupit shows no deviation from the analysis (Figure 5.65a, cf. Figure 5.62a). All scenarios show the initial ridgebuilding over and downstream of Japan between 120°E and 160°E. The amplitude of the ridge is weaker in the scenario without Lupit (red). Two days later all scenarios with Lupit reflect the strong amplification of the first downstream ridge-trough couplet (Figure 5.65b, cf. Figure 5.62b). However, the amplitude is a little less than in the analysis. The NOTC scenario (red) exhibits a much less amplified first downstream ridge-trough couplet. Also note, that the scenarios show a cut-off (not linked to the Rossby wave train considered) in the 1200 gpdm contour over the southern USA whereas the analysis has a trough.

At 00 UTC 29 October 2009, the first downstream trough transforms into a PV streamer at around 165°W which is reflected in the analysis (Figure 5.65c, cf. Figure 5.62c). Some scenarios (IL22, IL23, CL20) develop a broad cut-off that is most distinct in CL20. The other scenarios develop a trough, that is broader than in the analysis. The control scenario IL25 (blue) exhibits a trough that is closer to the analysis (in shape and position) than in the other scenarios. The NOTC scenario has only a weak trough in the 1200 gpdm contour due to the weaker amplification of the second downstream ridge. In all scenarios a broad second downstream ridge (155°W-125°W) and a broad second downstream trough (125-95°W) developed. In IL25 the second ridge is closest to the analysis. The other scenarios exhibit partly less amplification of the second ridge. However, all scenarios predict a similar evolution of the broad second downstream trough over the western USA. At this time also the third downstream ridge forms over the eastern USA. Despite small differences from the analysis, this is similar amongst the scenarios with Lupit.

On 31 October 2009 (Figure 5.65d, cf. Figure 5.62d) stronger deviations from the analysis occur. This is reflected in the ACC reaching its lowest values in all scenarios (Figure 5.64). At this time the analysed second downstream ridge moved to the western USA and the second downstream trough transformed into a PV streamer east of the Rocky Mountains (Figure 5.62d). The COSMO simulations are not able to reproduce the shift of the ridge and consequently the location and width of the second downstream trough/PV streamer are different in the scenarios. Furthermore, the third downstream ridge-trough couplet developed over the eastern USA and the western North Atlantic. Again this is represented closest to the analysis in the control simulation IL25.

In summary, the sensitivity experiments showed that Lupit has a strong impact on the ridge and trough structure directly downstream. However, this impact did not continue downstream so that the evolution of the upper-level flow over North America turned out to be relatively insensitive on the ET of Lupit. In particular, the second downstream trough causing the outbreak of polar air in the western and central USA and the explosive cyclogenesis affecting the eastern USA, was present in all scenarios. It can be concluded that the evolution of the Rossby wave train is mainly driven by midlatitude forcing.

### 5.4.4 Detailed quantification of Lupit's downstream impact

In the following, the downstream impact of Lupit is quantified by comparing the control simulation IL25 with the scenario in which Lupit has been removed from the initial conditions (NOTC=NL20). In the control scenario the ridgebuilding downstream of Lupit and the associated jet streak were strongest at 18 UTC 26 October 2009 (Figure 5.66a). At this time Lupit is located at about 140°E, 33°N (light blue labelled IL5L in Figure 5.60) and undergoes strong reintensification (978 hPa, light blue in Figure 5.61). The outflow lead to an amplification of the downstream ridge and trough. However, the ridgebuilding is weaker compared to the analysis (Figure 5.62b). Consequently the jet streak exceeds 70 $ms^{-1}$ only in a small region (140°E-160°E, 46°N; Figure 5.66a). Ahead of the downstream trough the first downstream cyclone deepened to 988 hPa (blue dashed in Figure 5.61). In the NOTC scenario the downstream ridge-trough couplet is much less strongly amplified and the jet streak weaker (Figure 5.66b). However, in the region where Lupit reintensifies an extratropical cyclone forms at around 147°E, 38°N (1005 hPa; orange in Figures 5.60 and 5.61). Also, the first downstream cyclone developed at about the same location as in IL25 (177°W, 40°N). Given the different width of the larger scale trough, the fine scale PV structure upstream of the cyclogenesis, which is similar in both scenarios, must be important for the evolution of the first downstream cyclone. The advection of low PV air by Lupit's outflow is reflected in the 340 K difference PV (Figure 5.67a). Just to the South of the jet streak in IL25 the minimum negative difference PV ($< -5\,PVU$) occurs and extends downstream along the midlatitude wave guide. This advection of low PV air leads to an amplification of the first downstream ridge-trough couplet. The increase in the PV gradient $\nabla_h PV$ along the midlatitude wave guide is reflected in the 340 K difference of $\nabla_h PV$ (Figure 5.67b). A band of difference $\nabla_h PV > 2$ PVU (100 km)$^{-1}$ extends along the midlatitude wave guide. The maximum in difference wind speed of up to 40 $ms^{-1}$ occurs where the difference PV gradient has a maximum (>6 PVU (100 km)$^{-1}$). Thus, as in Jangmi and Hanna, Lupit has a crucial impact on the downstream ridgebuilding and the acceleration of the midlatitude jet streak. Despite this, an extratropical cyclone also develops in the NOTC scenario at about the loca-

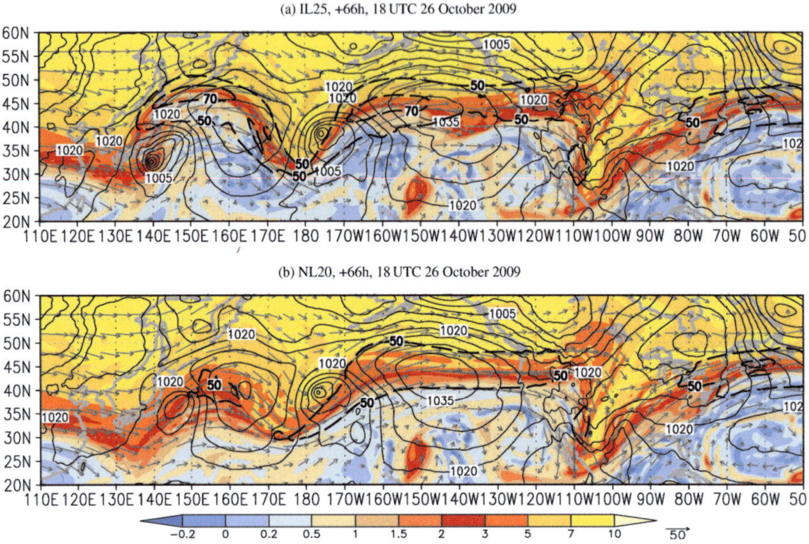


Figure 5.66: Similar to Figure 5.62b but the data is from the control scenario IL25 (a) and the NOTC scenario NL20 (b) after 66 hours forecast valid at 18 UTC 26 October 2009.

tion of Lupit and the downstream trough amplified, so that the first downstream cyclogenesis occurred.

For the subsequent evolution, only the transformation of the first downstream

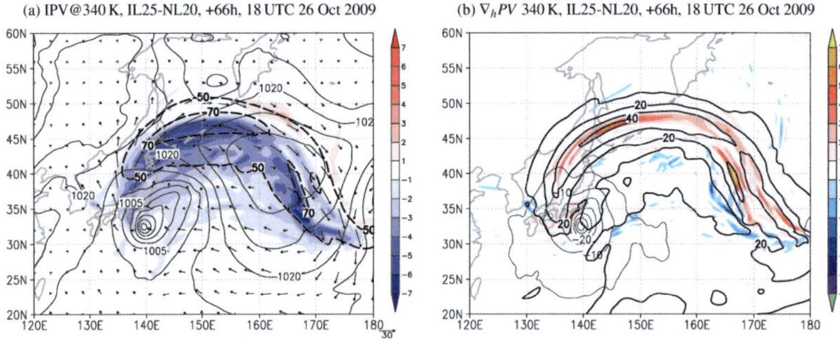


Figure 5.67: (a) Difference IL25 minus NL20 of PV (shaded) and wind vectors (grey) on the 340 K isentropic level, along with wind speed $|\vec{v}| > 50\,\mathrm{m\,s^{-1}}$ on 340 K (thick black dashed every $20\,\mathrm{m\,s^{-1}}$) and pmsl (black every 5 hPa) from the IL25 simulation. (b) Difference IL25 minus NL20 of the horizontal gradient of 340 K isentropic PV $\nabla_h PV$ (shaded every 1 PVU $(100\,\mathrm{km})^{-1}$), magnitude of the difference wind on 340 K (thick black dashed every $10\,\mathrm{m\,s^{-1}}$), and difference pmsl (black with a 2.5 hPa contour interval). Time is 18 UTC 26 October 2009.

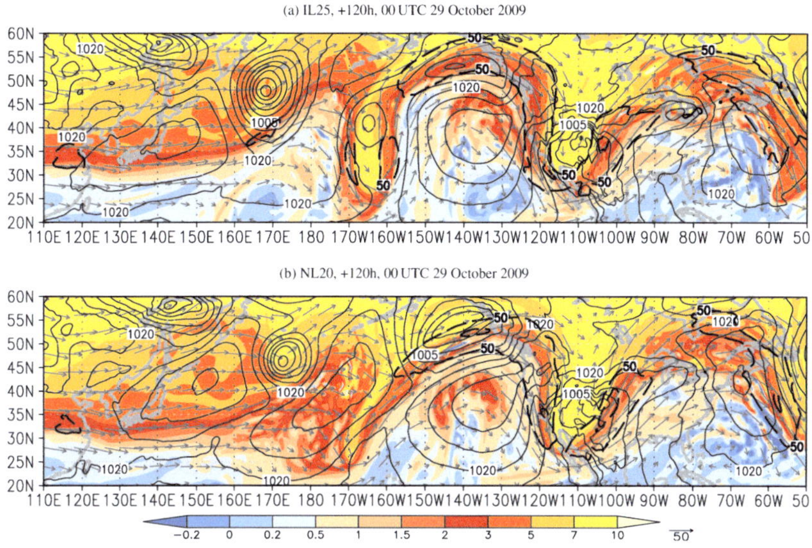


Figure 5.68: As Figure 5.66 but after 120 hours forecast valid at 00 UTC 29 October 2009.

trough and the amplification of the second downstream ridge-trough couplet at 00 UTC 29 October 2009 are considered. Afterwards the forecast skill significantly decreases (Figure 5.64). In the control simulation (Figure 5.68a), the first downstream trough transformed into a rather broad PV streamer. Lupit tracked further east- and southward than in the analysis (Figure 5.60), and its reintensification was not as intense (962 hPa, 18 UTC 27 October 2009, Figure 5.61). In the NOTC scenario (Figure 5.68b), the downstream trough begins to cut off, in contrast to the analysis. In the subsequent days this cut-off triggers a weak tropical transition, that undergoes ET (green, NL2T in Figures 5.60 and 5.61). The cyclone that forms instead of Lupit reached a minimum pmsl of 988 hPa at 12 UTC 28 October 2009 (Figure 5.61) and has a track similar to Lupit in IL25 (Figure 5.60). In both scenarios the first downstream cyclone exhibited a DRW-like propagation across the Pacific, as in the analysis (Figure 5.60). Also a second downstream ridge-trough couplet formed whether or not Lupit was present. However, the second downstream ridge is more amplified, in particular at the western edge in the presence of Lupit (cf. Figure 5.68 and analysis Figure 5.62c). The second downstream cyclone forms in both scenarios but its track and intensity shows differences to the analysis (Figures 5.60 and 5.61). In IL25 the second downstream cyclone merges with the first downstream cyclone on 31 October 2009.

The discussion of Lupit's downstream impact is summarised with Hovmöller diagrams of 250 hPa meridional wind in the control and no-TC scenarios (Figure 5.69). Although some differences occur between the analysis and the control simulation we concentrate on the differences between the scenarios.

The impact of Lupit on the evolution of the Rossby wave train is reflected in the difference between control and no-TC scenario (Figure 5.69b). In both scenarios the pronounced Rossby wave train emerges in the region where Lupit undergoes ET (120°E-180°E) during 24-26 October 2009 (Figure 5.69a cf. Figure 5.59). At 00 UTC 27 October 2009 the first downstream ridge-trough couplet is evident with axes at 160°E (ridge) and 180°E (trough) and the second downstream ridge with an axis at 140°W (Figure 5.69a cf. Figure 5.66). The positive difference wind of up to 20 m $s^{-1}$ emerging along the track of Lupit in IL25 (Figure 5.69b) reflects the amplification of the western part of the first downstream ridge and its more westward location when Lupit is present. The dipole of northerly and southerly difference wind starting at about 180°W 00 UTC 26 October indicates a stronger amplification of the midlatitude wave guide in IL25 between the first downstream ridge and trough. On 28 October the increasing

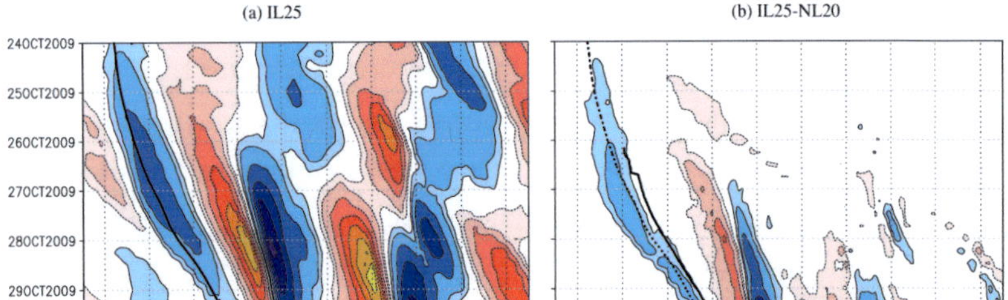


Figure 5.69: Hovmöller diagrams of the meridional wind component at 250 hPa (with a 5 m $s^{-1}$ contour interval) in the six-hourly control simulation IL25 (a) and IL25 and NL20 (b). The data is meridionally averaged from 25°N to 55°N. Forecast from 00 UTC 24 October 2009 (0 h) to 00 UTC 31 October 2009 (168 h) shown. The black solid line in (a) and dashed in (b) marks the track of Lupit in IL25, in (d) of cyclone NL2L in NOTC.

magnitude in the dipole of difference wind (Figure 5.69b) highlights the transformation of the first downstream trough into a PV streamer in IL25 whereas a cut-off forms in NOTC. At 00 UTC 29 October 2009 the first downstream ridge-trough couplet is still evident in both scenarios (Figure 5.69a cf. Figure 5.68) with axes at 180°E (ridge) and 165°W (trough/PV streamer). However, in the NOTC scenario the signal vanishes, whereas the first downstream PV streamer in the control scenario is evident until around 00 UTC 30 October 2009 centred at 160°W (Figure 5.69a). At 00 UTC 29 October 2009 the second downstream ridge-trough couplet is evident in both scenarios also (Figure 5.69a cf. Figure 5.68) with axes at 140°W (ridge) and 110°W (trough). The differences in meridional wind component (Figure 5.69b) east of 140°W after 00 UTC 29 October 2009 reflect the broader second downstream trough over the USA in the control experiment. Also the third downstream ridge-trough couplet appears in both scenarios after around 00 UTC 30 October 2009 with the ridge axis at 80°W and the trough axis at around 50°W and therefore the difference wind is weak (Figure 5.69b).

Moreover, several PV surgery experiments with an earlier version of the TC removal technique were performed (not shown). At this stage the sensitivity on the impact of the Hawaiian ET and on the definition of the moisture anomaly (smaller and much larger region for moisture removal) were tested. However, theses sensitivity experiments exhibit similar results about Lupit's impact on the evolution of the Rossby wave train as present before.

In summary, the detailed discussion of the control and no-TC scenarios showed that, in contrast to the Jangmi scenarios, a Rossby wave train would develop also if Lupit was not present. As in the previous case studies a distinct outflow-jet interaction occurred, resulting in a strengthening of the downstream ridgebuilding, an acceleration of the midlatitude jet streak and an amplification of the first downstream trough/PV streamer. Despite this modification of the detailed wave guide structure, Lupit was was not crucial for the evolution of the downstream Rossby wave train. In particular, the ET of Lupit only had weak impact on the evolution of the second downstream trough that caused high impact weather over the USA.

## 5.5 The ET of Typhoon Malakas (2010)

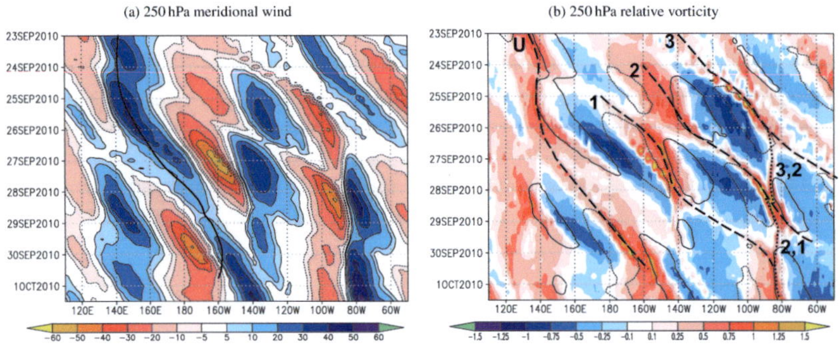


Figure 5.70: (a) Hovmöller diagram of meridional wind component at 250 hPa (shaded every 10 m s$^{-1}$) in the ECMWF analysis. Data meridionally averaged from **35°N to 55°N**. Track of Malakas in the analysis (dotted black line). (b) Hovmöller diagram of relative vorticity (shaded in $10^{-5}$s$^{-1}$) and meridional wind (black -20 and 20 m s$^{-1}$ contours) at 250 hPa in the ECMWF analysis. Data meridionally averaged from **40°N to 60°N**. The thick dashed lines labelled U,1,2,3 mark the maxima of relative vorticity indicating approximately the trough axes of trough U,1,2,3. The thick dotted lines labelled 3,2 and 2,1 mark the mergers of the troughs at 85°W.

Typhoon Malakas occurred in September 2010 in the western North Pacific. The midlatitude upper-level flow during and following ET was characterised by complicated Rossby wave activity. Two successive Rossby wave trains are evident in the Hovmöller diagram of 250 hPa meridional wind (Figure 5.70a). A first emerged on 23 September 2010 prior to recurvature with a first trough axis

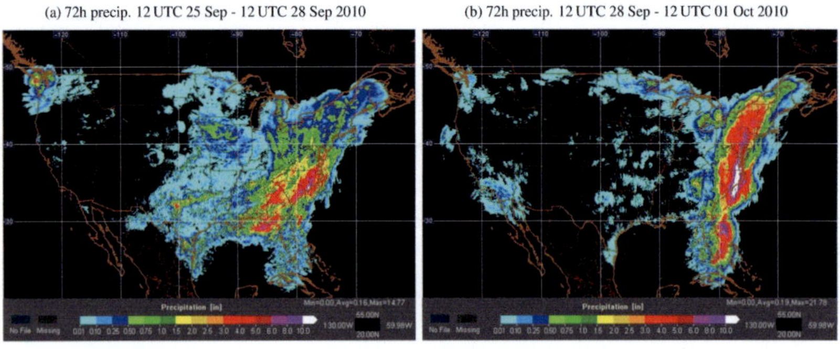


Figure 5.71: 72 h accumulated precipitation (shaded in inches) during the period (a) 12 UTC 25 September to 12 UTC 28 September 2010 and (b) 12 UTC 28 September - 12 UTC 01 October 2010. The data is taken from the NMQ project (`http://nmq.ou.edu/`) and a combination of radar composites and rain gauge data.

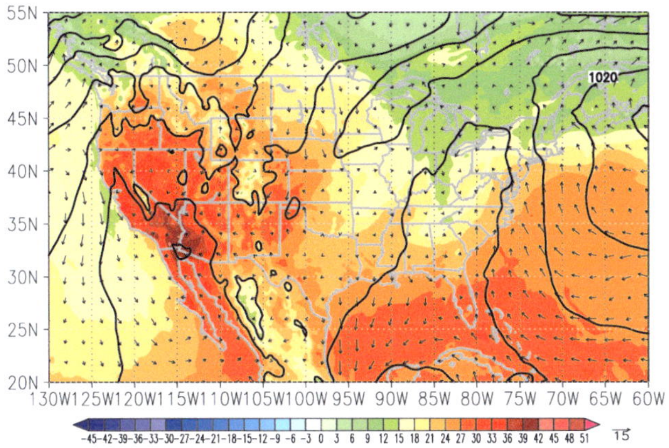


Figure 5.72: 2 m temperature (shaded every 3°C), 10 m wind vectors, and pmsl (black every 5 hPa) at 00 UTC 28 September 2010. Data from ECMWF analysis interpolated on COSMO grid, and 2 m temperature derived by COSMO.

at 130°E. Malakas recurved ahead of this approaching trough on 24 September 2010. At this time the ridgebuilding by Malakas outflow is reflected in an enhanced northward flow at 145°E. Downstream of this ridge a second Rossby wave train developed in the subsequent days. The first Rossby wave train caused a PV streamer over the eastern USA during 25-27 September 2010, reflected in the northerly flow around 100°W in Figure 5.70a. This PV streamer resulted in a first period of heavy precipitation with maxima of more than 380 mm in 72 h at the US East Coast (at around 78°W 34°N;Figure 5.71a). Following this event the second Rossby wave train also triggered a PV streamer over the eastern USA with an axis at around 85°W after 28 September 2010 (Figure 5.70a). This resulted in a second period of heavy precipitation of up to 560 mm at the US East Coast (at around 77°W 35°N; Figure 5.71b). Upstream of the PV streamer a ridge with an axis at around 115°W was located over western North America at 00 UTC 28 September 2010 (Figure 5.70a). At this time a heat wave struck the southwestern USA with an all time temperature record of 45°C in downtown Los Angeles, California, USA. In southern California, southwestern Arizona and northwestern Mexico the analysed 2 m temperature exceeded 36°C in a large region (Figure 5.72). Very high temperatures of more than 39°C occurred in the Los Angeles region along the southwestern Californian Coast and exceeded 42°C at the border to Arizona.

In the following, an overview of the large-scale flow evolution over the Pacific and North America is given (Section 5.5.1) and the interaction between the two Rossby wave trains leading to the high impact weather is explained. Then COSMO sensitivity experiments using the TC removal technique are introduced (Sections 5.5.2 and 5.5.3) and the impact of Malakas on the evolution of the Rossby wave train over the North Pacific and North America and on the extreme weather is quantified (Sections 5.5.4 and 5.5.5).

### 5.5.1 Synoptic overview

Typhoon Malakas formed on 21 September 2010 at around 145°E, 18°N. It headed slowly westward until 23 September 2010 and minimum pressure dropped by 25 hPa to 985 hPa according to JMA. On 23 September the track was more northward and Malakas recurved on 24 September 2010. Interestingly, the maximum intensity (945 hPa, maximum sustained winds 44 m $s^{-1}$) was reached during 24 September 2010 when the storm started to interact with a narrow midlatitude trough.

In the following the flow evolution in the ECMWF analysis is discussed starting on 23 September 2010 with reference to Malakas and four distinct upper-level troughs that we name upstream trough (U), first downstream trough (1), second downstream trough (2) and third downstream trough (3) (Figure 5.73). Although a signature of all of these troughs is present on 23 September 2010 we only discuss significant changes following the ET of Malakas.

At 12 UTC 23 September 2010 Malakas is located at 141°E, 21°N with a central pressure of 977 hPa (Figure 5.73a). It is well south of the midlatitude wave guide that is characterised by a broad trough at 130°E and a broad ridge at 160°E. Malakas starts to recurve ahead of this upstream trough with an axis at about 122°E.

One day later at 12 UTC 24 September 2010 the southern tip of the upstream trough forms a narrow PV streamer at around 132°E extending towards Malakas (Figure 5.73b). A strong northward flow is evident to the north of Malakas.

This reflects the midlatitude flow ahead of the trough enhanced by the strong outflow channel of Malakas. The midlatitude jet streak increases to more than $90\,\mathrm{m\,s^{-1}}$ between 150-160°E, 46°N. Very low PV air is advected northward by the outflow and starts building a ridge between the upstream trough and the 340 K PV maximum at 160°E (labelled "1" in Figure 5.73b). Malakas reaches peak intensity with a minimum pressure in the centre of 952 hPa at 00 UTC 25 September 2010 (in analysis data). Shortly after this time Malakas was officially declared extratropical.

On 25 September 2010 Malakas transitions into an extratropical cyclone. The narrow upstream PV streamer wraps cyclonically around Malakas and in its centre (Figure 5.73c). To the north of Malakas divergence in the outflow is enhanced compared to the previous day. Associated ridgebuilding continues and hinders a rapid eastward propagation of the upstream trough. Low PV air of the ridge also wraps cyclonically into the centre of Malakas similar to the wrapping-up during the ETs of Hanna and Choi-wan. The wrap-up of the PV streamer and of the downstream ridge reflect a poleward cyclonic wave breaking. The jet has a maximum at 180°W, 49°N. The 340 K PV maximum has moved to 173°W (labelled "1" in Figure 5.73c) and ahead of it in the poleward jet exit region an extratropical cyclone has developed ($<$ 995 hPa).

On 26 September 2010 Malakas is an intense extratropical cyclone (Figure 5.73d). It has slightly reintensified to a central pressure of 962 hPa. A very pronounced ridge is present downstream of Malakas. The ridgebuilding amplified the first downstream trough (1) which emerged from the 340 K PV maximum and appears with a clear axis at around 155°W. Ahead of trough 1 the extratropical cyclone has intensified and has merged with an extratropical cyclone that has developed ahead of the second downstream trough (2) during the previous days. The newly formed extratropical cyclone becomes the major low in the eastern North Pacific and initiates ridgebuilding between trough 1 and the second downstream trough (with the trough axis at about 125°W, labelled "2" in Figure 5.73d). The ridge that has been built during the previous days over North

America between trough 2 and the third downstream trough (labelled "3" in Figure 5.73d) experiences an equatorward anticyclonic wave breaking. In this wave breaking trough 3 becomes a distinct tilted PV streamer at 85°W and at its southern end a cut-off has formed (110°W, 31°N, labelled "3C2"). Ahead of the PV streamer (trough 3) southerly flow starts to advect subtropical moist air into the eastern USA.

At 12 UTC 27 September 2010 Malakas is a mature extratropical cyclone over the central North Pacific at 174°E, 53°N (Figure 5.73e). The first downstream trough (1) has become a narrowing PV streamer at 145°W. Downstream the extratropical cyclone has produced a very pronounced ridge over western North America, filled with low PV air. In its centre a surface anticyclone has formed at around 115°W, 42°N indicating large-scale descending air masses. In the southern part of the ridge the cut-off 3C2 of trough 3 has merged with a cut-off from trough 2 and easterly flow is evident between the cut-off and the ridge to the north. The northern part of trough 2 is transforming into a PV streamer at 95°W (Figure 5.73e). Slightly further downstream the southern portion of trough 3 repeatedly cut off (labelled "3S") and starts to merge with trough 2 over the eastern USA. To the east of this elongated cut-off strong southerly flow advects subtropical moist air from the Gulf of Mexico along the US East coast. The northern portion of trough 3 (labelled "3N") has moved into the North Atlantic.

During the next two days Malakas is decaying in the central North Pacific near 160°W, 55°N (Figure 5.73f). The northern part of trough 1 has moved to 100°W, but is linked with a narrow PV filament to the southern portion that cut off near 140°W. Trough 2 has become a narrow PV streamer at 85°W and has merged with the cut-off (3S) of trough 3 (Figure 5.73f). On its western side cool air is advected to the Gulf and to its eastern side subtropical moist air is advected northward. The upper-level low over southwestern North America resulting from the merger of the earlier cut-off of trough 2 and cut-off 3C2 has moved eastward and is located over southern California.

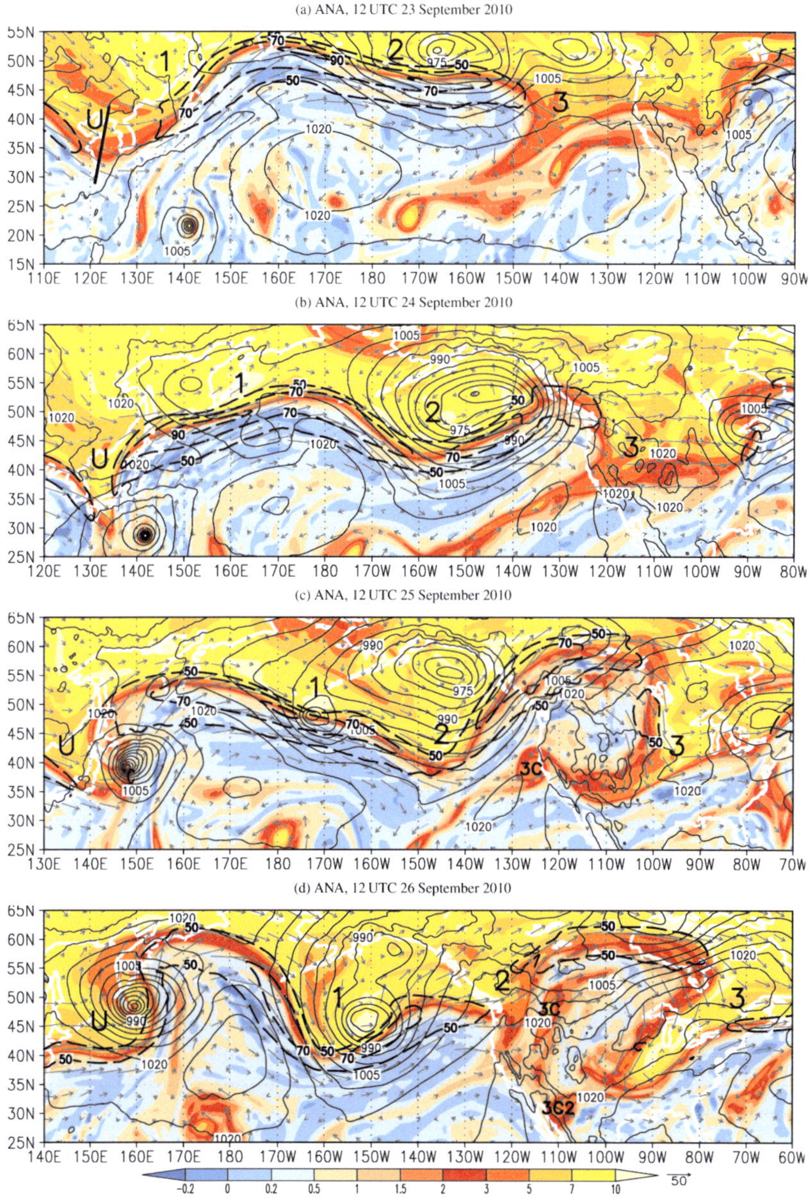


Figure 5.73: PV (in PVU, shaded), wind (grey vectors and black dashed with a $20\,\mathrm{m\,s^{-1}}$ contour interval for $|\vec{v}| > 50\,\mathrm{m\,s^{-1}}$) at the isentropic level of $\Theta$=340K, and pmsl (black contours every 5 hPa) shown for the ECMWF analysis... *Continued on page 253.*

At 12 UTC 01 October 2010 Malakas has dissolved. The northern portion of trough 1 has merged with trough 2 at 85°W (Figure 5.73g) and maintains the southerly advection of subtropical moist air at the US East Coast.

The evolution of the upper-level flow is briefly summarised with the help of Hovmöller diagrams of 250 hPa meridional wind and relative vorticity (Figure 5.70). Trough (ridge) axes also depict high (low) values of relative vorticity and are therefore occasionally better suited to track the trough axes, when they tilt or move out of the averaging band.

At 00 UTC 23 September 2010, the upstream trough (U) is evident with an axis at around 130°W (Figure 5.70). It remains almost stationary for about two days and two interacting Rossby wave trains emerge downstream of the trough. The first Rossby wave train emerges on 23 September 2010 and triggers a downstream trough at 160°W, 00 UTC 24 September 2010 (2, Figure 5.70b) and another downstream trough is already present at 140°W, 00 UTC 23 September 2010 (3, Figure 5.70b). Malakas recurves ahead of the upstream trough (Figure 5.70a) during 23-25 September 2010. On 24 September 2010 the ridgebuilding downstream of Malakas significantly enhances the southerly flow on the eastern side of the upstream trough (145°W, Figure 5.70a) and is likely responsible for the quasi-stationarity of the upstream trough. At 00 UTC 25 September 2010, the first downstream trough emerges at 180°E (Figure 5.70b), reflecting the triggering of the second Rossby wave train. On 27 September the ridgebuilding downstream of trough 1 is reflected in negative relative vorticity centred at 110°W over western North America (Figure 5.70b). Along with this, trough 2 becomes the second downstream trough of the Rossby wave train emerging downstream of Malakas ET (Figure 5.70a,b) and is merging with the southern portion of trough 3 (Figure 5.70b). The merger of both troughs results in a PV streamer at 85°W. On 29 September 2010 the PV streamer is regenerated by the merging of trough 1 and trough 2 (Figure 5.70b). This results in a rather stationary PV streamer at 85°W from 26 September to 1 October 2010 (Figure 5.70b). As a consequence subtropical, moist air is continuously advected along the US East Coast, contrasting the cooler air advected to the west of the PV streamer (Figure 5.70a).

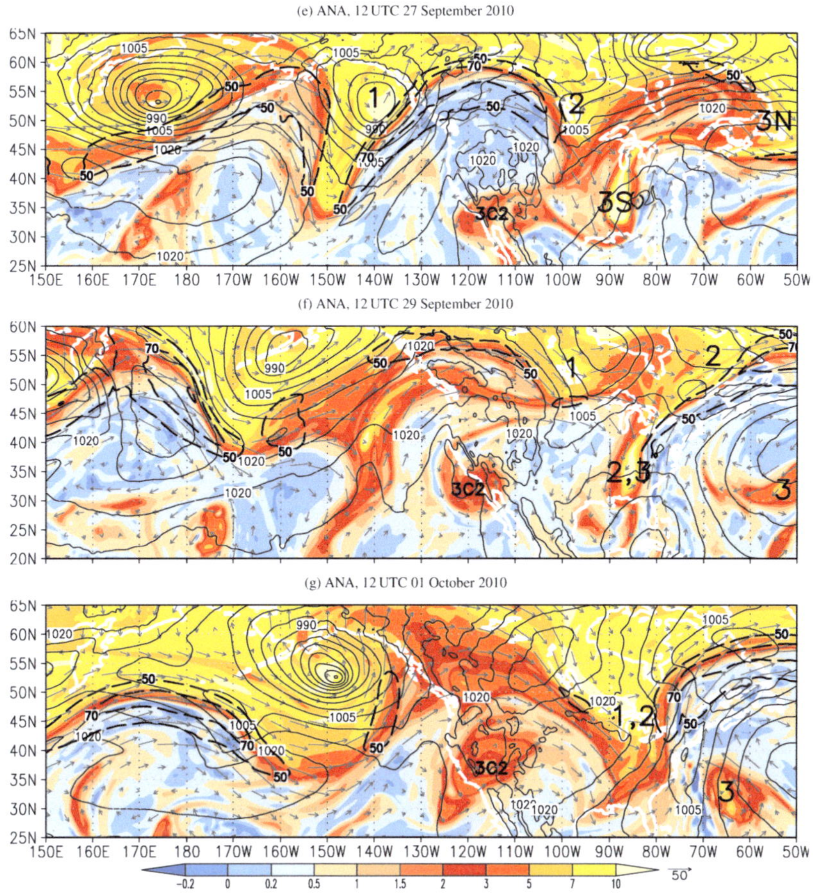


Figure 5.73: Continuation of Figure 5.73 ... The time is indicated above each panel. The section shown in (a-e) is shifted 10° east between two times and differs from (e-g).

The large-scale flow configuration gives an explanation for the high impact weather events that occurred in North America after the ET of Malakas. At 27 September 2010, when the heat wave occurred in southwestern North America, the second downstream ridge established over western North America accompanied by large-scale descending air masses (Figure 5.73e). At the same time the merger cut-off 3C2 is located south of the ridge. Easterly descending flow must be expected in the entire troposphere between the cut-off and the ridge and south of the anticyclone at the surface. This leads to the production and advec-

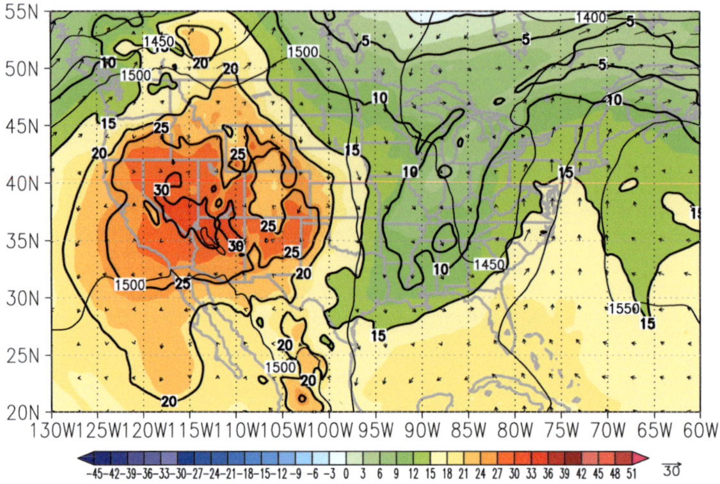


Figure 5.74: 850 hPa temperature (shaded every 3°C), 850 hPa wind vectors, and 850 hPa geopotential (thin black contours every 50 gpm) at 00 UTC 28 September 2010.

tion of a very hot continental air mass into the southwestern corner of the USA and explains the heat wave recorded there. Archambault (2011) has investigated the source of this hot air mass in more detail. She showed that the hot air originated from large-scale descent in the ridge and was advected within the anticyclonic circulation enhanced at the southern side of the ridge by the cut-off (Figure 4.12 in Archambault 2011). The 850 hPa temperature at 00 UTC 28 September 2010 shows the hot air mass centred with maxima above 30°C over the elevated terrain in southern Nevada and northern Arizona (Figure 5.74 cf. Figure 5.72). In southern California the 850 hPa temperature exceeds 27°C. The 850 hPa level is at around 1500 m amsl in this region (Figure 5.74). Assuming a dry adiabatic temperature gradient of $1\,\mathrm{K}(100\mathrm{m})^{-1}$, the temperature in the Los Angeles region at sea level must peak at 42°C. With low surface wind speed (cf. Figure 5.72) and strong insulation in an urban environment the measured temperature of 45°C is in agreement with the large-scale circulation.

At 27 September 2010 the advection of subtropical, warm air to the east of the PV streamer is seen with temperatures above 15°C at 850 hPa along the US East Coast (Figure 5.74). In and to the west of the PV streamer a cooler air mass (5-10°C) is advected southward to the Gulf of Mexico. The first heavy precipitation period detailed in the introductory section (cf. Figure 5.71a) occurs at the frontal zone between these contrasting air masses. At the frontal zone extratropical cyclones form repeatedly accompanied by heavy precipitation. Until 12 UTC 29 September 2010 (Figure 5.73f) a first cyclone has moved northward to 73°W, 49°N and a second cyclone forms at 80°W, 39°N. Further south tropical storm Nicole is evident at 82°W, 22°N. The advection of subtropical moist air continues at the US East Coast as the PV streamer is regenerated by the merger of trough 2 and 1. This merging inaugurated the second period of heavy precipitation at the US East Coast (cf. Figure 5.71b).

In summary, the upper-level flow configuration downstream of the ET of Malakas gives an explanation for the high impact weather in the USA observed at the end of September 2009. In the following we aim to quantify the impact of Malakas on this evolution with the help of COSMO simulation with and without the ET system.

## 5.5.2 Technical details of the PV surgery

The impact of Typhoon Malakas on the Rossby wave train and the high impact weather downstream is investigated with the help of PV surgery experiments. For Malakas, two sets of control and no-TC simulations are used. Reference simulation CM22 and no-TC simulation NM22 are initialised at 00 UTC 23 September 2010 and the simulations CM23 and NM23 one day later at 00 UTC 24 September 2010. The purpose of this approach is two-fold. The set CM22 and NM22 is initialised when Malakas is well away from the midlatitude trough. However, the high impact weather at the end of September 2010 occurs during the 4- to 10-day forecast range which in the previous studies has turned out to be a problematic in COSMO simulations. In the ECMWF no upper-level low PV air advection by the TC outflow and deflection of the midlatitude wave guide is detected until 00 UTC 24 September 2010, so that this time is also a suitable initialisation time. The set CM23 and NM23 thus should have a better forecast skill for the high impact weather and comparing both sets enables us to study the impact of the exact TC removal time. Apart from the different initialisation time, both sets are computed with the same configuration (Table A.11). Six-hourly ECWMF analyses (with 0.25° horizontal resolution) serve as boundary conditions. The horizontal model domain covers the entire North Pacific and North America. The control simulations CM22 and CM23 are initialised from the unmodified ECMWF analysis. Malakas is removed from the ECMWF analysis for the initial conditions of the no-TC runs NM22 and NM23 using the PV surgery technique (Section 3.2.4). Details of the TC removal of Malakas are summarised in the Appendix, Table A.12

### 5.5.3 Verification of the sensitivity experiments for Malakas

The anomaly correlation coefficient (ACC, Figure 5.75) of geopotential at 500 hPa and 200 hPa indicates a very good forecast skill for both control simulations (blue colours). Until 12 UTC 27 September 2010, the time when the heat wave and the first heavy precipitation event occurred, the ACC at mid and upper levels as well as low levels (not shown) is above 0.95. Subsequently the ACC decreases but, remains well above the mean IFS-ACC and above 0.6 - the threshold for a meaningful forecast. Interestingly, the skill of the simulation

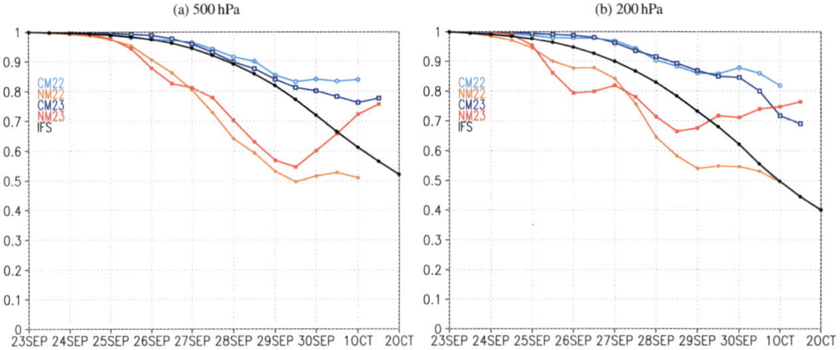


Figure 5.75: Anomaly correlation coefficient (ACC) of geopotential at (a) 500 hPa and (b) 200 hPa. Data is from the CM23 (blue) and NM23 (red) simulations initialised at 00 UTC 24 September 2010, and CM22 (light) blue) and NM22 (orange) simulations. The ACC is calculated on the subdomain 110-310°E, 20-60°N every 12 hours. The ECMWF analyses are used as the analysis data and the daily mean values of ERA-INTERIM interpolated on a 0.25°x0.25° horizontal grid is used as the climatology. Additionally the mean ACC for all deterministic ECMWF IFS forecasts during 1 August 2010 to 1 November 2010 calculated for the domain 110°E-300°E, 20°N-60°N is shown in black.

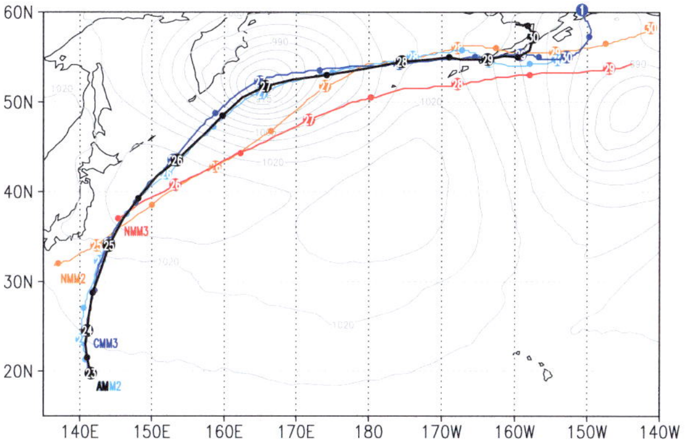


Figure 5.76: Tracks of Typhoon Malakas from analysis data (black), CNTRL simulations CM23 (blue) and CM22 (light blue), and NOTC simulations NM23 (red) and NM22 (orange). Thick points 00 UTC times, labelled with the day of September 2010, thin points 12 UTC times. Pressure at mean sea level (grey every 5 hPa) from the analysis data shown at 00 UTC 27 September 2010.

CM23 initialised one day later is lower than CM22 after 12 UTC 29 September 2010. Before this time both simulations exhibit a similar ACC and CM23 is slightly better at 1000 hPa (not shown). As expected, the ACC at mid and upper levels for the simulations without Malakas (red shades) is lower than for the control simulations and well below the mean forecast degradation in the IFS model. At 500 hPa the ACC is similar in both no-TC scenarios with the NM22 scenario (orange) a little closer to the analysis before 00 UTC 27 September 2010 and the NM23 (red) being a little closer to the analysis afterwards. The ACC for NM23 increases significantly after 12 UTC 30 September 2010.

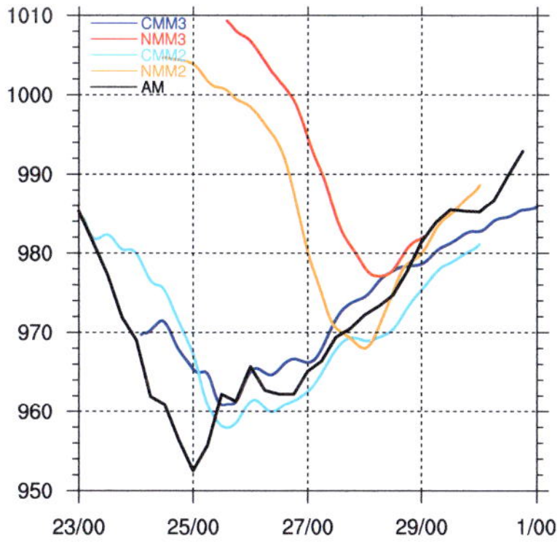


Figure 5.77: Minimum pmsl (in hPa) in the centre of Malakas. The same labels and colours as in Figure 5.76 are used.

The same pattern is evident at 200 hPa but the differences between the two no-TC scenarios are larger. At 1000 hPa (not shown) both scenarios exhibit a similar ACC. As will be shown later, an extratropical cyclone develops instead of Malakas in the no-TC scenarios.

The track of Malakas in control simulation CM23 is almost exactly the same as in the analysis until 00 UTC 29 September 2010 (cf. dark blue and black tracks in Figure 5.76). However,

COSMO needs about 36 h to (re-)intensify Malakas to the strength as in the analysis (Figure 5.77). Subsequently, the decay is similar to the analysis. The track of Malakas in CM22 (light blue) lags around 6 h behind the analysis. Also, it requires longer to (re-)intensify to the analysis intensity which is reached at about the same time as in CM23. Malakas in CM22 is slightly stronger but its subsequent decay is similar to the analysis. In both no-TC scenarios an extratropical cyclone develops and strongly intensifies ($>$30 hPa in 72 hours) ahead of the midlatitude trough in the region where Malakas (re-)intensifies (red colours in Figures 5.76 and 5.77). In NM22 this cyclone (NMM2, orange) forms at 09 UTC 24 September and starts to intensify after 00 UTC 25 September 2010. It reached the strength of Malakas in the corresponding reference simulation CM22 at about 18 UTC 27 September 2010. Afterwards the decay is similar to the analysis (cf. light blue and orange lines in Figure 5.77).At this time the track becomes identical to the track of Malakas, but the cyclone NMM2 is about 12 h in advance (orange in Figure 5.76). In NM23 the evolution of the extratropical cyclone (NMM3, red) is delayed about 12 h compared to NM22. It reaches the intensity of Malakas in CM23 at around 06 UTC 28 September 2010 and subsequently decays (red in Figure 5.77). However, the track of NMM3 is about 7° south of Malakas' original track (red in Figure 5.76). The strong extratropical development in the no-TC scenarios contrasts the simultaneously steady decay of Malakas in the control simulations. In terms of diabatic processes the extratropical development must have a different impact on the upper-levels than the decaying mature ET system in the control simulations. We also note that at the end of the simulation period scenarios CM22 and NM22 develop and maintain tropical storm Nicole in the Caribbean. In scenarios CM23 and NM23 Nicole also develops more than in the analysis, but it does not become as intense as in CM22/NM22.

The CM22/NM22 sensitivity experiments allow us to quantify the upper-level impact of Malakas at 00 UTC 24 September (+24h), the initialisation time of CM23/NM23 (Figure 5.78). The 340 K isentropic PV does not show significant differences in the upstream trough at 130°W, the ridge that is evident over the western North Pacific, the second downstream trough at 165°W or the third

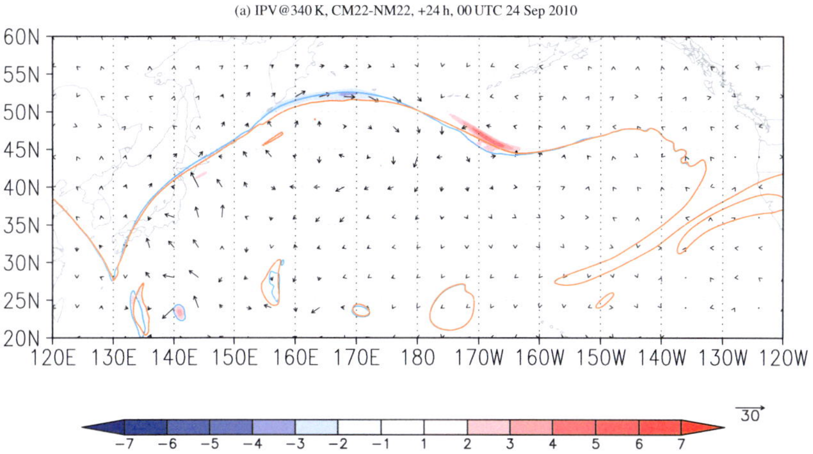


Figure 5.78: Difference CM22 minus NM22 of PV (shaded) and wind vectors (grey) on the 340 K isentropic level. Additionally the 340 K isentropic 2 PVU contour in CM22 (light blue) and NM22(orange) is shown. Time is 00 UTC 24 September 2010 (+24 h) for the sensitivity experiment initialised at 00 UTC 23 September 2010.

downstream trough at 140°W. We conclude that an initialisation time at 00 UTC 24 September 2010 in CM23/NM23 is equally suitable for the investigation of Malakas' downstream impact. We thereby assume that the outflow anomaly produced in the precedent 24 hours does not interact with the upper-level flow after 00 UTC 24 September 2010 which is confirmed by the subjective investigation of the upper-level flow evolution in the analysis.

The control simulations CM22 and CM23 represents well the structure of the upper-level Rossby waves seen in the analysis (cf. Figures 5.79a,b and 5.70a). Differences occur mainly after 00 UTC 27 September 2010 when the ACC started to decrease. The most significant deviation is in the merging of trough 2 and 3 at around 90°W on 27 September and trough 1 and 2 near 90°W around 30 September 2010 (Figure 5.79c,d). Both mergers appear slightly more eastward than in the analysis and the northward flow to the east of the PV streamer is not as intense as in the analysis.

The scenario with Malakas removed at 00 UTC 23 September 2010 (NM22) exhibits a slightly faster propagation of the Rossby wave train that occurs downstream of cyclone NMM2 developing instead of Malakas (Figure 5.79e). The downstream ridgebuilding at 150°E, 25 September 2010, is not as intense as with Malakas and particularly the first downstream trough (1) propagates farther east on 28 September 2010 (Figure 5.79e). A major difference in the simulation with and without Malakas initialised at 00 UTC 23 September 2010 is the lack of the merger of trough 1 and 2 at 80°W after 00 UTC 29 September 2010. As this was important for the second heavy precipitation period at the US East Coast one might expect a less severe downstream impact of the Rossby wave train in NM22.

When Malakas is removed at 00 UTC 24 September 2010 (NM23, Figure 5.79f) the Rossby wave train downstream of cyclone NMM3 occurs slightly later and is much weaker than in NM22 (Figure 5.79f cf. Figure 5.79e). The difference in the outflow anomaly at 150°E, 25 September 2010 is larger than in CM22/NM22, indicating that the diabatically enhanced ridgebuilding downstream of the extratropical cyclogenesis in NM23 is weaker than in NM22 due to the 24 hour later initialisation time. This is consistent with the lower final intensity of cyclone NMM3 compared to NMM2 (Figure 5.77). This also explains the lower ACC in NM23 at upper levels compared to NM22 before 27 September 2010 (Figure 5.75). Due to the earlier initialisation time the ridgebuilding downstream of the extratropical cyclogenesis in NM22 is closer to the evolution with Malakas than in NM23. Afterwards albeit in both no-TC simulations faster than in the control simulations, the Rossby wave train in NM22 propagates faster and earlier eastward than in NM23 leading to a larger forecast degradation in NM22. Similar to NM22, in particular, the first downstream trough (1) in NM23 extends farther eastward and NM23 also lacks of the merger of trough 1 and 2. In summary both sets of sensitivity experiments are found to be useful for the investigation of Malakas' downstream impact.

### 5.5.4 Quantification of Malakas' downstream impact

When Malakas interacted with the approaching midlatitude trough a pronounced ridgebuilding occurred to the north and downstream of the transitioning storm. At 12 UTC 25 September 2010 a poleward cyclonic wave breaking occurred as the trough and downstream ridge wrapped cyclonically into the centre of Mala-

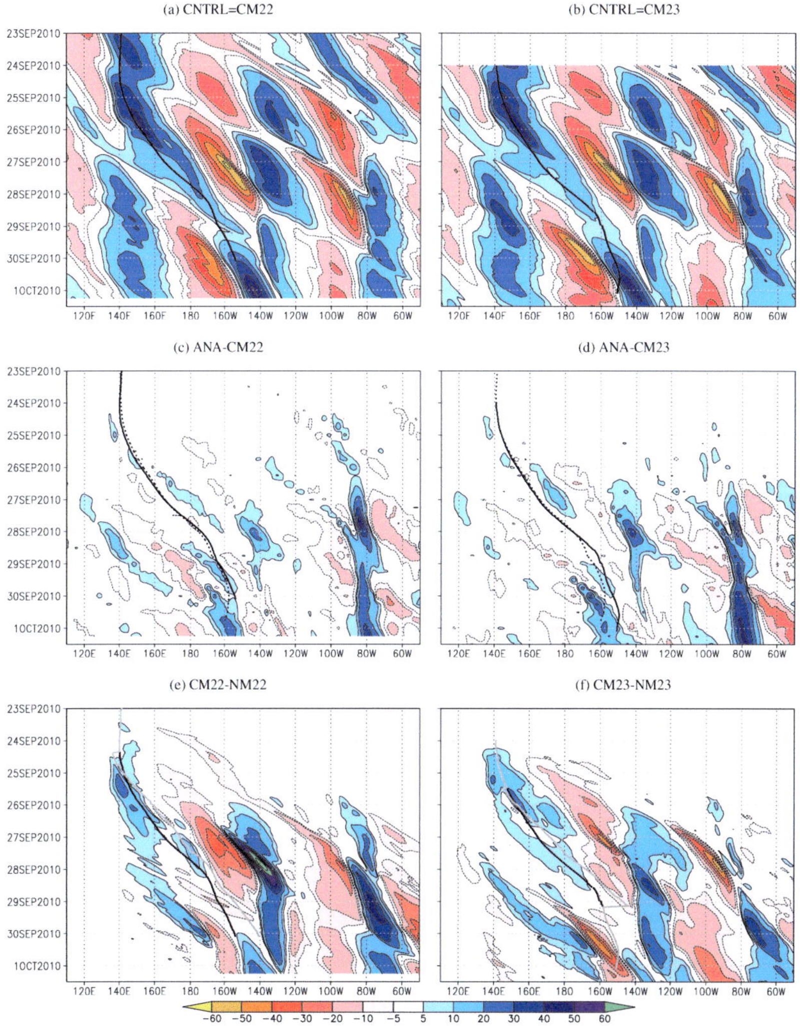


Figure 5.79: Hovmöller diagrams of the meridional wind component at 250 hPa (with a 5 m s$^{-1}$ contour interval) control simulation CM22,CM23 (a),(b) (one-hourly). Hovmöller diagrams of the six-hourly meridional wind component difference at 250 hPa (with a 5 m s$^{-1}$ contour interval) between ECMWF analysis and CM22,CM23 (c),(d) and CM22,CM23 and NM22,NM23 (e),(f). The data is meridionally averaged from 35°N to 55°N. Forecast from 00 UTC 23,24 September 2010 (0 h) to 06 UTC,12 UTC 1 October 2010 (198 h,180 h) shown in (a,c,e),(b,d,f). The black solid line in (a,c,e),(b,d,f) marks the track of Malakas in CM22,CM23; in (e),(f) in grey of cyclone NMM2,NMM3 in NM22,NM23 and the dashed black line in (c,d) marks the track of Malakas in the analysis.

kas (Figure 5.73c). Whereas this is reflected in both control simulations (blue colours in Figure 5.80a) CM23 (blue) is a little closer to the analysis (black) than CM22 (light blue) initialised one day earlier. At this time an extratropical cyclone formed in the early TC removal scenario NM22 leading to ridgebuilding (orange in Figure 5.80a) that occurs later than with Malakas, is much less amplified, and does not exhibit the wave breaking and cyclonic wrap-up. In NM23 (red) only weak ridgebuilding is evident in this region as an extratropical cyclone has not yet developed. One day later in the control simulations the ridge and first downstream trough (1) are close to the analysis (Figure 5.80b cf. Figure 5.73d).

Ahead of the downstream trough ridgebuilding begins. As the ridgebuilding downstream of Malakas and ahead of the downstream trough is much stronger in the control simulations, the eastward propagation of the upstream trough-ridge couplet and of the first downstream trough is delayed in the presence of Malakas compared to the trough-ridge-trough couplet that formed with the extratropical cyclone in NM22 (orange) and is shifted about 5° farther east. A weak ridge and broad downstream trough (1) is forming in the NM23 scenario (red) as extratropical cyclone NMM3 forms. The wave pattern has about the same phase as in the scenarios with Malakas as the ongoing (weak) ridgebuilding downstream of NMM3 is hindering an eastward propagation at this time.

The strong ridgebuilding due to the advection of low PV air by Malakas' outflow is reflected in the 340 K difference PV at 00 UTC 26 September 2010 (Figure 5.81a,b). South of the ridge in CM23 (as represented by the cyan 2 PVU contour in Figure 5.81b) a minimum of $< -7$ PVU occurs. To the North of the minimum the difference wind speed has a maximum of more than $60\,\mathrm{m\,s^{-1}}$. Thus, as in to the previous cases Jangmi, Hanna, Choi-wan, and Lupit, a strong acceleration of the jet core wind speed in the ridge downstream of the ET system occurs. Although the cyclonic wrap-up of low PV air occurred prior to the time shown, it is still reflected in low PV air east of 150°E. Along with the ridgebuilding the positive anomaly of $> 7$ PVU at 160°W indicates the amplification of the first downstream trough (1) when Malakas is present. The differences are less pronounced between simulations CM22 and NM22 initialised one day

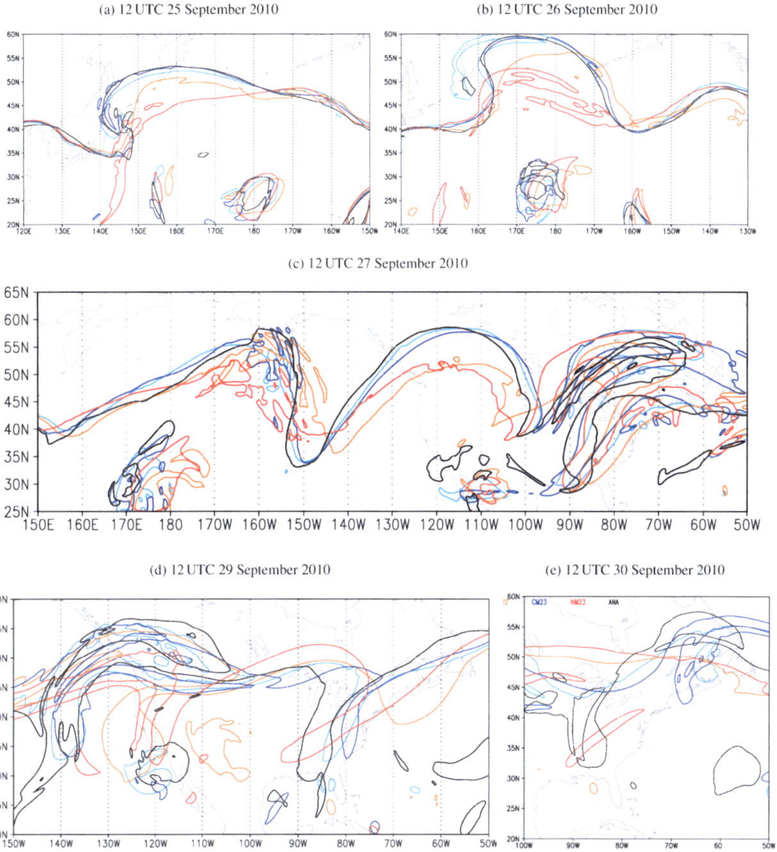


Figure 5.80: 340 K isentropic PV (2 PVU contour in (a-d), 3 PVU contour in (e)) in the different scenarios and ECMWF analysis (black). The scenarios are CM23 (blue) and NM23 (red) initialised at 00 UTC 24 September 2010, and NM22 (light blue) and NM22 (orange) initialised at 00 UTC 23 September 2010. Time is indicated above each panel.

earlier (Figure 5.81a) as the extratropical cyclone NMM2 lead to a ridgebuilding closer to Malakas than in NM23.

At 12 UTC 27 September 2010 (Figure 5.80c), the time when the heat wave struck the southwestern USA and heavy precipitation occurred at the East Coast, the signal of the second Rossby wave train triggered during Malakas' ET propagated downstream to eastern North America (80°W, Figure 5.79(a,b) and (e,f)). Consequently, the control and no-TC scenarios exhibit differences from the central Pacific to North America. Starting at the western end of the wave train the upstream trough and first downstream ridge east of 160°W are similar amongst the control simulations and the analysis. However, the first downstream ridge (axis at around 160°W) in the simulations without Malakas (orange, NM22; red NM23) is less amplified in NM22 and much less amplified in NM23 when smoothing the contour visually. Also the eastern part of the ridge and the first downstream trough at 145°W are much more amplified in the analysis and control simulations (Figure 5.80c cf. Figure 5.73e). The strong ridgebuilding ahead of the first downstream trough in the presence of Malakas delays the eastward propagation and consequently the first downstream trough is shifted 10° eastward in NM22 (orange) and 5° eastward and broader in NM23 (red). This has implications for the second downstream ridge over western North America, which is amplified much more strongly in the analysis and control simulations than in the simulations without Malakas. Under the upstream ridgebuilding the second downstream trough (2) narrows near 95°W and is shifted 5-10° eastward compared to the no-TC simulation. Over eastern North America the anticyclonic wave breaking results in a complicated PV structure (Figure 5.80c), that is discussed with the help of the difference PV (Figure 5.81c,d).

The PV streamer at around 85°W resulting from trough 3 is up to 5° further east in the no-TC simulations (orange, Figure 5.81c,d) compared to the control simulations (cyan). Together with the more westward trough 2 (at around 95°W) this hinders the merging of both troughs in the no Malakas simulations during the subsequent days. Thus in the control simulations as well as in the analysis trough 2 and trough 3 approached closer to each other over eastern North America than when Malakas is not present. In the CM23/NM23 experiment (Figure 5.81d) a strong dipole of negative difference PV to the west and positive to the east at 98°W reflects the farther east location of trough 2 in the control simulation. This is due to the enhanced ridgebuilding over western North America downstream of trough 1 and slightly less pronounced in the CM22/NM22 experiment (Figure 5.81c). The narrow tilted band of positive difference PV to the west and negative to the east at around 85°W (Figure 5.81d) reflects the westward location of trough 3 in the control simulation. This is more pronounced in the CM22/NM22 experiment (Figure 5.81c). With regard to the southwestern USA no strong differences are seen amongst the COSMO scenarios at the

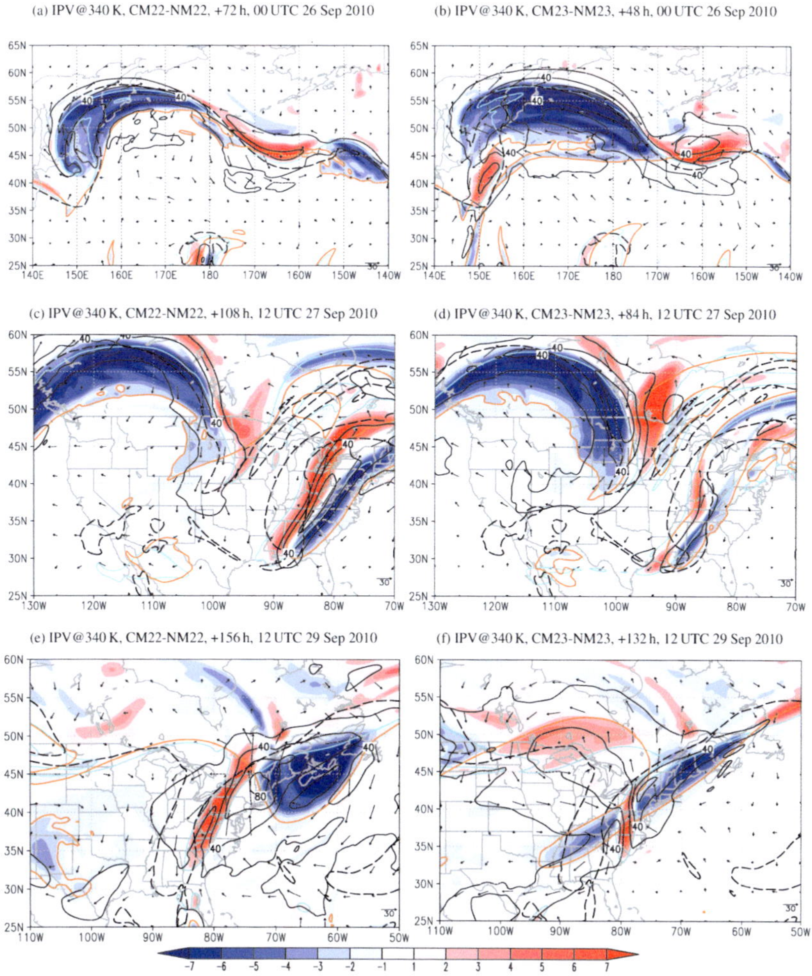


Figure 5.81: Difference control minus no-TC simulation (CM22/NM22, left; CM23/NM23, right) of PV (shaded) and wind (vectors and magnitude $>20\,\mathrm{m\,s^{-1}}$ black contours every $20\,\mathrm{m\,s^{-1}}$) and 2 PVU contours (cyan control, orange no-TC, black dashed analysis) on the 340 K isentropic level. Data shown at (a,b) 00 UTC 26 September 2010, (c,d) 12 UTC 27 September 2010, and (e,f) 12 UTC 29 September 2010.

340 K level (Figures 5.80c and 5.81c,d), however the cut-off 3C2 appears more southward in the COSMO scenarios centred at 110°W than in the analysis.

Two days later at 12 UTC 29 September 2010, the northern portion of trough 1 moved to about 100°W in the analysis and control simulations (5.80d cf. Figure 5.73f). The cut-off and PV filament of the southern portion of trough 1 occurs at about the same location at 137°W in the control simulations. The heat wave cut-off moved west to about 120°W. However, the pattern is farther east in the no-TC simulations. Over the eastern USA trough 2 and trough 3 merged. Although this occurs in the control simulations farther east (80°W) than in the analysis (85°W) the meridional orientation of the axis of the resulting PV streamer is favourable for the advection of subtropical moist air from the south. In contrast, the PV streamer in NM23 (red) appears much more strongly tilted and does not reach into the Gulf region. The northern part of the trough is farther east (70°W) than in the scenarios with Malakas. In NM22 (orange) the PV streamer from trough 3 was completely absorbed by trough 2 that has an axis at 65°W. The more meridional oriented and westward located merger from trough 2 and 3 in the control simulations is reflected in positive difference PV at around 80°W (5.81e,f). Trough 2 in NM22 appears with a strong negative difference PV centred around 65°W (5.81e) and the more tilted PV streamer in NM23 as an elongated band of negative difference PV (5.81f).

On 30 September 2010 the no-TC scenarios lack the merger of trough 1 and 2 reflected in an almost zonal orientation of the 3 PVU contour (orange, red in 5.80e). In the control simulations trough 1 is much broader and less strongly amplified than in the analysis (Figure 5.73f,g) and trough 2 is rather absorbed near 65°W by the broad trough 1 than forming a PV streamer at 85°W as in the analysis.

In summary, the COSMO sensitivity experiments revealed that Malakas' ET causes significant modifications to the Rossby wave train that develops downstream over the Pacific. The outflow of Malakas advects low PV air towards the midlatitudes, enhances the tropopause PV gradient, resulting in an accelerated jet streak. The ridgebuilding halts the eastward propagation of Malakas upstream trough. This diabatically enhanced modification of the upper-level flow, albeit present as an extratropical cyclone develops, is weaker and temporally delayed in the no-TC simulations. Therefore the amplification of the downstream Rossby wave train is weaker and the propagation faster without Malakas, resulting in an important shift of the ridge-trough patterns over the North Pacific and North America. The first downstream trough amplifies strongly in the presence of Malakas and ahead of the trough an intense extratropical cyclone develops. The ridgebuilding associated with this cyclone likewise halts the eastward propagation of the first downstream trough but amplifies and shifts eastward the second downstream trough over western North America. This yields in a narrowing of the second downstream trough over Central North America. In turn it approaches the third downstream trough and both merge into a PV streamer over the eastern USA. This chain of events is less distinct without Malakas and a marked PV streamer resulting from the merger of downstream trough 2 and 3 is missing. However this PV streamer was essential for regenerating the frontal zone at the US East Coast. In the next section the implications of this modification for the extreme weather events in the USA are explored.

An important mechanism describing the modified Rossby wave train propagation could be distilled: Initial ridgebuilding hinders an eastward propagation of the upstream trough, meanwhile amplifies the downstream trough and pushes it further east. Ridgebuilding ahead of the first downstream trough results in a repetition of this process. The eastward propagation of the first downstream trough is delayed, but the ridgebuilding amplifies and pushes eastward the second downstream trough.

### 5.5.5 Malakas' role in the North American high impact weather

Finally, the role of Malakas in the high impact weather in North America is investigated. In all sensitivity experiments the second downstream trough is accompanied with a hot air mass over the western USA reflected in the 850 hPa temperature at 00 UTC 28 September 2010 (20°C contours in Figure 5.82a,b cf. with analysis Figure 5.74). The warm air expands over a larger area in the simulations with Malakas (solid black contours), reflecting the stronger amplification of the ridge compared to the simulation without Malakas. The third downstream trough near 85°W is accompanied by 850 hPa temperature below 10°C and the eastward shift of this trough in the no-Malakas experiments is particularly distinct in NM22 (Figure 5.82a). In the simulation CM23 initialised one day after CM22 the warm air at 850 hPa expands over a larger region and exhibits stronger differences to the no-TC simulation NM23 than CM22 and NM22. This is consistent with the prior finding, that CM22/NM22 exhibits an extratropical cyclogenesis that is closer to the actual ET of Malakas, than CM23/NM23.

Albeit the control simulation CM23 exhibits a warm 2 m temperature anomaly of 0-4 K within the more pronounced ridge over western North America (shading in Figure 5.82b), the temperature differences of more than 6 K at the eastern

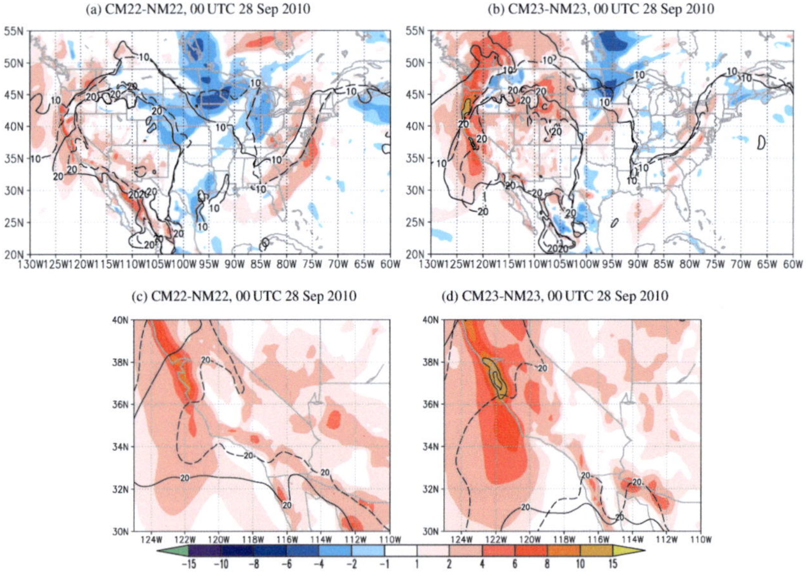


Figure 5.82: Difference of 2 m temperature between control and no-TC simulation (shaded, and contours every 2 K for $\Delta T \leq 10$ K) and 10, 20, 30°C contour of 850 hPa temperature from control (black, solid) and no-TC (black, dashed) simulation at 00 UTC 28 September 2010. In (c,d) focus is on southwestern North America. In (a,c) data from the CM22/NM22 experiments is used and in (b,d) from CM23/NM23. Cf. with analysis data in Figures 5.72 and 5.74.

and north eastern edges are striking, peaking at more than 10 K at the Coasts of northern California and Washington. A negative temperature anomaly of up to -6 K near 95°W, 50°N reflects the more eastward position of trough 2 in the CM23. Along the US East Coast a warm anomaly of up to 4 K reflects the advection of subtropical warm air ahead of trough 3 that is slightly more eastward in the control simulation.

The impact of Malakas in southwestern North America may partly explain the heat wave (Figure 5.82d). The 20°C contour of 850 hPa temperature expands further west in CM23 than NM23 (Figure 5.82b,d) which may reflect an enhanced easterly flow between the second downstream ridge and the cut-off to the south resulting in enhanced westward advection of the hot continental air mass. Over southern California the 2 m temperature differences are at about 2-4 K (up to 6 K near San Diego and in Mexico) and align along the coast. A more striking warm anomaly ($>$12 K) occurs near 122°W, 37°N south of the San Francisco Bay region. Here the 10 m wind indicates cool onshore winds ($<$15°C) in the NM23 simulation, while the continental hot air is advected eastward to the coast in CM23 (not shown). The alignment of the maximum temperature anomalies along the US West Coast indicates that the flow is more off-shore in the control simulation and may indicate that the rather small differences are not due to random numerical noise.

Regarding the Californian heat wave it may be concluded that the modification of the Rossby wave train by the ET of Malakas slightly enhanced the maximum temperatures making the heat wave a record weather event. However, the overall midlatitude flow configuration was favourable for a heat wave and the modification yields only in 2-4 K higher 2 m temperatures in southern California. The eastward shift of the hot continental air mass exhibits a larger impact at the West Coast and in the entire western half of North America (Figure 5.82d).

Qualitatively similar results are found when comparing the simulations CM22/NM22 (Figure 5.82a,c). The differences within the hot air mass associated with the anticyclone are weaker, downstream trough 2 exhibits a larger negative temperature anomaly and the more distinct shift of trough 3 results in a more pronounced positive anomaly along the US East Coast (Figure 5.82a).

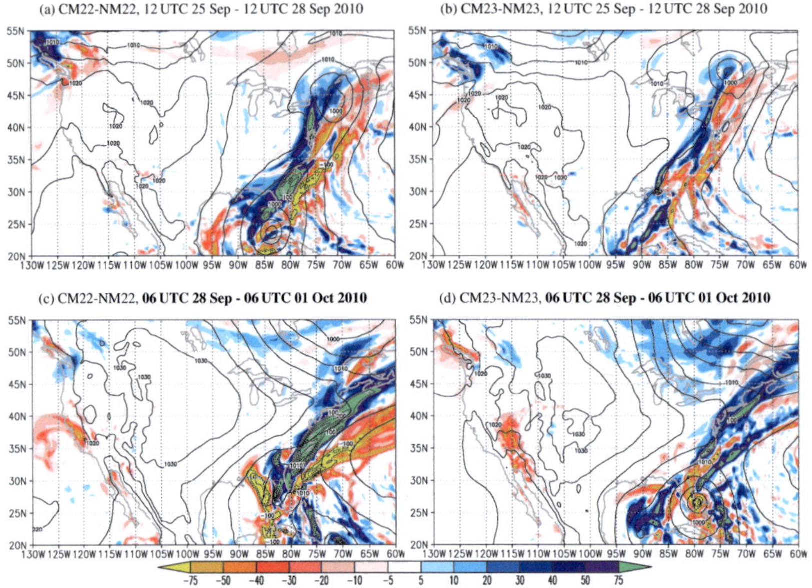


Figure 5.83: Difference in 72 hour accumulated precipitation between control and no-TC simulations (shaded, and contours every 25 $kgm^{-2}$ starting at 100 $kgm^{-2}$) along with pmsl in the control simulation. The periods are 12 UTC 25 September to 12 UTC 28 September 2010 in (a,b) and 06 UTC 28 September to 06 UTC 1 October 2010 in (c,d). Data is from the CM22/NM22 simulations (a,c) and CM23/NM23 (b,d) scenarios. The same domain as in Figure 5.71 is shown.

The two subsequent heavy precipitation periods at the US East Coast at the end of September 2010 are simulated in the COSMO scenarios with similar intensities (>350 mm in 72 h, not shown), although the most intense rainfall is slightly shifted east in the COSMO scenarios compared to the measurements (Figure 5.71a).

For the first precipitation period from 25-28 September 2010, the precipitation is shifted eastward over the Atlantic when Malakas is not present. This is reflected in the difference of precipitation between control and no-TC simulation being positive along the US East Coast and negative in a band off the coast (Figure 5.83a,b). Thereby the differences are stronger in CM22/NM22 compared to CM23/NM23 consistent with the more eastward location of trough 3 in NM22. It has to be noted that scenarios CM22/NM22 develop tropical storm Nicole in the Caribbean and therefore exhibit more precipitation over the southeastern USA.

For the second period of heavy precipitation from 28 September to 1 October 2010 the forecast skill in all COSMO simulations is reduced. Thus we can only get a qualitative picture of a possible scenario with and without the impact of Malakas. In particular an unrealistic diabatic Rossby wave occurs along the US East Coast in scenarios CM22/NM22 and scenarios CM23/NM23 maintain tropical storm Nicole during this period, while it actually had already dissolved (see contours of pmsl in Figure 5.83d). Thereby CM23 exhibits a more realistic track of Nicole to the east of Florida. Without considering the precipitation due to Nicole in the Caribbean, CM23 exhibits much more precipitation along and off the East Coast than NM23. Similarly more precipitation is evident in CM22 along the US East Coast. The higher rainfall amounts in the control simulations north of 30°N are due to the fact, that, in contrast to the simulations without Malakas, the PV streamer of the merger from trough 1 and 2 was present. We also note that in the no-TC simulations precipitation occurs in southwestern North America in the region of the more eastward upper-level cut-off low.

In summary we could link the upper-level flow modification during the ET of Typhoon Malakas in the western North Pacific to some hight impact weather events in North America. The interaction of Malakas and the midlatitude flow lead to a complicated shift of the Rossby wave pattern in the North Pacific North American region. As a result a pronounced hot continental air mass formed over western North America. This air mass was about 2-4 K warmer in the presence of Malakas enhancing the heat wave in the southeastern USA. Further downstream a stationary PV streamer developed from the successive mergers of the third and second, and thirdsecond and first downstream troughs. The merging was more distinct and occurred further west over land when the Rossby wave train was modified by the ET of Malakas. This yield in an aggravation of the heavy precipitation period at the US East Coast.

# 6 Implications for the downstream impact of ET

The detailed discussion of the individual ET cases revealed common features and general physical processes that determine the midlatitude impact of ET. In this chapter we highlight the outflow-jet interaction directly downstream of an ET event (Section 6.1), the downstream impact beyond the first downstream ridge (Section 6.2), and the general implications of ET for the reduction in predictability (Section 6.3).

## 6.1 Outflow-jet interaction

The initial interaction of the TC outflow with the upper-level midlatitude wave guide is a common feature of all ET cases presented in this study. The TC out-

Table 6.1: Characteristics of the outflow-jet interaction in the different case studies as depicted from data shown in Figures 6.1, 6.2, and 6.3. Case name, forecast hour, maximum magnitude of horizontal difference wind in m s$^{-1}$ ($|\vec{v}_h^c - \vec{v}_h^n|_{max}$), maximum and minimum PV difference control-no-TC ($\Delta PV_{max,min} = (PV^c - PV^n)_{max,min}$) in PVU, maximum of wind speed difference in m s$^{-1}$ between the control and no-TC simulation ($(|\vec{v}_h^c| - |\vec{v}_h^n|)_{max}$), sum of maximum and minimum difference in PV gradient ($\nabla PV^c - \nabla PV^n$) close to the jet streak ($\Delta\nabla PV = (\nabla PV^c - \nabla PV^n)_{max} + (\nabla PV^c - \nabla PV^n)_{min} \approx \nabla PV^c_{max} - \nabla PV^n_{max}$) in PVU(100 km)$^{-1}$ (represents the maximum difference in $\nabla PV$ between control and no-TC). The following variables are not directly seen in Figures 6.1, 6.2, 6.3: maximum PV gradient in control run ($\nabla PV^c$) in PVU(100 km)$^{-1}$, maximum wind speed of the first downstream jet streak in control and in no-TC run ($|\vec{v}_h^c|_{max}, |\vec{v}_h^n|_{max}$) in m s$^{-1}$. The reference level for all cases and variables is **340 K**.

| **case** | $t$ | $\|\vec{v}_h^c - \vec{v}_h^n\|_{max}$ | $(\Delta PV)_{max}$ | $(\Delta PV)_{min}$ | $(\|\vec{v}_h^c\| - \|\vec{v}_h^n\|)_{max}$ | $\Delta\nabla PV$ | $(\nabla PV^c)_{max}$ | $\|\vec{v}_h^c\|_{max}$ | $\|\vec{v}_h^n\|_{max}$ |
|---|---|---|---|---|---|---|---|---|---|
| Jangmi | 66 | 45.0 | 6.0 | -5.4 | 39.9 | 1.4 | 5.5 | 81.1 | 59.9 |
| Hanna | 54 | 55.7 | 7.1 | -10.6 | 52.0 | 2.1 | 7.7 | 75.4 | 65.9 |
| Choi-Wan | 60 | 54.9 | 6.8 | -8.3 | 33.9 | 4.6 | 9.4 | 70.3 | 85.3 |
| Choi-Wan | 102 | 104.3 | 8.5 | -8.6 | 58.8 | 1.2 | 9.3 | 79.9 | 76.3 |
| Lupit | 66 | 42.4 | 2.3 | -7.2 | 42.3 | 3.8 | 8.5 | 81.2 | 53.6 |
| Malakas | 45 | 74.5 | 10.1 | -9.8 | 73.7 | 3.9 | 10.1 | 92.2 | 75.5 |
| Malakas | 66 | 75.9 | 7.4 | -10.1 | 66.9 | -0.2 | 9.0 | 81.6 | 67.8 |

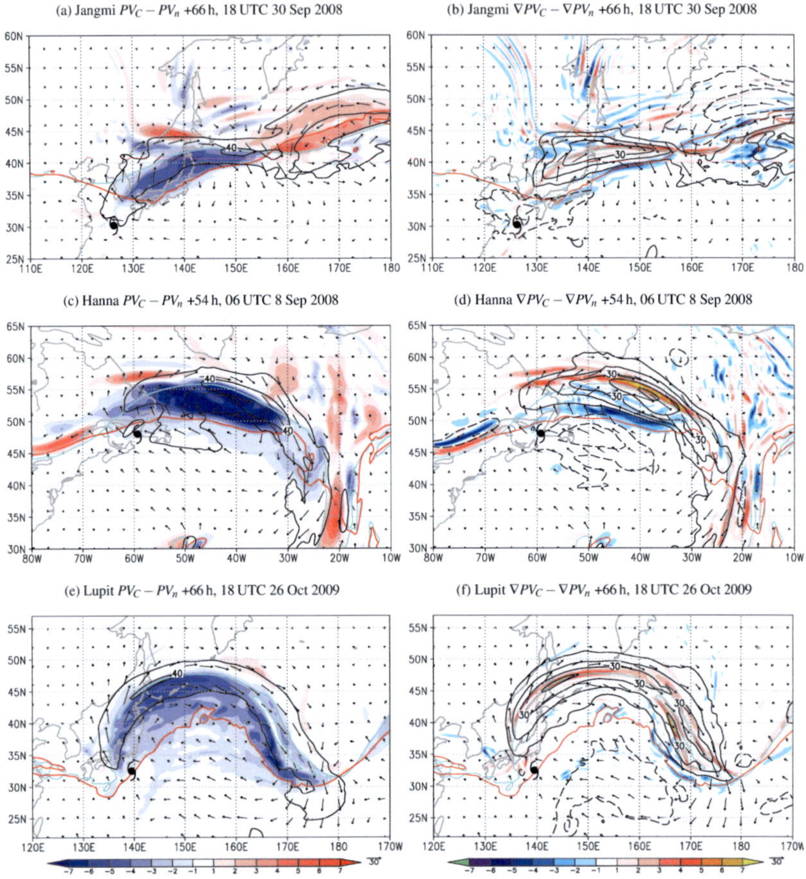


Figure 6.1: (a,c,e) Difference PV (shaded) and difference wind vectors (vectors), along with magnitude of difference wind speed $|\vec{v}_h^c - v_h^n| \geq 20\,\mathrm{m\,s^{-1}}$ (black contours every $20\,\mathrm{m\,s^{-1}}$) on the 340 K isentropic level. (b,d,f) Difference of the horizontal gradient of 340 K isentropic PV ($\nabla PV^c - \nabla PV^n$, shaded every 1 PVU $(100\,\mathrm{km})^{-1}$), absolute wind speed difference $|\vec{v}_h^c| - |v_h^n|$ on 340 K (black contours every $10\,\mathrm{m\,s^{-1}}$), and difference wind vectors on 340 K (as in (a,c,e)). The difference is taken between the control and no-TC simulations. These are in (a,b) Jangmi S20-N20, Table 5.1; in (c,d) Hanna SH20-NH30, and in (e,f) Lupit IL25-NL20, Table 5.3. The time shown exhibits the minimum 340 K difference PV and is indicated above each panel. Additionally, the 3 PVU contour for the control (cyan) and no-TC (red) simulation is shown in each plot. The black TC symbol marks the position of the TC in the control simulation. The extent of the domain is the same for each case in Figures 6.1, 6.2, and 6.3.

flow advects low PV air at upper levels towards the midlatitude wave guide. This leads to an anticyclonic deflection and an increase in the tropopause PV gradient resulting in ridgebuilding and an enhanced upper-level jet streak downstream of the ET system. As the low-level cyclonic TC circulation advects warm moist air with typically low PV values northwards, this process is enhanced in the presence of a baroclinic zone. When the air impinges on the baroclinic zone buoyant convective ascent ahead of the baroclinic zone and dynamically forced slantwise ascent along the tilted (moist-) isentropes occurs. The PV of the air parcels increases below the level of maximum condensational heating above the PV reduces to the low values of the parcel's starting level or below. Thus as a net effect of diabatic PV modification low PV air is transported to jet level, while a lower and mid tropospheric PV maximum develops at the baroclinic zone. The cyclonic circulation induced by this PV maximum acts to sustain the interaction with the baroclinic zone and thus to sustain a continuous diabatically enhanced transport of low PV air to upper levels. In addition to the low PV air of the TC outflow this low PV air is advected towards the midlatitude wave guide by the TC outflow which results in an increased tropopause PV (and temperature) gradient, enhancing the jet streak. While all ET cases studied here exhibit this physical mechanism the cases differ in the downstream propagation of this upper-level flow modification.

In the following a joint quantification of the outflow-jet interaction is presented for all cases at the time when the minimum difference PV occurred downstream of the ET system on the 340 K isentropic level (Figures 6.1, 6.2, and 6.3 and Table 6.1).

In the case of Jangmi (Figure 6.1a,b) negative difference PV of below -5 PVU over Japan downstream of the decaying storm reflects weak ridgebuilding. The associated difference wind speed peaks at 45 m s$^{-1}$ near 150°E (Figure 6.1a). In the no-TC run a broad trough is evident in this region and the wave pattern is shifted eastward and much wider. A jet streak is located east of 160°E near the axis of a broad ridge. The absolute wind speed difference over Japan has a maximum of around 40 m s$^{-1}$ in the same region where the maximum in the difference of the PV gradients occurs (Figure 6.1b, Table 6.1). The absolute wind speed maximum of the newly developed jet streak over Japan in the control simulation is $|\vec{v}_h^c|_{max} = 81.1$ m s$^{-1}$ (Table 6.1). This is about 20 m s$^{-1}$ more than the wind speed at the same longitude but slightly further south in the tail of the jet streak in the no TC simulation ($|\vec{v}_h^n|_{max} = 59.9\,\mathrm{m\,s^{-1}}$). The PV gradient is about 1.4 PVU (100 km$^{-1}$) higher in the presence of Jangmi, reflecting the warm frontogenesis at the baroclinic zone.

During the ET of Hurricane Hanna a strong amplification of the downstream ridge occurred, reflected in negative difference PV extending over a larger region than in Jangmi and reaching a much lower minimum of -10.6 PVU (Figure 6.1c, Table 6.1). The magnitude of the difference wind is higher also. The lowest difference PV is in the region where warm frontogenesis occurs. However, the dipoles of difference PV gradient (Figure 6.1d) and wind speed difference reveal a northern band of positive and a southern band of negative values, reflecting a northward shift of the jet stream in the control run which would be also present without Hanna. Although the PV gradient increases by 2.1 PVU (100 km$^{-1}$), the jet core wind speed is only about 10 m s$^{-1}$ higher.

In the case of Lupit, the widespread amplification of the first downstream ridge is reflected in the negative difference PV (Figure 6.1e). A strong PV gradient of up to 8.5 PVU (100 km$^{-1}$) is evident along the midlatitude wave guide in the control simulation which is about 3.8 PVU (100 km$^{-1}$) higher than without Lupit. This results in a jet streak of 81.2 m s$^{-1}$ in the simulation with Lupit which is almost 30 m s$^{-1}$ faster than in the simulation without the storm (Table 6.1). The increase of the jet-level wind speed expands over a much larger area than in the cases of Jangmi and Hanna. Despite the strongly modified jet the impact further downstream was relatively weak.

The outflow-jet interaction during the ET of Choi-wan and of Malakas is twofold (Figures 6.2 and 6.3). A first minimum in difference PV occurs, as in the previous cases, when the storms approach the midlatitudes. However, a second, stronger minimum follows when the storms become a mature extratropical cyclone.

As Typhoon Choi-wan approaches the midlatitudes the initial outflow-jet interaction leads to a north and westward extension of the downstream ridge (Figure 6.2a). East of 170°E, where a jet streak is evident with and without Choi-wan, this shift and amplification of the first downstream ridge results in positive difference PV. Consequently, the jet streak wind speed in the control simulation

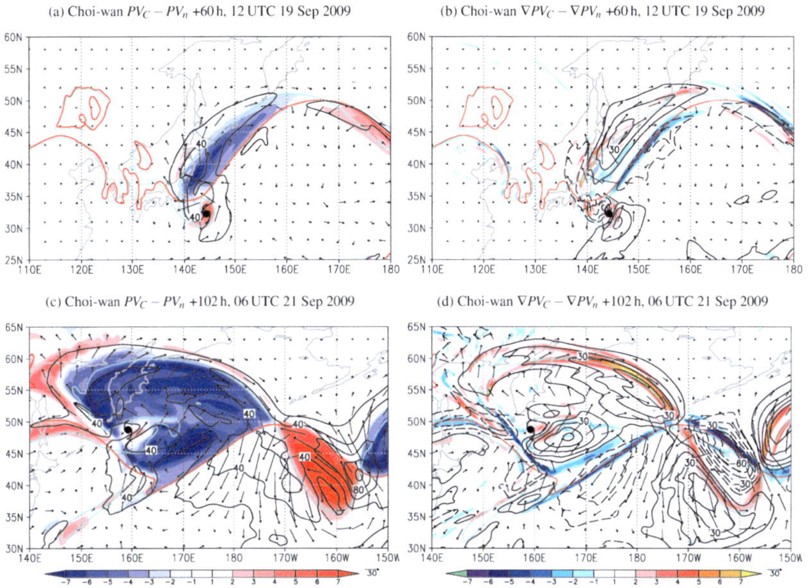


Figure 6.2: As Figure 6.1 but for Typhoon Choi-wan CNTRL-NOTC at 12 UTC 19 September 2009 (a,b) and 06 UTC 21 September 2009 (c,d).

is 15 m s$^{-1}$ less than in the no-TC simulation (Table 6.1). However, the high difference in the PV gradient and the positive wind speed difference north of Choi-wan (Figure 6.2b) indicate the extension of the jet streak in the control run into the region of enhanced ridgebuilding and warm frontogenesis. Two days later very low difference PV of -8.6 PVU northwest and south of Choi-wan reflects the cyclonic wrap-up of low PV air in CNTRL and the 20° westward shift of the first downstream ridge in CNTRL compared to NOTC (Figure 6.2c). The very high values of difference PV gradient and of absolute wind speed difference downstream of the reintensified storm (Figure 6.2d) reflect the enhancement of the jet streak and the strong differences in position and amplification of the first downstream trough. Thus Choi-wan significantly hindered an eastward propagation of the Rossby wave packet resulting in an westward shift of the first downstream ridge by 20° and of the first downstream trough by 10°. The jet core wind speed in the control run increased to around 80 m s$^{-1}$ which is slightly higher than in the no-TC simulation (Table 6.1).

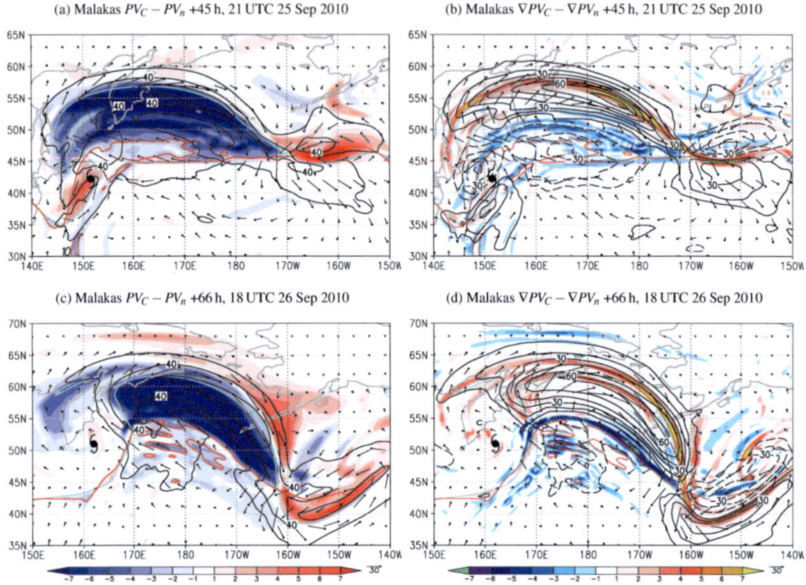


Figure 6.3: As Figure 6.1 but for Typhoon Malakas CM23-NM23 at 21 UTC 25 September 2012 (a,b) and 06 UTC 27 September 2010 (c,d).

In the case of Malakas, strong initial downstream ridgebuilding is reflected in a large area of negative difference PV (Figure 6.3a). The ridgebuilding also triggers a downstream trough reflected in a positive difference PV at around 165°W. The very high values of difference wind speed (Figure 6.3a) and of absolute wind speed difference (Figure 6.3b) reflect the significantly northward shifted jet streak in the simulation with Malakas. The PV gradient in the control simulation has a maximum of 10.1 PVU ($100\,km^{-1}$) which is 3.9 PVU ($100\,km^{-1}$) higher than in the no-TC simulation. Consequently, the jet core wind speed in the presence of Malakas peaks at $92\,m\,s^{-1}$ which is almost $17\,m\,s^{-1}$ faster than in the no-TC simulation. One day later the region of negative difference PV has propagated downstream (Figure 6.3c). In addition, negative difference PV to the northwest of Malakas indicates the cyclonic wrap-up of low PV air around the centre of the maturing extratropical cyclone. Downstream of Malakas, at around 155°W positive difference PV indicates the shift and amplification of the first downstream trough when Malakas is present. Also, the maxima in difference

PV gradient and absolute wind speed difference have propagated downstream (Figure 6.3d). Despite the lower jet core wind speed of 82 m $s^{-1}$ compared to the previous time shown it is still about 14 m $s^{-1}$ faster with the storm than without Malakas (Table 6.1).

In all cases, the minimum in difference PV occurs two to three days after the initialisation of the sensitivity experiments, which was mostly during peak intensity of the tropical stage at a time before recurvature and before interaction with the midlatitudes was seen subjectively in the ECMWF analysis. The lowest values of 340 K difference PV ($< -8.6$ PVU, Table 6.1) occur in the cases with a significant modification of the flow further east of the first downstream ridge (Hanna, Choi-wan, Malakas). In these cases also the 340 K difference wind exceeds a magnitude of 50 m $s^{-1}$. It is also striking that in these cases the spatial extent of difference PV $|\Delta PV| > 1$PVU is much larger than in the case of Jangmi and Lupit, in which no significant downstream impact was observed despite the modification of the first downstream ridge and jet streak. The cases of Choi-wan and Malakas exhibit a modification of the second and third downstream ridge-trough couplets. In both cases the highest maximum PV gradients in the control simulation and high differences between the maximum PV gradients in the control and no-TC simulations occur. However, Lupit has similar values for these variables, although no further downstream flow modification occurred.

In summary, the outflow-jet interaction is manifested in all cases by strong diabatically enhanced negative 340 K difference PV directly downstream of the ET system and south of the midlatitude wave guide, an associated strong anticyclonic difference wind circulation enhancing the jet core wind speed, and an enhanced PV gradient at the midlatitude wave guide. In the cases in which the midlatitude flow was characterised by a preexistent Rossby wave train or a pronounced upstream trough a westward shift of the downstream ridge occurs. The magnitude of these anomalies is mostly larger if a downstream propagation of this modification occurs.

## 6.2 Modification of the midlatitude wave guide

In some of the cases, ET had a downstream impact beyond the ridgebuilding directly downstream. This is summarised for all cases with the temporal evolution of the 340 K meridional difference wind component between the control and no-TC scenarios ($v^c - v^n$) meridionally averaged around the midlatitude wave guide represented by the 340 K 2 PVU contour (Figure 6.4). In all cases the downstream ridgebuilding is reflected by a southerly difference wind of $v^c - v^n > 10\,\mathrm{m\,s^{-1}}$ during the first 96 h of simulation. This southerly wind is mostly more pronounced to the east of the TC track. An exception is Choi-wan (Figure 6.4c) where the northerly flow in the eastern part of the ridge is slightly more pronounced than the southerly flow in the western part.

In the case of Jangmi (Figure 6.4a, see also Section 4.4.3) a modification of the upper-level midlatitude wave guide is manifested in a weak ($10\,\mathrm{m\,s^{-1}} < |v^c - v^n| < 20\,\mathrm{m\,s^{-1}}$) first downstream trough (axis at 170°E, 12 UTC 30 September 2008) and second downstream ridge (axis at 150°W, 00 UTC 01 October 2008). However, a dramatic downstream impact and triggering of a Rossby wave train occurs, when Jangmi is relocated more than 2.5° to the East or North of its initial position (Figure 6.4b, cf. Section 5.1.3, Figure 5.11, page 159 and Figure 5.12, page 160).

During the ET of Hanna (Figure 6.4c) the most striking feature is the downstream ridgebuilding that lasts for almost 6 days. No Rossby wave train occurs in the difference wind. This is reflected in a dipole of southerly and northerly $v^c - v^n$ difference wind. The maximum of $|v^c - v^n| > 30\mathrm{ms}^{-1}$ on 9 September 2008 reflects the far southward extension of the eastern part of the ridge and of the first downstream trough (see also Section 5.2.4). After 00 UTC 10 September 2008 with the beginning wrap-up of the downstream ridge, the dipole decays and the southerly wind component becomes located to the west of Ex-Hanna. This is accompanied by a northerly difference wind to the west after 00 UTC 11 September 2008 both reflecting the transformation of Ex-Hanna's trough into a PV streamer.

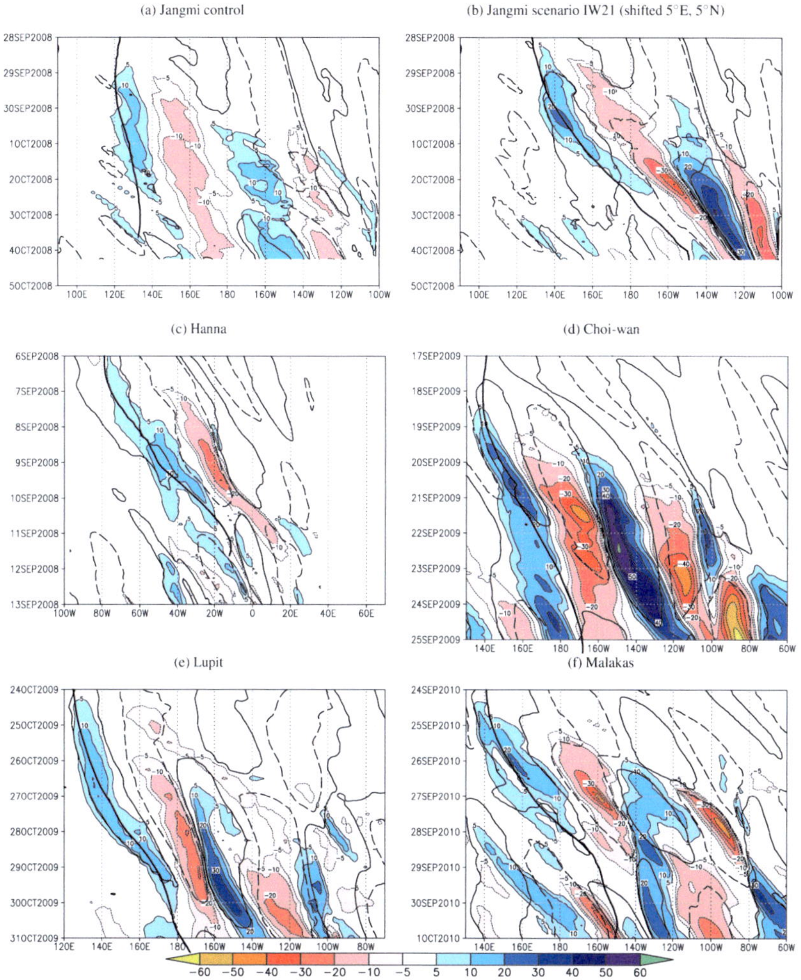


Figure 6.4: Hovmöller diagrams of meridional difference wind component ($v^c - v^n$) at the 340 K isentropic level (shaded every $10\,\mathrm{m\,s^{-1}}$). (a) Jangmi control simulation S20 minus N22 averaged from 30-50°N, (b) Jangmi scenario with a 5°E and 5°N shifted initial storm position IW21 minus N22 averaged from 35-55°N, (c) Hanna SH20 minus NH30 averaged from 35-55°N, (d) Choi-wan CNTRL minus NOTC averaged from 35-55°N, (e) Lupit IL25 minus NL20 averaged from 30-50°N, (f) Malakas CM23 minus NM23 averaged from 35-55°N. Track of the TCs in the control simulation/scenario in (b) is given by the black line. The abscissa and ordinate of all plots span the same longitudinal and time ranges (except for (d) showing 1 day more). Three-hourly (six-hourly in (e)) data is used. Additionally the -10 and $10\,\mathrm{m\,s^{-1}}$ contours of meridional wind speed in the control simulation is shown.

Downstream of Typhoon Choi-wan a very pronounced Rossby wave train emerges with a strong first and second downstream ridge-trough couplet (see also Section 5.3.3). This Rossby wave train exhibits a phase shift of around 10° to the west in the presence of Choi-wan. This is reflected in the dipoles of northerly and southerly difference wind (Figure 6.4d) which are shifted westward compared to the dipoles in the CNTRL run (black contours). Also the downstream ridge-trough couplets are more amplified in the presence of Choi-wan, which is reflected in the increase of absolute difference wind with time.

During the ET of Lupit the downstream impact was limited to the modification of the first downstream ridge-trough couplet (see also Section 5.4.4). The long-lasting amplification of the downstream ridge is reflected in the southerly difference wind component east and along the track of Lupit (Figure 6.4e). Starting at 180°E, 00 UTC 26 October 2009, the first downstream trough amplifies and subsequently transforms into a PV streamer. This is less pronounced when Lupit is not present, and thus reflected in $v^c - v^n$. As the forecast skill for the COSMO sensitivity experiments were poor after 00 UTC 29 October 2009 we omit a further discussion of (Figure 6.4e).

Typhoon Malakas triggered a Rossby wave train in a midlatitude environment that was already characterised by another Rossby wave train (Section 5.5, Section 5.5.4). This lead to a complicated interaction of both Rossby wave trains and a significant modification of the first, second and third downstream ridge-trough couplets. The triggering of the Rossby wave train downstream of Malakas is clearly reflected in the dipoles of southerly and northerly meridional difference wind emerging to the east of Malakas after 25 September 2010 (Figure 6.4f). Differences in $v^c - v^n$ after 28 September 2010 and east of 160°W reflect the modification of the initial Rossby wave train by the interaction with the one triggered by Malakas.

In summary, the meridional difference wind highlights that in the strong ET scenarios (Jangmi IW21, Choi-wan, Malakas), in which the TC transitions in an intense extratropical cyclone, a significant modification of the midlatitude wave guide occurs beyond a modification of the first downstream ridge-trough couplets. One exemption is Lupit, which reintensified to an intense extratropical cyclone, but the downstream impact was mainly limited to the first ridge-trough couplet. In the weaker ET cases (Jangmi, Hanna) the downstream ridgebuilding and ridge amplification constitute the major downstream impact.

## 6.3 The role of ET in reducing the midlatitude forecast skill

The reduction in predictability in downstream regions is the biggest challenge in the operational numerical weather prediction of ET. The understanding of the physical and dynamical processes that govern the interaction of the TC and midlatitudes help to address this problem. The methods developed in this study for investigating these processes allow us for quantifying the reduction of predictability associated with ET.

### 6.3.1 Measures for quantifying the forecast skill modification

As detailed in Section 3.3.3 the forecast skill is quantified in terms of the anomaly correlation coefficient $ACC_f^a$ of geopotential (where $f$ stands for the considered simulation and $a$ for the reference analysis). As climatology ($c$ in Equation (3.18)) we use the daily mean values of ERA-INTERIM interpolated on a 0.25° x 0.25° horizontal grid. For the analysis $a$ we follow two approaches:

1. Using the operational ECMWF analysis ($a = ANA$ in Equation (3.18), as employed for the $ACC$ in the previous sections), allows us to quantify the forecast skill with respect to the actual evolution. We name this $ACC_f^{ANA}$.
2. Using the COSMO control simulation as analysis ($a = CNTRL$ in Equation (3.18)) allows us to quantify the forecast skill reduction with respect to the PV surgery (either TC removal or TC relocation) in the COSMO model framework. We name this $ACC_f^{CNTRL}$.

The COSMO simulations use the same model framework and have the same background flow. Therefore the change in $ACC_f^{CNTRL}$ should reflect the forecast

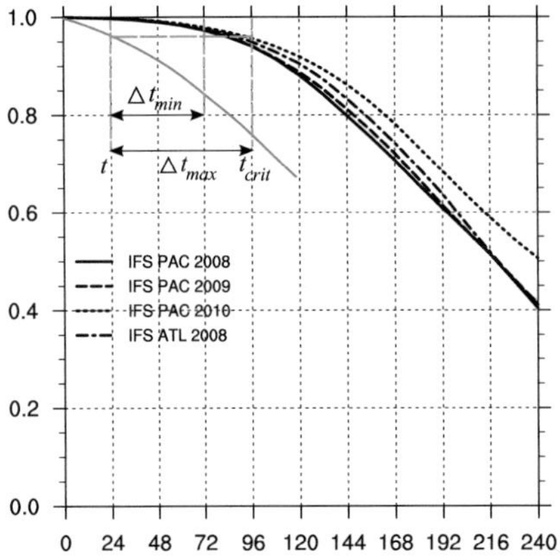


Figure 6.5: Mean anomaly correlation coefficient $ACC^{ANA}_{IFS}$ of geopotential at 200 hPa for all deterministic ECMWF IFS forecasts during 1 August to 1 November of 2008 (solid), 2009 (dashed), 2010 (dotted) for the North Pacific (IFS PAC yyyy) and 2008 (dash-dotted) for the North Atlantic (IFS ATL 2008). The abscissa indicates the forecast hour. Additionally the identification of the time $t_{crit}$ is illustrated with randomly chosen grey lines (see text for details).

skill reduction that is mainly due to the modification of the initial conditions.

In addition to the *ACC* for the COSMO simulations, we consider the mean *ACC* over the relevant ocean basin of all deterministic ECMWF IFS forecasts in August, September, and October (with 00 UTC and 12 UTC base times). These mean *ACCs* are named $ACC^{ANA}_{IFS}$ and were calculated [1] in the North Pacific for 2008, 2009 and 2010, and in the North Atlantic for 2008. An example is shown for geopotential at 200 hPa (Figure 6.5). Interestingly, the $ACC^{ANA}_{IFS}$ increased strongly in 2010 improving the predictability by one day. When comparing the $ACC^{ANA}_{IFS}$ against the *ACC*s for the COSMO simulations one has to consider that the COSMO simulations are actually hindcasts driven by the analysis at the lateral boundaries. Therefore - as seen before (Figures 5.13, 5.33, 5.43,, 5.64, 5.75) - the control runs mostly exhibit an *ACC* similar to $ACC^{ANA}_{IFS}$ in the beginning and larger than $ACC^{ANA}_{IFS}$ at later simulation times when the skill of the freely running IFS model decreases.

The foregoing *ACC* definitions are used to define a relative measure for the forecast skill modification during ET.

$$R_{CNTRL} = \frac{ACC^{ANA}_{NOTC} - ACC^{ANA}_{CNTRL}}{ACC^{ANA}_{CNTRL}}, \tag{6.1}$$

[1] kindly provided by Simon Lang (KIT)

where $CNTRL$ denotes the COSMO control simulation, and $NOTC$ denotes the COSMO simulation in which PV surgery has been applied to the initial conditions (either removing or relocating the storm). $R_{CNTRL}$ is a measure for the relative forecast skill modification in the COSMO model framework, i.e. it quantifies the forecast skill modification due to the PV surgery with respect to the COSMO control simulation.

A measure for the reduction in predictability due to ET is the difference $\Delta t$ between the time $t_{crit}$ and time $t < t_{crit}$ at which $ACC_{NOTC}^{CNTRL}(t)$ is larger than $ACC_{IFS}^{ANA}(t_{crit})$ for the first time. This is illustrated in Figure 6.5 and becomes clear in the discussion of Figure 6.7. As we have twelve-hourly $ACC$ values $t_{crit}$ is only given in twelve-hourly time steps. E.g. $t_{crit} = 96$ h and $t = 24$ h means that the predictability is reduced by at least $\Delta t_{min} = 48$ h and the actual reduction in predictability $\Delta t$ is within the range 48 h$= \Delta t_{min} < \Delta t < \Delta t_{max} = 72$ h. In the following we discuss the reduction in predictability with respect to the lower end $\Delta t_{min}$ of this time range.

The ACC is calculated on a region centred around the midlatitude wave guide in the respective ocean basin. As the COSMO model domain in the scenarios differs slightly, the ACC is calculated for a different region for each scenario. The region covers the midlatitude wave guide, while having enough distance[2] from the boundaries of the model domain. For the western North Pacific typhoons the ACC is calculated in the region extending from 20°N to 60°N in meridional direction. The region in the zonal direction extended for Jangmi from 100°E to 115°W, for Choi-wan from 120°E to 60°W, and for Lupit and Malakas from 110°E to 50°W. The ACC for the deterministic ECMWF IFS forecasts is calculated for the typhoons in the Northern Pacific in the band 110°E-60°W, 20°N-60°N. For Hurricane Hanna in the North Atlantic the region for the ACC calculation extends from 25°N to 55°N and 80°W to 20°E. The same region is used for the COSMO scenarios and deterministic ECMWF IFS forecasts. In the fol-

[2]at least 15° from the western and at least 9° from the eastern model domain boundary. As the flow is, in general, into the model domain from the west and out of model domain in the east, less impact of the eastern boundaries were found. The ACC domain is also chosen to cover the regions shown in the detailed discussion of the dynamical evolution in the previous chapter. Therefore we try to extend the ACC calculation domain over the whole zonal range of the model domain. We found the ACC relatively insensitive to the exact distance to the lateral boundaries as long as a distance of more than 5° was used.

lowing we discuss the forecast skill for geopotential at 200 hPa representing the upper levels (jet level). Qualitatively similar results are found at 500 hPa and 1000 hPa and these plots are given without further discussion in the Appendix (Figures A.2 - A.6).

### 6.3.2 Control and no-TC experiments

In all TC removal experiments $R_{CNTRL}$ indicates a reduction of the forecast skill for geopotential at 200 hPa during the first 96 h (Figure 6.6). The average ACC reduction (dashed in Figure 6.6) relative to the control simulation reaches 15% after 72 h. This is mostly due to the lack of the ridgebuilding by the TC outflow. The average skill reduction remains at 15% until 120 h forecast. After this time the cases depict different behaviour. For Jangmi, Hanna, and Malakas CM22/NM22 $R_{CNTRL}$ increases as the $ACC_{NOTC}^{ANA}$ of the NOTC simulation is higher than $ACC_{CNTRL}^{ANA}$ at the end of the calculation period. In the case of Choi-wan and Malakas CM23/NM23 a maximum ACC reduction of up to 40% is seen. In these scenarios strong modification and a shift of the Rossby wave train downstream of ET occurred.

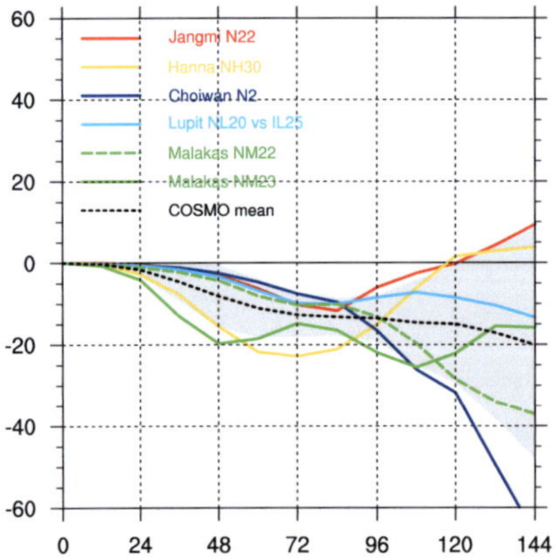


Figure 6.6: Relative forecast skill modification $R_{CNTRL}$ (in %) of geopotential at 200 hPa in the different COSMO TC removal experiments with respect to the corresponding COSMO control simulation. The abscissa indicates the forecast hour. The black short dashed line is the average, and the grey shading indicates the average ± 1 standard deviation.

To give an idea of the implications of the relative ACC reduction during ET for the predictability, we consider the $ACC_{NOTC}^{CNTRL}$ of the no-TC simulations relative to the corresponding control simulations (Figure 6.7a). For all scenarios the $ACC_{NOTC}^{CNTRL}$ decreases during the first 72 h of simulation. This reduction is larger than the reduction of the mean ACC in the IFS model. On average the $ACC_{NOTC}^{CNTRL}$

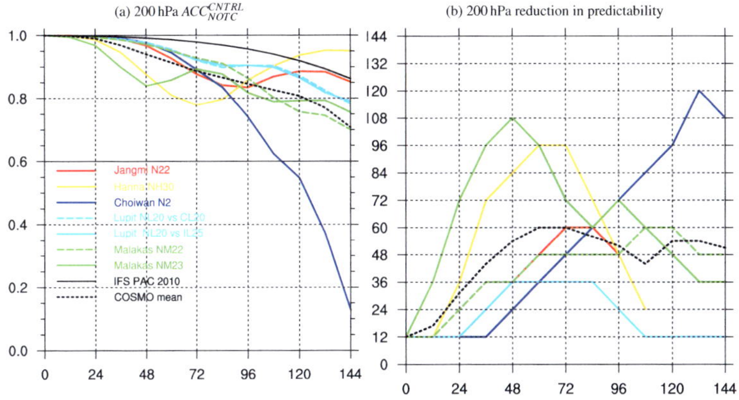


Figure 6.7: (a) Anomaly correlation coefficient $ACC_{NOTC}^{CNTRL}$ of geopotential at 200 hPa for the different COSMO sensitivity experiments. The corresponding control simulations are used as the analysis data. The mean $ACC_{IFS}^{ANA}$ for the Pacific in 2010 (black, solid, IFS PAC 2010) is given. (b) On the ordinate the time difference $\Delta t_{max} = t_{crit} - t$ (in h, but only in 12 hourly steps) between the simulation time $t$ and the time $t_{crit}$ at which $ACC_{NOTC}^{CNTRL}(t) > ACC_{IFS}^{ANA}(t_c rit)$ is valid for the first time is shown. The $ACC_{IFS}^{ANA}$ for the corresponding region and year is used to calculate $t_{crit}$. In (a) and (b) the abscissa indicates the forecast hour $t$, the black short dashed line is the average of the COSMO TC removal experiments, and the grey shading indicates the average $\pm$ 1 standard deviation.

dropped to about 0.87 after 72 h. In the IFS model the $ACC_{IFS}^{ANA}$ typically reaches this level after 5 (2008) to 6 (2010) days (cf. Figure 6.5). Thus a reduction in predictability by at least 2 days would occur in a forecast that misses the ET system completely. This measure for reduction in predictability is shown in Figure 6.7b. The reduction in predictability after 72 h forecast is on average at least 2 days and ranges from at least 1 day (Lupit) to at least 3.5 days (Hanna) within the scenarios (Figure 6.7b). After 96 h the average $ACC_{NOTC}^{CNTRL}$ reduction (Figure 6.7a) continues but the decrease is no longer as pronounced as for the $ACC_{IFS}^{ANA}$. However, this still means a reduction in predictability by on average at least 1.5 days (Figure 6.7b). The predictability reduction peaks by at least 4 days after 48 h forecast in the case of Malakas CM23/NM23 and on average by at least two days after 60 h. At later times Choi-wan is outstanding. The reduction in predictability exceeds at least 4.5 days after 132 h forecast. For both Choi-wan and Malakas a strong modification and shift of the Rossby wave train occurred.

In summary, we could show that on average the anomaly correlation coefficient at upper levels is reduced by 10-20% when the ET system is not present in the simulation (discussion of $R_{CNTRL}$, Figure 6.6). During the first 72 h forecast the $ACC_{NOTC}^{CNTRL}$ of the simulations without the TC relative to the control simulation showed a strong decrease, which was stronger than in the typical mean IFS forecasts (Figures 6.7a and 6.5). The decrease continued during the following forecast time and remained below the mean $ACC_{IFS}^{ANA}$ of the IFS model in scenarios Choi-wan and Malakas, that exhibited a strong modification of the Rossby wave train. This decrease in $ACC_{NOTC}^{CNTRL}$ is accompanied by a reduction in predictability peaking by at least 2 days during the first three days of simulation (6.7a). Afterwards it remains on average at a level of at least 1.5 days. We conclude that a first significant reduction in forecast skill is linked to the interaction of the TC outflow and the midlatitude wave guide. In the control simulations with TC this lead to downstream ridge building and an acceleration of the jet (see also Section 6.1), that is missing in the no-TC scenarios. The reduction in predictability further increased or remained at a high level in the scenarios in which the downstream Rossby wave train was significantly modified by the TC (Malakas, Choi-wan, see also Section 6.2). In the other scenarios the forecast skill reduction at later times was less significant compared to the mean reduction represented by the $ACC_{IFS}^{ANA}$. A similar behaviour of the forecast skill reduction was found at 500 hPa and 1000 hPa and the corresponding plots are given in the Appendix (Figures A.2 - A.3).

### 6.3.3 ET scenarios with relocated storms

In the Jangmi scenarios with relocated storms (Table 5.1 on page 147) the forecast skill reduction in terms of the ACC of 200 hPa geopotential is much more pronounced than in the TC removal case studies presented in the previous section. The average forecast skill reduction is about 20% after 72 h forecast ($R_{CNTRL}$, Figure 6.8). At this time a maximum reduction of 55% is reached in the early reintensifying scenarios IW21 and IW22 (5° east, 5° and 2.5° north). The scenario with Jangmi relocated east of Taiwan (IW27) and dissipating at the Chinese Coast depicts a small forecast improvement. After the initial peak in skill reduction in the other scenarios, $R_{CNTRL}$ increases but negative values indicate a forecast skill reduction until 120 h simulation. Interestingly, in the scenarios in which Jangmi reintensifies $R_{CNTRL}$ drops below -15% at some forecast time. The scenarios that produce a strong Rossby wave train at the end of the forecast period (IW22, IW23) indicate a reduction by more than 30% by the end of the calculation period.

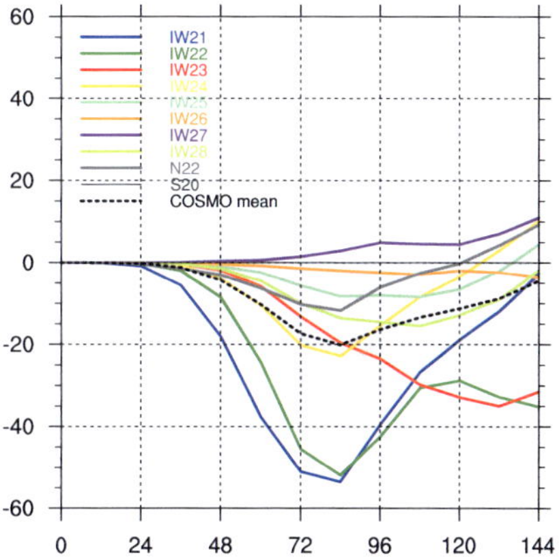


Figure 6.8: As Figure 6.6 but $R_{CNTRL}$ for the Jangmi scenarios with respect to the COSMO control simulation CNTRL=S20.

The two main scenarios - decaying Jangmi and reintensifying Jangmi - are also clearly separated in the $ACC_{NOTC}^{CNTRL}$ of the PV surgery experiments relative to the control simulations t (Figure 6.9a). The decay scenarios IW26 (1.25° east) and IW27 are very close to the control simulation. Decay scenario IW25 (2.5° east) is close to the mean $ACC_{IFS}^{ANA}$ of the IFS model. The intensification scenarios all

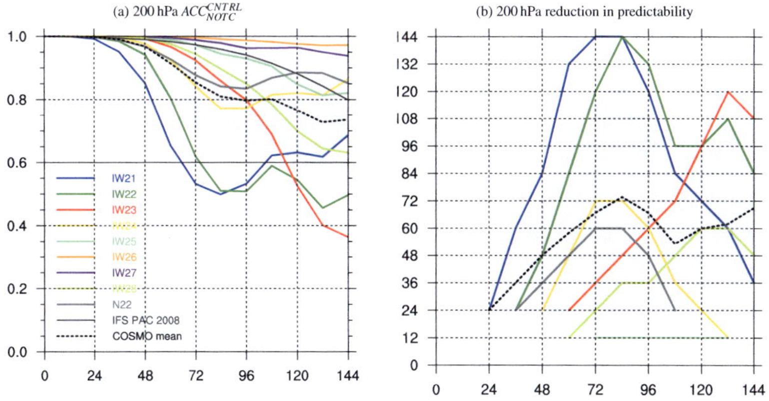


Figure 6.9: As Figure 6.7 but for the Jangmi scenarios with respect to the control simulation CNTRL=S20.

depict an $ACC_{NOTC}^{CNTRL}$ that is well below the mean $ACC_{IFS}^{ANA}$ (except at the end of the simulation for IW24 (2.5° north)). This is clearer in the predictability reduction $\Delta t$ (Figure 6.9b). The decay scenarios - if present at all - only depict a reduction in predictability by at most 12 h (IW25). If the storm is relocated by only 1.25° (128 km) further east the reduction in predictability increases to a maximum of at least 2 days (IW28). Thus the predictability in the reintensifying scenarios to the northeast of the bifurcation point is at least 1.5 days lower than for those in which Jangmi remains to the Southwest of the bifurcation point and decays (cf. Section 5.1.5). The predictability reduction in the decay scenarios is also less than the reduction in the no-TC experiment (at least 2 days after 72 h). The predictability reduction in the intensification scenarios is on average larger than 2 days after 48 h forecast. The strongly reintensifying scenarios IW21, IW22 have a peak predictability reduction of at least 5.5 days after 84 h forecast and IW23 of 4.5 days two days later, reflecting practically no forecast skill for the downstream flow, if the ET system is predicted 5° to far to the east.

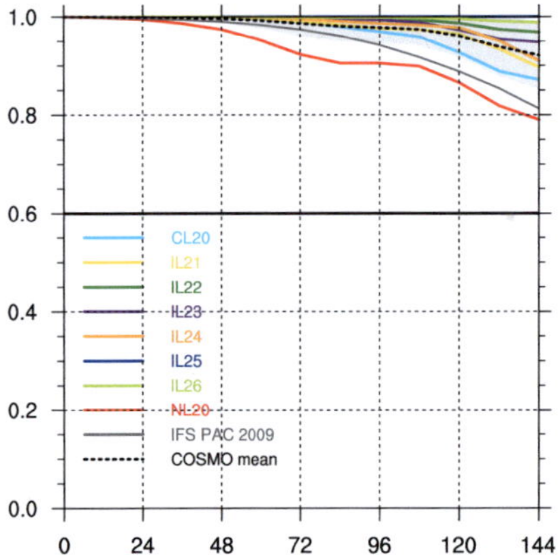


Figure 6.10: (a) asFigure 6.7a but $ACC_{NOTC}^{CNTRL}$ for the Lupit scenarios and with respect to the COSMO control simulation CNTRL=IL25.

In the case of Lupit the storms were only relocated about few grid points (Table 5.3 on page 236). The forecast skill reduction in these scenarios is very weak and strongest in the no-TC scenario (Figure 6.10). Although the $ACC_{NOTC}^{CNTRL}$ for the relocated storms decreases to the end of the calculation period it remains above the mean $ACC_{IFS}^{ANA}$ in the IFS model. Therefore the relocations exhibit no reduction in predictability ($\Delta t$=0). The scenario without Lupit has a reduction in predictability between of at least 1 day after 48 h forecast time (Figure 6.7b).

In summary, the Jangmi and Lupit scenarios confirm the qualitative picture of forecast skill reduction seen for the no-TC scenarios (Section 6.3.2). The interaction of the outflow with the midlatitudes yields a peak in forecast skill reduction during the early stages of ET. If the storms reintensify (reintensification scenarios Jangmi, Choi-wan, Malakas, Lupit) the forecast skill reduction persists or even increases. Otherwise, if the storms decays (decay scenarios of Jangmi) the forecast skill reduction weakens. A similar behaviour of the forecast skill reduction was found at 500 hPa and 1000 hPa and the corresponding plots are given in the Appendix (Figures A.4 - A.6).

# 7 Summary

This thesis presents a quantification of the midlatitude and tropical cyclone contributions to the downstream impact of extratropical transition (ET) in real case studies. Physical processes during ET are explored that are responsible for the modification of the midlatitude flow and the downstream impact in midlatitude regions remote from the ET system.

Typhoon Jangmi in September 2008 is used as an archetypal ET case for this investigation. Jangmi occurred during the THORPEX Pacific Asian Regional Campaign which provides a very good model analysis and observational data basis. Although Jangmi recurved and underwent the transformation stage of ET (Klein et al. 2002), it did not reintensify but decayed. As the midlatitude flow during recurvature was characterised by zonal flow with a primary low to the northeast, Jangmi underwent the northeast type of ET in the classification of Harr et al. (2000). However, a distinct outflow-jet interaction and an extratropical cyclogenesis directly downstream occurred. A PV surgery method and other analytical tools are developed and enable us not only to quantify Jangmi's role in the acceleration of a midlatitude jet over Japan but also to answer why Jangmi did not reintensify and what would have happened if Jangmi had phased differently relative to the midlatitude flow. Thus, despite the rather unusual actual ET, scenarios based on Jangmi allow us to conduct a more general investigation of the physical processes that govern the downstream impact of ET. The analytical tools developed for Jangmi are applied to four other real ET cases. The cases were selected with regard to a possible downstream impact of ET beyond the initial ridgebuilding in different midlatitude flow configurations. The impact of the ET system on the downstream midlatitude flow evolution is quantified for these cases. In the following, the main results are summarised, directions for future work given, and the research questions raised in the introduction are answered.

Jangmi was a very intense Typhoon in the western North Pacific in September 2008. It formed in the Philippine Sea and headed westnorthwestward whilst intensifying. Shortly after reaching peak intensity it made landfall at the northern tip of Taiwan. After landfall the symmetric TC structure collapsed, the storm started to recurve north of Taiwan and to transform into an extratropical system. During this transformation the sustained TC outflow was directed towards a jet streak over Japan. Weak ridgebuilding occurred downstream of Jangmi due to the advection of low PV air by the TC outflow. The jet became stationary over Japan and the jet core wind speed increased to more than $80\,\mathrm{m\,s^{-1}}$ at the $340\,\mathrm{K}$ isentropic level ($\approx 200\,hPa$). At the same time, a weak trough approached from the west. Subsequently Jangmi moved eastward phase-locked to the south of the trough axis (Section 4.1).

The verification of the ECWMF analysis against aircraft and satellite observations during the outflow-jet interaction confirmed a very good representation of the TC outflow, the midlatitude jet and the baroclinic zone in the model analysis (Section 4.2).

A complex interaction of the low-level TC circulation and the midlatitude baroclinic zone which is linked to the interaction of the upper-level TC outflow and the midlatitude jet was revealed by the detailed analysis of the three-dimensional PV, $\Theta_e$, and jet structure along with trajectory calculations (Section 4.3). This interaction governs the structural changes during the ET of Jangmi from the erosion of the TC inner core to an accelerated jet streak. The interaction also explains how Jangmi acts to transport tropical air to jet level and the role of the baroclinic zone thereby.

During the early stages of ET air parcels with initially low PV values from the TC environment experience convective buoyant ascent in the TC inner core. Condensational heating leads to a strong increase of the parcels' PV by diabatic PV production in the low and mid troposphere below the level of maximum heating. Above, PV steadily decreases and reaches values as in the lower troposphere or lower. Reaching the upper troposphere, the air parcels are advected by the TC outflow towards the jet streak following the (moist-)isentropes. Thus, as a net effect of diabatic heating, low PV air from the low levels is transported towards jet level. The Falcon dropsondes indicate that at its eastern side Jangmi advects tropical air poleward towards the baroclinic zone. This air mass exhibits strongly unstable stratification of $\Theta_e$ and high values of CAPE. At the baroclinic zone rather neutral stratification and no CAPE are evident. Within the unstable air mass, convective bursts occur ahead of the midlatitude baroclinic zone and lead to additional upright ascent within a new PV tower supporting the transport of low PV air to jet level. Also the low-level cyclonic flow induced by this new PV tower and the low-level cyclonic TC circulation advect tropical air poleward on their eastern side. When the air impinges on the baroclinic zone, dynamically forced slantwise ascent along the tilted (moist-)isentropes occurs and likewise transports low PV air to jet level. Similar isentropic upgliding was reported by Atallah and Bosart (2003) during the ET of Hurricane Floyd (1999).

Both the upright ascent within the TC inner core and within new convection ahead of the baroclinic zone and the slantwise ascent along the tilted (moist-) isentropes zone have a dramatic impact on the midlatitude upper-level flow. Low PV air is transported to jet level elevating the tropopause and increasing the tropopause PV gradient. As a consequence, weak ridgebuilding occurs directly downstream of Jangmi and the jet core wind speed increases. This transport of low PV air to upper levels is confirmed quantitatively by a statistical analysis of variables traced along the trajectories. Similar changes of the upper-level flow during ET were observed by Agustí-Panareda et al. (2004) without giving a detailed explanation. The results presented here add some more physical background on the dynamical processes that determine the modification of the upper-level flow during ET.

The strengthening of the jet by the transport of low PV air to upper levels, enhances the conditions favourable for extratropical cyclogenesis in the equatorward jet entrance region which has implications for the downstream flow evolution. Quasi-geostrophic diagnostics reveal that the position of Jangmi relative to the weak midlatitude trough was unfavourable for extratropical reintensification. Jangmi missed the region of upper-level forcing ahead of the midlatitude trough and in the equatorward jet entrance region. Consequently, Jangmi did not reintensify and subsequently decayed. However, slightly downstream of Jangmi a new extratropical cyclone developed in that region.

When the air parcels that ascend slantwise along the tilted (moist-)isentropes pass through a region of condensational heating, positive PV anomalies are produced in low and mid levels at the baroclinic zone. Above the region of condensational heating the air parcels ascend almost isentropically to jet level. The positive PV anomalies at the baroclinic zone likewise induce a cyclonic flow that sustains the advection of tropical air towards the baroclinic zone further downstream. This triggers a process of diabatic Rossby wave-like downstream propagation (Parker and Thorpe 1995; Böttcher and Wernli 2011) of the PV anomaly along the baroclinic zone. At later stages of ET the tropical PV towers collapsed and a complete transition from upright ascent within the TC inner core and new convection ahead of the baroclinic zone to (moist-)isentropic upgliding along the baroclinic zone occurred. The new extratropical cyclone developed from the positive PV anomaly below the equatorward jet entrance region and propagated quickly over the Pacific as a frontal wave. However, due to the slower propagation speed of the midlatitude trough, the cyclone lacked upper-level forcing from the trough and did not further intensify. It became secondary to a steering low over Alaska and dissipated at the North American coast.

A PV surgery method based on PV inversion (Section 3.2.4) is employed to remove Jangmi in the initial conditions of a COSMO simulation. Comparing this no TC run against a control simulation with unmodified initial conditions allows us to quantify the impact of Jangmi on the midlatitude jet streak and the development of the extratropical cyclone (Section 4.4). Comparing the jet-level wind in both simulations revealed that the interaction of the outflow of Jangmi and the midlatitude jet resulted in an acceleration of the jet core wind speed of about $25\,\mathrm{m\,s^{-1}}$. In conjunction with the northward shift of the jet streak over Japan an absolute wind speed difference of more than $40\,\mathrm{m\,s^{-1}}$ occurred. At the same time, a weak upper-level ridgebuilding occurred downstream of Jangmi whereas without Jangmi a broad trough developed. As a consequence, the midlatitude jet is rather zonally oriented in the presence of Jangmi while a wave pattern with a broad trough over Japan and a broad ridge to the east is present without Jangmi. Thus the complex interaction of Jangmi with the baroclinic zone and the midlatitude jet acts to zonalise the midlatitude upper-level flow. The flow of the subtropical anticyclone to the east of Jangmi was enhanced with a magnitude of up to 20 m $s^{-1}$ by the TC outflow. In the presence of Jangmi the flow between the subtropical anticyclone and a TUTT cell further east was significantly accelerated also, the TUTT cell was more pronounced and located further westward at the end of ET. This may constitute a feedback of the ET of Jangmi to the tropics.

Trajectory calculations revealed that the upright ascent in the region of Jangmi and ahead of the baroclinic zone is completely missing without Jangmi. The process of sustained advection towards the baroclinic zone and subsequent development and propagation of a diabatic PV anomaly along the baroclinic zone is absent also. Thus the no TC run lacks the sustained transport of low PV air to upper levels. In the control simulation this transport and the northward shift of the midlatitude wave guide due to the weak ridgebuilding are reflected in the 340 K difference PV which is more than 6 PVU lower south of the jet streak than in the no TC run. As a consequence, there is no enhanced equatorward jet entrance region without Jangmi. However, upper-level forcing by the broader midlatitude trough - albeit not very strong - triggers low-level cyclogenesis at the baroclinic zone which results in a frontal wave-like cyclone. As the trough exhibits a slower propagation than the jet streak in the control run, the weak extratropical cyclone has a significantly slower propagation speed and decays earlier than the DRW-like cyclone in the control run.

The importance of the relative position of a tropical cyclone and the midlatitude flow during ET is investigated in realistic scenarios (Section 5.1). Therefore Jangmi is relocated at initialisation time with the help of the PV surgery method. A first set of experiments is designed so that Jangmi becomes located ahead of the approaching midlatitude trough. In these scenarios Jangmi reintensifies and triggers a pronounced downstream Rossby wave train that propagates across the Pacific. It is remarkable that the ET of these relocated storms results in a highly amplified midlatitude wave guide, whereas during the actual ET rather zonal flow dominates. Motivated by the idealised studies of Scheck et al. (2011b) the impact of the longitudinal relocation was addressed. It turned out that similar to their findings a critical eastward relocation point exists. If Jangmi was moved east by 2.5° or less it would have decayed and the weak frontal wave-like extratropical cyclone would have developed. A shift by more than 2.5° would have lead to a reintensification of Jangmi and the triggering of a Rossby wave train. Beyond that critical eastward shift, the more Jangmi was relocated to the east the more pronounced and amplified the Rossby wave train was. The Rossby wave train triggering occurred earlier if Jangmi was additionally shifted northward. The relocation scenarios were designed so that the initial position of Jangmi varies in the order of typical track errors in current numerical weather prediction models, e.g. the ECWMF ensemble prediction system. This indicates that the position of Jangmi relative to the midlatitude trough and to the critical bifurcation point explains the uncertainty seen in the operational ensemble forecasts.

During the tropical stage until around recurvature the tropical cyclone motion was dominated by the environmental background flow, represented by the deep layer mean wind of the no TC experiment. However, we found that during the transformation stage of ET only the trough-relative non-divergent environmental background flow represents the TC motion correctly. In this frame of reference solely the reintensification scenarios recurve. It turned out that a critical bifurcation point in the trough-relative stream function exists, that determines whether the storm recurves or not. Thus, the theoretical existence of the bifurcation point shown by Scheck et al. (2011b) is proven in a realistic scenario. This bifurcation point is defined as a saddle point in the trough-relative stream function south of the axis of the approaching midlatitude trough. In theory, all storms to the northeast of this bifurcation point must recurve and reintensify whereas all storms to the southeast move back to the tropics and decay. The sum of the trough-relative wind components being zero ($u_r + v_r = 0\,\mathrm{m\,s^{-1}}$) defines a critical line that separates the region to the northeast ($u_r + v_r > 0\,\mathrm{m\,s^{-1}}$) and southwest ($u_r + v_r < 0\,\mathrm{m\,s^{-1}}$) of the bifurcation point. However in the case of Jangmi, we found the vertical average from 100 to 300 hPa of $u_r + v_r = 10\,\mathrm{m\,s^{-1}}$ as a criterion to discriminate decay from reintensification scenarios, which is simpler to derive than the discriminating parameters defined by Ritchie and Elsberry (2007). For all reintensification scenarios $u_r + v_r$ was larger than $10\,\mathrm{m\,s^{-1}}$, for all decay scenarios $u_r + v_r$ was smaller than $10\,\mathrm{m\,s^{-1}}$. Provided that the environmental background flow is known this allows for determining regions favourable for extratropical reintensification and may give a better forecast guidance in operational prediction of the downstream impact of ET. Forecasts in which a TC recurves close to a bifurcation point should be considered to exhibit more forecast uncertainty than forecasts in which the TC moves well away from a bifurcation point.

In the remainder of this thesis the downstream impact of ET in the midlatitudes has been quantified in four other ET case studies selected to cover a variety of different midlatitude flow configurations (Sections 5.2-5.5).

In the case of Hurricane Hanna over the Atlantic in September 2008 repeated diabatic processes associated with the ET system and an extratropical cyclone upstream had a crucial impact in modifying the upper-level Rossby wave guide (Grams et al. 2011, Section 5.2). In addition to the findings of Grams et al. (2011), the focus here lays on the downstream impact of Hanna. During ET, Hanna moved into a preexistent ridge over the western North Atlantic. The PV surgery experiment revealed that downstream ridgebuilding lead to a stronger amplification of this ridge and an amplification of the first downstream trough. A downstream cyclone associated with this trough moved closer to the British Isles and up to 50 mm more precipitation occurred in the UK. The southern part of the amplified downstream trough cut off off the Moroccan coast in the presence of Hanna and triggered precipitation in the Madeira region. The cyclonic wrap-up of Ex-Hanna's downstream ridge and ridgebuilding by an extratropical cyclone upstream lead to an elongation of Ex-Hanna's trough. This trough formed a PV streamer and cut off over the western Mediterranean region. The ridgebuilding due to Hanna's outflow lead to a more eastward position of this cut-off. This resulted in an important shift of heavy precipitation patterns associated with a cold frontal structure ahead of Ex-Hanna's PV streamer and a Mediterranean cyclone that developed along with the cut-off.

The investigation of the ET of Typhoon Choi-wan over the western Pacific in September 2009 was motivated by the presence of a preexistent Rossby wave train that became much more pronounced following the ET (Section 5.3). The PV surgery experiment showed, that Choi-wan lead to a significant westward shift of the Rossby wave pattern, which had crucial impact on the triggering of high impact weather in North America. A pronounced second downstream ridge developed over the eastern North Pacific and western North America and a PV streamer formed from the second downstream trough. Between the second downstream ridge and the PV streamer northerly advection of cool air over the mountainous regions of western North America lead to a drop of the maximum 2 m temperature by up to 20 K within one day. Further west the warm and dry continental air mass shifted westward in the centre of the ridge and lead to unusual warmth along the US West Coast. The PV streamer subsequently cut off over the central USA and triggered heavy precipitation there. The shift in the Rossby wave train also lead to a northward deviation of an extratropical cyclone that developed over the Pacific ahead of the first downstream trough. Without Choi-wan this cyclone would have hit Vancouver Island and would have caused heavy precipitation. Instead an early autumn heat wave occurred.

Typhoon Lupit one month later in October 2009 recurved ahead of a strongly amplified midlatitude trough and a pronounced Rossby wave train occurred downstream of the ET system (Section 5.4). This Rossby wave train lead to the formation of a PV streamer that advected tropical moist air to the US East Coast resulting in a heavy precipitation and flooding event. Although modifying the first downstream ridge and trough, Lupit had no further downstream impact and the Rossby wave train developed with and without the ET system as revealed by PV surgery. We also tested the sensitivity on different regions for the removal of moisture associated with Lupit. Further the additional or solely removal of a weak ET system that occurred close to Hawaii was tested. None of these sensitivity experiments showed an impact on the evolution of the strongly amplified PV streamer over North America. Thus in this case, the evolution of a Rossby wave train was completely forced by the midlatitude flow and the ET of Lupit did only play a role in modifying the first downstream ridge.

Lastly the ET of Typhoon Malakas in the western North Pacific in September 2010 was investigated (Section 5.5). Malakas recurved into a broad ridge over the Pacific and interacted with a narrow PV streamer. As in the previous typhoon cases a pronounced Rossby wave train occurred downstream of the ET system. Following the ET, a very unusual heat wave occurred in southern California and northwestern Mexico. The US East Coast experienced two subsequent periods of heavy precipitation with accumulated precipitation in excess of 500 mm in six days at some places. It turned out that a Rossby wave train preexisted in the midlatitude flow. However, Malakas triggered a second Rossby wave train which lead to a complicated interaction of the troughs, cut-offs and ridges of both Rossby wave trains. This interaction resulted in a small enhancement of the heat wave in southern California, which made it a record event, and a westward shift of the PV streamer in the eastern USA. As a consequence, the region of heavy precipitation was located along the US East Coast in the presence of Malakas rather than over the Atlantic when Malakas would not have been present.

Finally more general aspects of ET based on the cases studied here are elucidated (Chapter 6). The complex interaction of the TC circulation with the baroclinic zone and the midlatitude jet identified and discussed in detail for the ET of Jangmi was proven to exist during the ET of Hanna with the help of trajectory calculations. Although a Lagrangian analysis was not done for the other cases, in all cases upper-level 340 K PV was 5.4 to 10.6 PVU lower over a vast region in the downstream ridge in the presence of the ET system compared to the simulation without the ET system. This suggests that low PV air is not only advected from the outflow anomaly but also transported to upper levels by diabatic processes as shown for Jangmi and Hanna.

The transport of low PV air to upper levels results in an elevated tropopause and enhanced ridgebuilding directly downstream of the ET system. Thus an important result of this study is that the upper-level outflow-jet interaction can only be explained in conjunction with the interaction of the low-level TC circulation and the baroclinic zone. The transport of low PV air towards the upper-levels also leads to an enhancement of the tropopause PV gradient by 1.4 to 3.9 PVU$(100\text{km})^{-1}$ in all cases. Consequently, the jet core wind speed was accelerated by up to 28 m s$^{-1}$ in the presence of the ET system. The spatial extent of the upper-level ridgebuilding was much larger in the cases which exhibited a strong modification of the downstream flow (Hanna, Choi-wan, Malakas). In these cases the strong amplification of the first downstream ridge lead to an amplification of the first downstream trough. Extratropical development ahead of this trough lead to subsequent triggering of a Rossby wave train. Thus a strong downstream impact beyond the initial amplification of the first downstream ridge occurred.

In all case studies an extratropical cyclone (the new cyclone in the case of Jangmi) developed in the no TC runs at about the region of the reintensifying ET system in the control runs. We showed that despite this a significant modification of the midlatitude wave guide occurred beyond the first trough-ridge couplet in the cases that exhibit a strong downstream impact of ET. This suggests that ET alters the timing and location of extratropical development. In the cases of Choi-wan and Malakas this was expressed in a strongly amplified Rossby wave train and a more westward location of the second and third downstream ridge-trough couplet which was due to the different timing of the diabatic PV modification at upper-levels in the (re-)intensifying extratropical cyclones. Thus the initial ridge amplification retards the downstream propagation of Rossby waves, in agreement with the findings of Riemer et al. (2008). The ET of Hanna does not fit into this picture as here a more eastward location of a cut-off occurred. However, this was linked to the interaction of the ET system and the upstream trough. Interestingly, in the case of Lupit which exhibited a strong reintensification, the downstream impact was limited to the first downstream ridge. Thus the Rossby Wave train emerging downstream of the ET system was primarily midlatitude forced and may have favoured the recurvature of the TC. The Jangmi scenarios with relocated initial storm position showed that ET can trigger a pronounced Rossby wave train in an otherwise unperturbed midlatitude flow.

The PV surgery experiments enable us to elucidate the forecast skill reduction that is attributable to ET. Therefore the anomaly correlation coefficient of geopotential of the no TC simulation with respect to the control simulation has been investigated. This quantifies the forecast skill reduction in a forecast that completely misses the ET event with respect to a correct simulation of the ET. We could show that in all cases an initial reduction in forecast skill for the upper levels of on average 15% is linked to the first downstream ridgebuilding which resulted from the interaction of the TC with the baroclinic zone and the midlatitude jet. Compared to the typical anomaly correlation coefficient in the operational ECMWF IFS model this means a reduction of predictability by on average at least 2 days. The predictability further decreased dramatically in the scenarios that exhibit a strong modification of the downstream Rossby wave train. On average the reduction in predictability remained at about 1.5 days.

This study opens various directions for future research. Some recommendations on the technical side are made. The vast region affected by ET requires the model domain to cover at least an entire ocean basin. Furthermore, the impact of western North Pacific ET may propagate downstream into the Atlantic (Martius et al. 2008). Thus for future studies quantifying the downstream impact of ET in real case studies we recommend the use of a global or at least a hemispheric or channel model. Boundary conditions, if needed, have to be selected carefully to avoid having Rossby wave signals via the model boundaries in the no TC simulation.

A main result of this study is that the outflow-jet interaction and initial downstream ridgebuilding during ET can be explained only with the contribution of the interaction with the midlatitude baroclinic zone. Details of the condensational processes at the baroclinic zone should be investigated in numerical simulations that resolve convection explicitly. This gives more insight in the transition from upright ascent in the TC inner core and convective bursts ahead of the baroclinic zone to slantwise ascent along the tilted isentropes. This may also help to assess the relative contribution of upright and slantwise ascent to the transport of low PV air to jet level.

In all the cases presented here a weaker extratropical cyclone developed also if the TC was removed and weak Rossby wave activity occurred. This raises the question if an approaching midlatitude trough is necessary to attract the TC towards the midlatitudes and for ET to occur. The existence of the critical bifurcation point in the Jangmi scenarios gives some evidence for this assumption.

The work of Archambault (2011) also points in that direction. Future research should investigate if such a bifurcation point explains the reintensification in other ET cases and thus if it is a common feature of ET. However, all the cases studied here are undergoing a more or less "baroclinic mode" of reintensification ahead of a midlatitude trough. Therefore midlatitude forcing for extratropical development must be expected. Future studies should also quantify the downstream impact of ET in the rarer "tropical mode" of reintensification at the poleward jet exit region in which the ET system is supposed to play a major role (Thorncroft and Jones 2000; McTaggart-Cowan et al. 2004). In this type of ET the existence of a bifurcation point is not expected.

Archambault (2011) opens a climatological perspective on western North Pacific ET using a composite technique. We plan to quantify the downstream impact of ET in this composite using the PV surgery method. This allows for a quantification of the midlatitude and TC contribution to the downstream impact of ET in a more climatological view than from individual case studies. This also helps to assess if a midlatitude trough is necessary for ET to occur and if in that case a bifurcation point governs the reintensification. Furthermore, the quantification of the downstream impact of ET in a composite framework could be expanded to North Atlantic ET.

This thesis is finished by summarising the main results with specific answers to the research questions raised in the introduction (Chapter 1).

*Which physical processes govern the structural changes during the ET of Jangmi from the erosion of the tropical cyclone inner core to the interaction with a zonal jet stream?*

> In this study we reveal the joint interaction of both the low-level TC circulation with the midlatitude baroclinic zone *and* the upper-level TC outflow with the midlatitude jet as the crucial process governing the transformation into an extratropical cyclone and the interaction with the jet. The differential advection of drier, cooler air to the west and moister, warmer air to the east of Jangmi by the low-level

TC circulation results in the formation of a frontal system at the baroclinic zone. Moreover, buoyant upright ascent in convective bursts in the unstable stratified warm air and dynamically forced slantwise ascent along the baroclinic zone lead to a diabatically enhanced net transport of low PV air from lower levels to jet level and a lifting of the tropopause. The TC outflow advects this low PV air together with environmental low PV air poleward, enhancing the midlatitude PV gradient.

*Which contribution did Jangmi have to the acceleration of the jet streak over Japan? Which role did the interaction of Jangmi with the baroclinic zone play?*

PV surgery experiments revealed that PV is up to 6 PVU lower to the south of the jet streak and the PV gradient around 1.4 PVU $(100\,\mathrm{km})^{-1}$ higher in the presence of Jangmi. As a consequence, the jet core wind speed increased by up to $25\,\mathrm{m\,s^{-1}}$. As outlined in the answer to the previous question, the interaction with the baroclinic zone was crucial for the transport of low PV air to jet level. This physical process is completely absent when Jangmi is not present.

*What are the implications of the modified jet for the evolution of the downstream flow?*

The PV surgery experiment also showed that a weak upper-level ridgebuilding occurred downstream of Jangmi whereas a broad trough and ridge were present without Jangmi. Jangmi accelerated the flow around the subtropical anticyclone to its east and thereby indirectly influenced the track of an upper-level TUTT cell. This may constitute a feedback mechanism to the tropics.

*Why did Jangmi decay and an extratropical cyclone develop directly downstream?*

The QG-diagnostics showed that Jangmi misses the region favourable for extratropical reintensification ahead of the approaching midlatitude trough and decays. The scenarios with relocated initial position of Jangmi revealed that the ET of Jangmi crucially depends upon the phasing of Jangmi and the approaching midlatitude trough. When Jangmi becomes located ahead of the midlatitude trough it reintensifies and becomes a pronounced extratropical cyclone.

*Which role does the phasing of Jangmi and the midlatitude flow play for the uncertainty in the track forecast prior to recurvature?*

The scenarios further allowed to detect a critical bifurcation point south of the trough axis in which the track of Jangmi diverges. As Jangmi moves very close to the bifurcation point this gives an explanation for the uncertain track forecast prior to recurvature.

*Which role does the position of the tropical cyclone relative to the midlatitude flow play for the reintensification of the ET system and its downstream impact?*

In addition to the bifurcation point, the Jangmi scenarios with relocated storm position revealed a critical line defined by the sum of the trough-relative non-divergent wind components of the background flow. The region beyond this critical line is favourable for poleward motion and reintensification. If Jangmi moved in this region it would reintensify otherwise it would decay. The reintensification scenarios exhibit a strong downstream impact by triggering a Rossby wave train or a PV streamer remote from the ET system. Moreover the ET case studies for Hanna, Choi-wan, Lupit and Malakas showed that the interaction with an upper-level trough is crucial for the downstream impact of ET.

*What are the physical processes involved in the interaction with the baroclinic zone and what implications do they have for the modification of the midlatitude flow?*

The interaction of the low-level TC circulation with the midlatitude baroclinic zone associated with an upper-level interaction of the TC outflow and the midlatitude jet was detected in all case studies.

Therefore the crucial net transport of low PV air to upper levels, detailed for the case of Jangmi, is a common process in the ET cases studied here. Furthermore, the low- and mid-level diabatic PV production at the baroclinic zone can trigger a propagation along the baroclinic zone, that has characteristics of diabatic Rossby waves.

*Is an acceleration of the midlatitude jet streak a common feature of ET? What governs the outflow-jet interaction? What are the implications for the downstream flow?*

A lifting of the tropopause, ridgebuilding and an associated acceleration of the midlatitude jet was seen in all our ET case studies. The interaction of the low-level TC circulation and the baroclinic zone together with the interaction of the TC outflow and the upper-level flow govern this outflow-jet interaction. The ridgebuilding directly downstream of the ET system hinders the eastward propagation of upper-level Rossby waves. In the cases of strong ridgebuilding (Hanna, Choi-wan, Malakas) an amplification of the midlatitude wave guide beyond the first downstream ridge occurred, associated with a strong modification of the downstream flow.

*Which contribution does the tropical cyclone remnants have to the midlatitude flow evolution beyond the ridgebuilding directly downstream of the ET system?*

In the cases of Choi-wan and Malakas the interaction of the ET system with the midlatitudes results in a phase shift and an amplification of the downstream Rossby wave train. This alters the location of weather systems that leads to high impact weather in downstream regions.

*Does a Rossby wave train or uncertainty downstream of an ET system originate from ET or is it preexistent in the midlatitude flow?*

In the no TC run of all ET cases studied here an extratropical cyclone developed in the region of the removed storm and also a Rossby wave train occurred. This suggests that the midlatitude flow plays a primary role in the downstream evolution. However, the scenarios with relocated position of Jangmi indicate that the TC remnants can be crucial for the downstream flow evolution. Also in the other scenarios the TC remnants caused a significant modification of the downstream flow. Thus both the midlatitude flow and the TC remnants are important for the downstream flow evolution in our case studies. Small shifts in the relative position of the TC undergoing ET and the midlatitude flow can result in a different evolution of the downstream flow, explaining the increased forecast uncertainty during ET.

*To what extent does ET contribute to reduced predictability and which are the physical processes primarily involved?*

The TC removal experiments allowed to quantify the reduction of predictability due to ET in our case studies to on average 2 days. An initial reduction in predictability occurs due to the ridgebuilding directly downstream of the ET system. In the cases that exhibit a strong amplification of the downstream Rossby wave train (Choiwan, Malakas) this reduction persists or increases.

# Acknowledgements

Acknowledgements - often the most read part of a thesis, but also the hardest, as they are written in the last minute and I risk to forget some of the people who supported me. I will try anyway...

I am deeply grateful to my supervisor, Prof Sarah Jones, for the exciting topic, the mentoring, the introduction into the science community, enlightening and motivation, and enabling me to participate in T-PARC and the research stays in Monterey and Boulder. I also thank Sarah, her family and friends, for the occasional joyful weekend excursions.

I am very thankful to Prof Klaus Dieter Beheng for motivating conversations, advice and the kind acceptance to do the proofreading.

This work benefits from discussions with Dr Chris Davis, Prof Pat Harr, Dr Michael Riemer and Prof Heini Wernli and I am much obliged for sharing ideas and their prompt response to various research questions. I want to thank Dr Chris Davis and Dr Michael Riemer for their support with the PV inversion. I am grateful to Dr Chris Davis for enabling me to come to Boulder. The visit at NCAR essentially pushed forward my research. I thank Prof Pat Harr for enabling me to participate in T-PARC and being such a cordial host during the enriching research stays in Monterey. And I thank Prof Heini Wernli for being a scientific mentor in PANDOWAE and for support in the coordination of our joint case study.

I am happy to have met my working group scientifically and personally. Doris, Julia, Simon, Leonhard, Juliane, thank you for not only being colleagues but good friends. In particular I owe gratitude to my fellow PhD students Julia Keller and Simon Lang who shared every moment of joy and desperation and always gave me new motivation to continue.

I thank Dr Maxi Böttcher, Jana Čampa, and Prof Heini Wernli for enabling access to and giving advice for the QG-diagnostics, LAGRANTO, and other tools, and for sharing ideas and advice. Many thanks to University of Wisconsin Senior Scientist Christopher Velden and CIMSS for providing and assisting with the AMV data. I am grateful to Max Bangert, Leonhard Gantner, Vera Klüpfel, Dr Juliane Schwendike and Dr Heike Vogel for assistance with COSMO.

I thank my PANDOWAE fellow PhD students Maxi Böttcher, Jana Čampa, Ilona Glatt, Florian Harnisch, Julia Keller, Simon Lang, Andreas Schäfler, Arwen Twitchett, and Lars Wiegand for our inspiring Young Scientist Summer/Winter Schools and excellent cooperations.

I am happy to have met some international young scientists, namely Dr Heather Archambault, Dr Clark Evans, Dr Stephanie Leroux, Andrew Penny, Emily Riley, whose contact is both personal and scientific enrichment.

I would like to thank Olgica Birnmeier for always bringing sunshine to IMK. I likewise thank Friederike Schönbein and Doris Stenschke for their kind and professional assistance with various administrative tasks. Many thanks also to Gerhard Brückel for his support with IT and data storage, to Bernhard Mühr for weather discussions and being a good friend, the people of the 13th floor and to all other colleagues who supported me throughout the last years.

This work has been done within the Research Unit PANDOWAE (Forschergruppe - FOR896) sponsored by the German Research Council (Deutsche Forschungsgemeinschaft, DFG). This study benefits a lot from the inspiring discussions and collaborations within the PANDOWAE projects. Every single workshop gave me new ideas and motivation. I thank Aurelia Müller for the excellent organisation of the communication and teamwork within PANDOWAE. I am especially grateful to the PIs for their outstanding support of us PANDOWAE Young Scientists and to Sarah for being the heart of PANDOWAE.

I thank the German Weather Service (Deutscher Wetterdienst, DWD) for support with COSMO and data provision. Also I thank the Steinbuch Centre for Computing at KIT for providing super-computing facilities.

I am very grateful to the Karlsruhe House of Young Scientists at KIT for the financial support of a 3-month research stay in the USA (Auslandsaufenthalt wurde vom Karlsruhe House of Young Scientists (KHYS) gefördert). The code to derive the upper- and lower-level forced quasi-geostrophic omega fields was derived from code in ND-DIAG, a diagnostics package developed and supported by NCAS-CMS (National Centre for atmospheric science computational modelling services, UK). The DLR-Falcon data were collected as part of the THORPEX Pacific Asian Regional Campaign (T-PARC). In the United States, the primary sponsors of T-PARC are the U.S. National Science Foundation (NSF) and the Office of Naval Research. The involvement of the NSF-sponsored National Center for Atmospheric Research (NCAR) Earth Observing Laboratory (EOL) is acknowledged. The data were provided from the T-PARC Data Archive, which is maintained by NCAR's Earth Observing Laboratory (EOL). The acquisition of various T-PARC observations via the T-PARC support page of the Japan Meteorological Agency (JMA) is acknowledged. The German Aerospace Center (DLR) is acknowledged for providing Falcon data.

I thank Kiki and Mareike for their absolute support, understanding and help with the corrections. I'm grateful to Ehrfried and Petra for always listening and giving advice. Lastly I am deeply thankful to my parents, Ira and Warnfrid, my brother Anting, and Simon Z. for their support and understanding, particularly during the last weeks.

# A Appendix

## Technical details of the PV surgery

Table A.1: Details of the model domain and simulation period for all COSMO runs of **Jangmi**.

| | | |
|---|---|---|
| **model domain** | | |
| longitudinal extent | 83.0°E - 259.5°E (83.0°E - 100.5°W) | i=707 |
| latitudinal extent | 4.0°N - 63.5°N | j=239 |
| model half levels | z(1)=28377 m, z(80)=0 m, $p_0(1)$=2.7 hPa, $p_0(80)$=1000.0 hPa | n+1=80 |
| **simulation period** | | |
| dates | 00 UTC 28 September 2008 - 06 UTC 04 October 2008 | |
| output | hourly model output for forecast 0 h - 150 h | |
| **resources** | | |
| CPUs | 128 CPUs, 16 in zonal and 8 in meridional direction | |
| numerical timestep | 60s | |

Table A.2: Details of the parameters for the PV surgery of **Jangmi** performed on the ECMWF analysis at 00 UTC 28 September 2008. See also Figures in Section 3.2.4.

| | | |
|---|---|---|
| **PV inversion domain** | | |
| longitudinal extent | 82.0°E - 260.5°E (82.0°E - 99.5°W) | i=715, i/3=239 |
| latitudinal extent | 3.0°N - 64.5°N | j=247, j/3=83 |
| vertical extent | 900 hPa - 100 hPa, every 25 hPa | |
| horizontal bnd. cond. | $\Theta$ from model data | |
| **PV removal domain** | | |
| domain | 118.0°E - 127.0°E, 18.0°N - 26.0°N | |
| vertical extent | 900 hPa - 100 hPa, every 25 hPa | |
| $PV_{thres}$ | 0.3 PVU constant threshold | |
| **PBL treatment** | | |
| method | 30% linear wind speed reduction, 30° directional wind change | |
| affected levels | $k_{PBL} = 65 < k < 79 = n$ | |
| **anomaly definition $\mathbf{T^m, p*^m, u^m, v^m}$** on model full levels | | |
| modification region | 85.0°E - 145.0°E, 5.0°N - 40.0°N, k=25-79 | ($\approx$ 100-1000 hPa) |
| mask criteria | $p*_{crit} = -0.5$ hPa, $\Psi_{crit} = -2.0 \cdot 10^{-6} m^2 s^{-1}$ | |
| smoothing | 10 grid points $(T, p*)$, 15 grid points $(u, v)$ | |
| **anomaly definition $\mathbf{w^m}$** on model half levels | | |
| modification region | 118.0°E - 127.0°E, 18.0°N - 26.0°N, k=25-79 | ($\approx$ 100-1000 hPa) |
| replaced by | level average $w^m$ of total domain | |
| **anomaly definition $\mathbf{q_x^m}$** on model full levels | | |
| modification region | 118.0°E - 136.0°E, 14.0°N - 31.0°N, k=25-79 | ($\approx$ 100-1000 hPa) |
| mask criterion | $q_x^m > q^{thres} = 0.9\, \bar{q}_x^m$, $qv_{fact}$=0.9, $\bar{q}_x^m$: level average modification region | |

Table A.3: Details for the TC relocation of Typhoon **Jangmi** at 00 UTC 28 September 2008 based on the anomaly definition in Table A.2.

**relocation of anomaly in $\mathbf{T^m, p^{*m}, u^m, v^m}$**
anomalies as defined in Table A.2 added at new location
**relocation of anomaly in $\mathbf{w_x}$**
original vertical velocity in the box as defined in Table A.2 added at new location
**relocation of anomaly in $\mathbf{q_x^m}$**
original values within the $\mathbf{q_x^m}$ masks as defined in Table A.2 replace $\mathbf{q_x^m}$ at the new location.

Table A.4: Details of the model domain and simulation period for the COSMO runs of **Hanna**.

| | | |
|---|---|---|
| **model domain** | | |
| longitudinal extent | 97.0°W - 36°E | i=533 |
| latitudinal extent | 6.0°N - 67.0°N | j=245 |
| model half levels | z(1)=28377 m, z(80)=0 m, $p_0(1)$=2.7 hPa, $p_0(80)$=1000.0 hPa | n+1=80 |
| **simulation period** | | |
| dates | 00 UTC 6 September 2008 - 06 UTC 14 September 2008 | |
| output | hourly model output for forecast 0 h - 198 h | |
| **resources** | | |
| CPUs | 128 CPUs, 16 in zonal and 8 in meridional direction | |
| numerical timestep | 40s | |

Table A.5: Details of the parameters for the PV surgery of **Hanna** performed on the ECMWF analysis at 00 UTC 6 September 2008.

| | | |
|---|---|---|
| **PV inversion domain** | | |
| longitudinal extent | 98.0°W - 50.0°W | i=193, i/3=65 |
| latitudinal extent | 10.0°N - 55.0°N | j=181, j/3=61 |
| vertical extent | 900 hPa - 25 hPa, every 25 hPa | |
| horizontal bnd. cond. | $\Theta$ from model data | |
| **PV removal domain** | | |
| domain | 87.0°W - 72.0°W, 22.0°N - 38.5°N | |
| vertical extent | 900 hPa - 200 hPa, every 25 hPa | |
| $PV_{thres}$ | 0.3 PVU constant threshold | |
| **PBL treatment** | | |
| method | 30% linear wind speed reduction, 30° directional wind change | |
| affected levels | $k_{PBL} = 65 < k < 79 = n$ | |

| | | |
|---|---|---|
| **anomaly definition $\mathbf{T^m}, \mathbf{p^{*m}}, \mathbf{u^m}, \mathbf{v^m}$** on model full levels | | |
| modification region | 95.0°W - 68.0°W, 15.0°N - 45.0°N, k=38-79 | (≈ 200-1000 hPa) |
| mask criteria | $p^*_{crit} = -0.1$ hPa, $\Psi_{crit} = -2.0 \cdot 10^{-6} m^2 s^{-1}$ | |
| smoothing | 5 grid points $(T, p^*)$, 10 grid points $(u, v)$ | |

| | | |
|---|---|---|
| **anomaly definition $\mathbf{w^m}$** on model half levels | | |
| modification region | 90.0°W - 72.0°W, 22.0°N - 52.0°N, k=1-79 | (≈ 2-1000 hPa) |
| replaced by | level average $w^m$ of total domain | |

| | | |
|---|---|---|
| **anomaly definition $\mathbf{q_x^m}$** on model full levels | | |
| modification region | 90.0°W - 50.0°W, 22.0°N - 55.0°N, k=1-79 | (≈ 2-1000 hPa) |
| mask criterion | $q_x^m > q^{thres} = 0.9\ \bar{q}_x^m$, $qv_{fact}$=0.9, $\bar{q}_x^m$: level average modification region | |

Table A.6: Details of the model domain and simulation period for COSMO runs of Typhoon **Choi-wan**.

| | | |
|---|---|---|
| **model domain** | | |
| longitudinal extent | 92.0°E - 319.0°E (92.0°E - 41.0°W) | i=909 |
| latitudinal extent | 5.0°N - 78.0°N | j=293 |
| model half levels | z(1)=28377 m, z(80)=0 m, $p_0(1)$=2.7 hPa, $p_0(80)$=1000.0 hPa | n+1=80 |
| **simulation period** | | |
| dates | 00 UTC 17 September 2009 - 00 UTC 27 September 2009 | |
| output | hourly model output for forecast 0 h - 240 h | |
| **resources** | | |
| CPUs | 128 CPUs, 16 in zonal and 8 in meridional direction | |
| numerical timestep | 10s | |

Table A.7: Details of the parameters for the PV surgery of Typhoon **Choi-wan** performed on the ECMWF analysis at 00 UTC 17 September 2009.

| | | | |
|---|---|---|---|
| **PV inversion domain** | | | |
| longitudinal extent | 117.0°E - 165.0°E | | i=193, i/3=65 |
| latitudinal extent | 3.0°N - 42.0°N | | j=157, j/3=53 |
| vertical extent | 900 hPa - 100 hPa, every 25 hPa | | |
| horizontal bnd. cond. | $\Theta$ from model data | | |
| **PV removal domain** | | | |
| domain | 137.0°E - 145.0°E, 17.0°N - 24.0°N | | |
| vertical extent | 900 hPa - 100 hPa, every 25 hPa | | |
| $PV_{thres}$ | 0.3 PVU constant threshold | | |
| **PBL treatment** | | | |
| method | 30% linear wind speed reduction, 30° directional wind change | | |
| affected levels | $k_{PBL} = 65 < k < 79 = n$ | | |
| | | | |
| **anomaly definition $\mathbf{T^m}, \mathbf{p^{*m}}, \mathbf{u^m}, \mathbf{v^m}$** on model full levels | | | |
| modification region | 120.0°E - 160.0°E, 7.0°N - 34.0°N, | k=27-79 | ($\approx$ 100-1000 hPa) |
| mask criteria | $p^*_{crit} = -0.5$ hPa, $\Psi_{crit} = -2.0 \cdot 10^{-6} m^2 s^{-1}$ | | |
| smoothing | 5 grid points $(T, p^*)$, 20 grid points $(u, v)$ | | |
| | | | |
| **anomaly definition $\mathbf{w^m}$** on model half levels | | | |
| modification region | 136.0°E - 145.0°E, 16.0°N - 25.0°N, | k=1-79 | ($\approx$ 2-1000 hPa) |
| replaced by | level average $w^m$ of total domain | | |
| | | | |
| **anomaly definition $\mathbf{q_x^m}$** on model full levels | | | |
| modification region | 130.0°E - 160.0°E, 13.0°N - 40.0°N, | k=1-79 | ($\approx$ 2-1000 hPa) |
| mask criterion | $q_x^m > q^{thres} = 1.0\, \bar{q}_x^m$, $qv_{fact}$=1.0, $\bar{q}_x^m$: level average modification region | | |

Table A.8: Details of the model domain and simulation period for all COSMO runs of **Lupit**.

| | | |
|---|---|---|
| **model domain** | | |
| longitudinal extent | 92.0°E - 319.0°E (92.0°E - 41.0°W) | i=909 |
| latitudinal extent | 5.0°N - 70.0°N | j=261 |
| model half levels | z(1)=28377 m, z(80)=0 m, $p_0(1)$=2.7 hPa, $p_0(80)$=1000.0 hPa | n+1=80 |
| **simulation period** | | |
| dates | 00 UTC 24 October 2009 - 11 UTC 02 November 2009 | |
| output | hourly model output for forecast 0 h - 227 h | |
| **resources** | | |
| CPUs | 128 CPUs, 16 in zonal and 8 in meridional direction | |
| numerical timestep | 40s | |

Table A.9: Details of the parameters for the PV surgery of **Lupit** performed on the ECMWF analysis at 00 UTC 24 October 2009.

| | | |
|---|---|---|
| **PV inversion domain** | | |
| longitudinal extent | 100.0°E - 145.0°E | i=181, i/3=61 |
| latitudinal extent | 5.0°N - 50.0°N | j=181, j/3=61 |
| vertical extent | 900 hPa - 100 hPa, every 25 hPa | |
| horizontal bnd. cond. | $\Theta$ from model data | |
| **PV removal domain** | | |
| domain | 122.5°E - 126.0°E, 19.0°N - 24.0°N | |
| vertical extent | 900 hPa - 100 hPa, every 25 hPa | |
| $PV_{thres}$ | 0.1 PVU constant threshold | |
| **PBL treatment** | | |
| method | 30% linear wind speed reduction, 30° directional wind change | |
| affected levels | $k_{PBL} = 65 < k < 79 = n$ | |

| | | |
|---|---|---|
| **anomaly definition** $\mathbf{T^m}, \mathbf{p^{*m}}, \mathbf{u^m}, \mathbf{v^m}$ on model full levels | | |
| modification region | 110.0°E - 140.0°E, 12.0°N - 30.0°N, k=35-79 | (≈ 170-1000 hPa) |
| mask criteria | $p^*_{crit} = -0.1$ hPa, $\Psi_{crit} = -2.0 \cdot 10^{-6} m^2 s^{-1}$ | |
| smoothing | 5 grid points $(T, p^*)$, 15 grid points $(u, v)$ | |
| **anomaly definition** $\mathbf{w^m}$ on model half levels | | |
| modification region | 120.0°E - 128.0°E, 16.0°N - 24.0°N, k=27-79 | (≈ 100-1000 hPa) |
| replaced by | level average $w^m$ of total domain | |
| **anomaly definition** $\mathbf{q_x^m}$ on model full levels | | |
| modification region | 116.0°E - 180.0°E, 14.0°N - 37.0°N, k=27-79 | (≈ 100-1000 hPa) |
| mask criterion | $q_x^m > q^{thres} = 0.8\ \bar{q}_x^m$, $qv_{fact}$=0.8, $\bar{q}_x^m$: level average modification region | |

Table A.10: Details for the TC relocation of **Lupit** at 00 UTC 24 October 2009 based on the anomaly definition in Table A.9.

**relocation of anomaly in** $\mathbf{T^m}, \mathbf{p^{*m}}, \mathbf{u^m}, \mathbf{v^m}$
anomalies as defined in Table A.9 added at new location
**relocation of anomaly in** $\mathbf{w_x}$
original vertical velocity in the box as defined in Table A.9 added at new location
**relocation of anomaly in** $\mathbf{q_x^m}$
original values within the $\mathbf{q_x^m}$ masks as defined in Table A.9 replace $\mathbf{q_x^m}$ at the new location.

Table A.11: Details of the model domain and simulation period for COSMO runs of Typhoon **Malakas**.

| | | |
|---|---|---|
| **model domain** | | |
| longitudinal extent | 92.0°E - 319.0°E (92.0°E - 41.0°W) | i=909 |
| latitudinal extent | 5.0°N - 70.0°N | j=261 |
| model half levels | z(1)=28377 m, z(80)=0 m, $p_0$(1)=2.7 hPa, $p_0$(80)=1000.0 hPa | n+1=80 |
| **simulation period CM22 and NM22** | | |
| dates | 00 UTC 23 September 2010 - 06 UTC 01 October 2010 | |
| output | hourly model output for forecast 0 h - 198 h | |
| **simulation period CM23 and NM23** | | |
| dates | 00 UTC 24 September 2010 - 12 UTC 01 October 2010 | |
| output | hourly model output for forecast 0 h - 180 h | |
| **resources** | | |
| CPUs | 128 CPUs, 16 in zonal and 8 in meridional direction | |
| numerical timestep | 40 s | |

Table A.12: Details of the parameters for the PV surgery of Typhoon **Malakas** performed on the ECMWF analysis at 00 UTC 23 September 2010 for NM22 (middle) and at 00 UTC 24 September 2010 for NM23 (right)

| parameter | NM22, 00 UTC 23 Sep 2010 | NM23, 00 UTC 24 Sep 2010 |
|---|---|---|
| **PV inversion domain** | | |
| longitudinal extent | 119.0°E - 170.0°E; i=205, i/3=69 | 110.0°E - 170.0°E; i=241, i/3=81 |
| latitudinal extent | 5.0°N - 50.0°N; j=181, j/3=61 | 5.0°N - 56.0°N; j=205, j/3=69 |
| vertical extent | 900 hPa - 100 hPa, every 25 hPa | 900 hPa - 100 hPa, every 25 hPa |
| horizontal bnd. cond. | $\Theta$ from model data | $\Theta$ from model data |
| **PV removal domain** | | |
| domain | 138.0 - 146.0°E, 16.0 - 23.0°N | 138.0 - 145.0°E, 22.0 - 29.0°N |
| vertical extent | 900 - 100 hPa, every 25 hPa | 900 - 100 hPa, every 25 hPa |
| $PV_{thres}$ | 0.3 PVU constant threshold | 0.3 PVU constant threshold |
| **PBL treatment for NM22 and NM23** | | |
| method | 30% linear wind speed reduction, 30° directional wind change | |
| affected levels | $k_{PBL} = 65 < k < 79 = n$ | |
| **anomaly definition $\mathbf{T^m}, \mathbf{p*^m}, \mathbf{u^m}, \mathbf{v^m}$** on model full levels | | |
| modification region | 130.0 - 155.0°E, 12.0 - 30.0°N, k=37-79 ($\approx$ 200-1000 hPa) | 125.0 - 157.0°E, 12.0 - 36.0°N, k=38-79 ($\approx$ 200-1000 hPa) |
| mask criteria NM22 and NM23 | $p*_{crit} = -0.5$ hPa, $\Psi_{crit} = -2.0 \cdot 10^{-6} m^2 s^{-1}$ smoothing 5 grid points $(T, p*)$, 15 grid points $(u, v)$ | |
| **anomaly definition $\mathbf{w^m}$** on model half levels | | |
| modification region | 135.0 - 150.0°E, 14.0 - 26.0°N, k=25-79 ($\approx$ 90-1000 hPa) | 135.0 - 155.0°E, 15.0 - 33.0°N, k=1-79 ($\approx$ 2-1000 hPa) |
| | replaced by level average $w^m$ of total domain | |
| **anomaly definition $\mathbf{q_x^m}$** on model full levels | | |
| modification region | 132.0 - 170.0°E, 13.0 - 37.0°N, k=25-79 ($\approx$ 90-1000 hPa) | 125.0 - 202.0°E, 15.0 - 55.0°N, k=1-79 ($\approx$ 2-1000 hPa) |
| mask criterion | $q_x^m > q^{thres} = 0.8\, \bar{q}_x^m$, $qv_{fact}$=0.8 | $q_x^m > q^{thres} = 1.0\, \bar{q}_x^m$, $qv_{fact}$=1.0 |
| | $\bar{q}_x^m$: level average modification region | |

# Steering of Jangmi

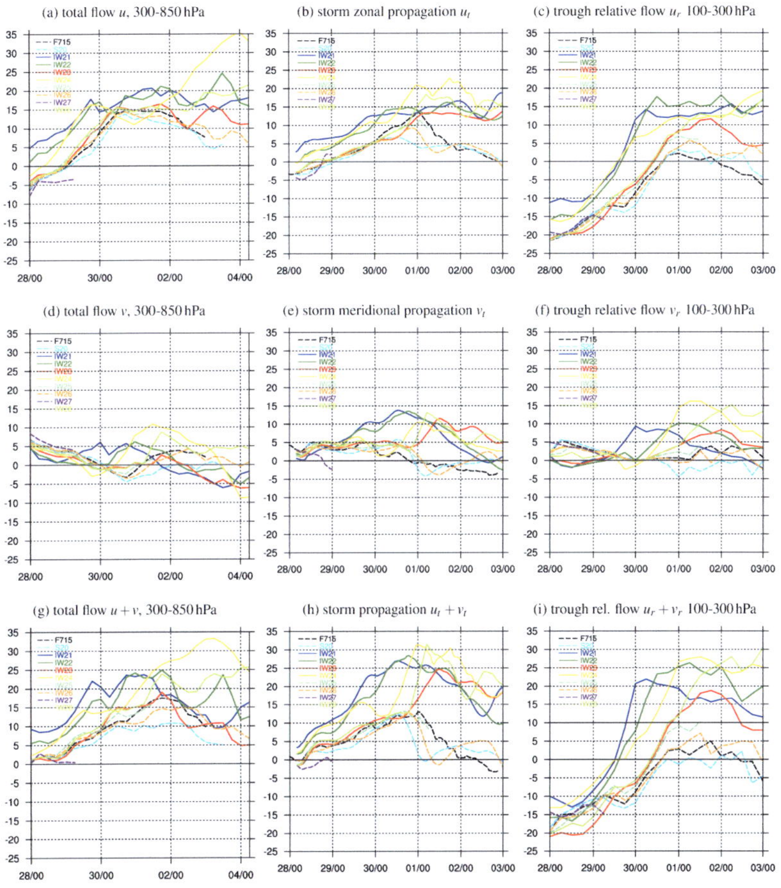


Figure A.1: Wind components (in $m\,s^{-1}$) along the tracks of Jangmi in the scenarios. (a),(d),(g) (left column) total background flow $u, v$ vertically averaged from 300-850 hPa. (c),(f),(i) (right column) non-divergent flow $u_r, v_r$ in the frame relative to the trough moving with constant zonal propagation speed of $c = 0.5^\circ\,h^{-1}$ vertically averaged from 300-850 hPa. The data in (a),(d),(g) and (c),(f),(i) is six-hourly from NOTC and the average of a 2°x2° box centred on the storm location. (b),(e),(h) (middle column) components of the storm propagation speed ($u_t + v_t$, in $m\,s^{-1}$) calculated from one-hourly track data, smoothed with a 10 point running mean. Reintensifiers are shown with solid contours, decayers dashed. The top, middle, bottom rows show the zonal, meridional and zonal+meridional components. X-axis indicates DD/HH day (DD) and hour (HH) in UTC of September/October 2008.

# Complementary ACC plots

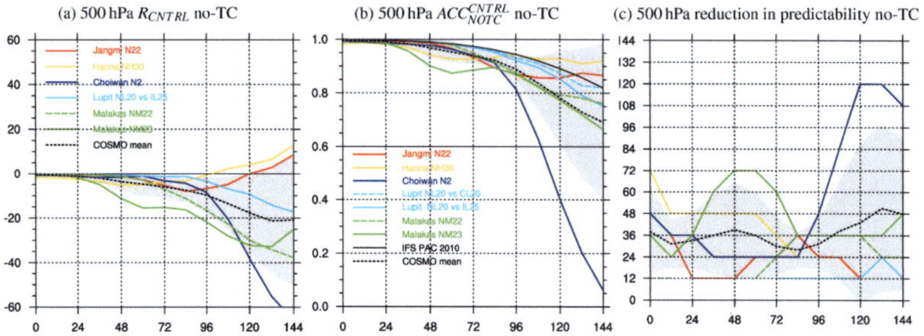


Figure A.2: (a) as Figure 6.6 but at 500 hPa. (b,c) as Figure 6.7 but at 500 hPa.

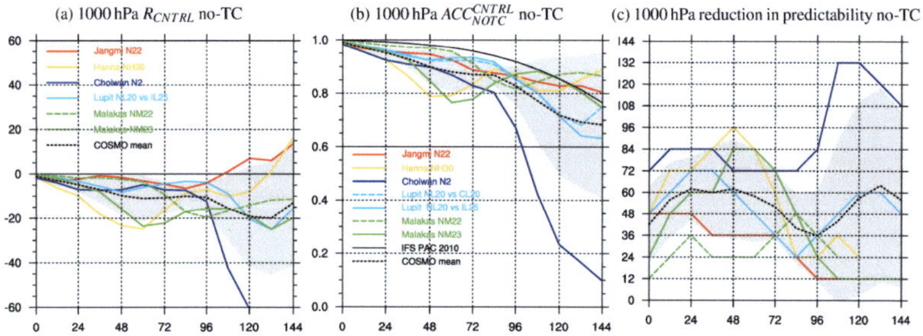


Figure A.3: As Figure A.2 but at 1000 hPa

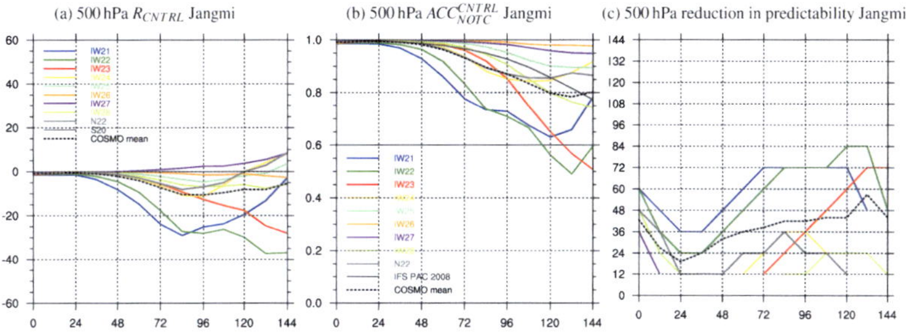


Figure A.4: (a) as Figure 6.8 but at 500 hPa. (b,c) as Figure 6.9 but at 500 hPa

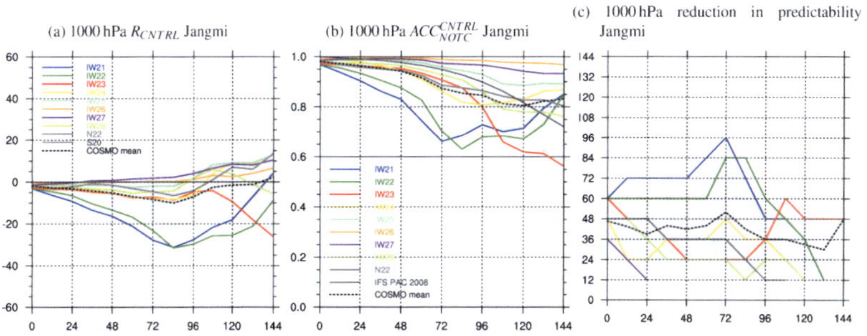


Figure A.5: As Figure A.4 but at 1000 hPa.

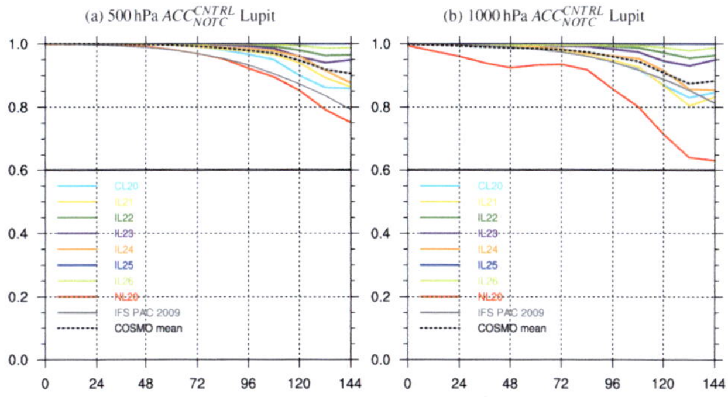


Figure A.6: As Figure 6.10 but at 500 hPa and 1000 hPa.

# Bibliography

Agustí-Panareda, A., 2008: The Contribution of Ex Tropical Cyclone Gert (1999) toward the Weakening of a Midlatitude Cyclogenesis Event. *Monthly Weather Review*, **136**, 2091–+, doi:10.1175/2007MWR1637.1.

Agustí-Panareda, A., S. L. Gray, G. C. Craig, and C. Thorncroft, 2005: The Extratropical Transition of Tropical Cyclone Lili (1996) and Its Crucial Contribution to a Moderate Extratropical Development. *Monthly Weather Review*, **133**, 1562–+, doi:10.1175/MWR2935.1.

Agustí-Panareda, A., C. D. Thorncroft, G. C. Craig, and S. L. Gray, 2004: The extratropical transition of hurricane Irene (1999): A potential-vorticity perspective. *Quarterly Journal of the Royal Meteorological Society*, **130**, 1047–1074, doi:10.1256/qj.02.140.

Ahmadi-Givi, F., G. C. Graig, and R. S. Plant, 2004: The dynamics of a mid-latitude cyclone with very strong latent-heat release. *Quarterly Journal of the Royal Meteorological Society*, **130**, 295–323, doi:10.1256/qj.02.226.

Anwender, D., P. A. Harr, and S. C. Jones, 2008: Predictability Associated with the Downstream Impacts of the Extratropical Transition of Tropical Cyclones: Case Studies. *Monthly Weather Review*, **136**, 3226–+, doi: 10.1175/2008MWR2249.1.

Arakawa, A. and V. R. Lamb, 1981: A potential enstrophy and energy conserving scheme for the shallow water equations. *Monthly Weather Review*, **109**, 18–36, doi:10.1175/1520-0493(1981)109<0018:APEAEC>2.0.CO;2.

Archambault, H. M., 2011: The Downstream Extratropical Flow Response to Recurving Western North Pacific Tropical Cyclones. Ph.D. thesis, University of Albany, New York, (USA).

Atallah, E. H. and L. F. Bosart, 2003: The Extratropical Transition and Precipitation Distribution of Hurricane Floyd (1999). *Monthly Weather Review*, **131**, 1063–+, doi:10.1175/1520-0493(2003)131<1063:TETAPD>2.0.CO;2.

Bennetts, D. A. and B. J. Hoskins, 1979: Conditional symmetric instability - a possible explanation for frontal rainbands. *Quarterly Journal of the Royal Meteorological Society*, **105**, 945–962, doi:10.1002/qj.49710544615.

Berrisford, P., 1988: Potential vorticity in extratropical cyclones. Ph.D. thesis, University of Reading (United Kingdom).

Bolton, D., 1980: The Computation of Equivalent Potential Temperature. *Monthly Weather Review*, **108**, 1046–+, doi:10.1175/1520-0493(1980)108<1046:TCOEPT>2.0.CO;2.

Bosart, L. F. and G. M. Lackmann, 1995: Postlandfall tropical cyclone reintensification in a weakly baroclinic environment: A case study of Hurricane David (September 1979). *Monthly Weather Review*, **123**, 3268–+, doi:10.1175/1520-0493(1995)123<3268:PTCRIA>2.0.CO;2.

Böttcher, M., 2010: Dynamik diabatischer Rossby-Wellen. Ph.D. thesis, Universität Mainz (Germany).

Böttcher, M. and H. Wernli, 2011: Life cycle study of a diabatic rossby wave as a precursor to rapid cyclogenesis in the north atlantic - dynamics and forecast performance. *Monthly Weather Review*, **139 (6)**, 1861–1878, doi:10.1175/2011MWR3504.1.

Brand, S. and C. P. Guard, 1979: An Observational Study of Extratropical Storms Evolved from Tropical Cyclones in the Western North Pacific. *Journal of the Meteorological Society of Japan*, **57**, 479–482.

Bretherton, F. P., 1966: Critical layer instability in baroclinic flows. *Quarterly Journal of the Royal Meteorological Society*, **92**, 325–334, doi:10.1002/qj.49709239302.

Brown, D. P. and T. Kimberlain, 2008: Tropical Cyclone Report: Hurricane Hanna 28 August - 7 September 2008. Tech. rep., National Hurricane Center.

Browning, K. A., 1990: Rain, rainclouds and climate. *Quarterly Journal of the Royal Meteorological Society*, **116**, 1025–1051, doi:10.1002/qj.49711649502.

Browning, K. A., P. Panagi, and G. Vaughan, 1998: Analysis of an ex-tropical cyclone after its reintensification as a warm-core extratropical cyclone. *Quarterly Journal of the Royal Meteorological Society*, **124**, 2329–2356, doi:10.1002/qj.49712455108.

Chan, J. C. L. and W. M. Gray, 1982: Tropical Cyclone Movement and Surrounding Flow Relationships. *Monthly Weather Review*, **110**, 1354–+, doi:10.1175/1520-0493(1982)110<1354:TCMASF>2.0.CO;2.

Chan, J. C. L. and J. D. Kepert, 2010: *Global Perspectives on Tropical Cyclones - From Science to Mitigation*, World Scientific Series on Asia-Pacific Weather and Climate, Vol. 4. World Scientific Publishing Co. Pte. Ltd., Singapore, Hackensack, London.

Charney, J. G., 1955: The use of primitive equations of motion in numerical prediction. *Tellus*, **7**, 22–26.

Clough, S. A., C. S. A. Davitt, and A. J. Thorpe, 1996: Attribution concepts applied to the omega equation. *Quarterly Journal of the Royal Meteorological Society*, **122**, 1943–1962, doi:10.1002/qj.49712253610.

Cordeira, J. M. and L. F. Bosart, 2010: The Antecedent Large-Scale Conditions of the "Perfect Storms" of Late October and Early November 1991. *Monthly Weather Review*, **138**, 2546–2569, doi:10.1175/2010MWR3280.1.

Davis, C. A., 1992: Piecewise Potential Vorticity Inversion. *Journal of Atmospheric Sciences*, **49**, 1397–1411, doi:10.1175/1520-0469(1992)049<1397:PPVI>2.0.CO;2.

Davis, C. A. and K. A. Emanuel, 1991: Potential Vorticity Diagnostics of Cyclogenesis. *Monthly Weather Review*, **119**, 1929–+, doi:10.1175/1520-0493(1991)119<1929:PVDOC>2.0.CO;2.

Davis, C. A., E. D. Grell, and M. A. Shapiro, 1996: The Balanced Dynamical Nature of a Rapidly Intensifying Oceanic Cyclone. *Monthly Weather Review*, **124**, 3–+, doi:10.1175/1520-0493(1996)124<0003:TBDNOA>2.0.CO;2.

Dee, D. P., S. M. Uppala, A. J. Simmons, P. Berrisford, P. Poli, S. Kobayashi, U. Andrae, M. A. Balmaseda, G. Balsamo, P. Bauer, P. Bechtold, A. C. M. Beljaars, L. van de Berg, J. Bidlot, N. Bormann, C. Delsol, R. Dragani, M. Fuentes, A. J. Geer, L. Haimberger, S. B. Healy, H. Hersbach, E. V. Hólm, L. Isaksen, P. Kållberg, M. Köhler, M. Matricardi, A. P. McNally, B. M. Monge-Sanz, J.-J. Morcrette, B.-K. Park, C. Peubey, P. de Rosnay, C. Tavolato, J.-N. Thépaut, and F. Vitart, 2011: The era-interim reanalysis: configuration and performance of the data assimilation system. *Quarterly Journal of the Royal Meteorological Society*, **137 (656)**, 553–597, doi:10.1002/qj.828,

Deveson, A. C. L., K. A. Browning, and T. D. Hewson, 2002: A classification of FASTEX cyclones using a height-attributable quasi-geostrophic vertical-motion diagnostic. *Quarterly Journal of the Royal Meteorological Society*, **128**, 93–117, doi:10.1256/00359000260498806.

Dixon, M. A. G., A. J. Thorpe, and K. A. Browning, 2003: Layer-wise attribution of vertical motion and the influence of potential-vorticity anomalies on synoptic development. *Quarterly Journal of the Royal Meteorological Society*, **129**, 1761–1778, doi:10.1256/qj.02.83.

Doms, G., J. Förstner, E. Heise, H.-J. Herzog, M. Raschendorfer, T. Reinhardt, B. Ritter, R. Schrodin, J.-P. Schulz, and V. G., 2007: *A description of the nonhydrostatic Regional Model LM.*

*Part II: Physical Parametrization.* Consortium for Small-Scale Modelling (COSMO), URL http://www.cosmo-model.org, available at http://www.cosmo-model.org.

Doms, G. and U. Schättler, 2002: *A description of the nonhydrostatic Regional Model LM. Part I: Dynamics and Numerics.* Consortium for Small-Scale Modelling (COSMO), URL http://www.cosmo-model.org, available at http://www.cosmo-model.org.

Dong, K. and C. J. Neumann, 1986: The Relationship between Tropical Cyclone Motion and Environmental Geostrophic Flows. *Monthly Weather Review*, **114**, 115–+, doi:10.1175/1520-0493(1986)114<0115:TRBTCM>2.0.CO;2.

Dudhia, J., 1989: Numerical Study of Convection Observed during the Winter Monsoon Experiment Using a Mesoscale Two-Dimensional Model. *Journal of Atmospheric Sciences*, **46**, 3077–3107, doi:10.1175/1520-0469(1989)046<3077:NSOCOD>2.0.CO;2.

Eckhardt, S., A. Stohl, H. Wernli, P. James, C. Forster, and N. Spichtinger, 2004: A 15-Year climatology of warm conveyor belts. *Journal of Climate*, **17**, 218–237, doi:10.1175/1520-0442(2004)017<0218:AYCOWC>2.0.CO;2.

ECMWF, 2010: *IFS Documentation Cy36r1 - Part II: Data Assimilation.* European Centre for Medium-range weather forecast (ECMWF), available at http://www.ecmwf.int/research/ifsdocs/CY36r1/ASSIMILATION/IFSPart2.pdf.

Emanuel, K., 2003: Tropical Cyclones. *Annual Review of Earth and Planetary Sciences*, **31**, 75–104, doi:10.1146/annurev.earth.31.100901.141259.

———, 2005: *Divine Wind - The History and Science of Hurricanes.* Oxford University Press.

Ertel, H., 1942: Ein neuer hydrodynamischer Erhaltungssatz. *Meteorologische Zeitschrift*, **59**, 277–281.

Evans, J. L. and R. E. Hart, 2003: Objective Indicators of the Life Cycle Evolution of Extratropical Transition for Atlantic Tropical Cyclones. *Monthly Weather Review*, **131**, 909–925, doi:10.1175/1520-0493(2003)131<0909:OIOTLC>2.0.CO;2.

Grams, C. M., H. Wernli, M. Böttcher, J. Čampa, U. Corsmeier, S. C. Jones, J. H. Keller, C.-J. Lenz, and L. Wiegand, 2011: The key role of diabatic processes in modifying the upper-level tropospheric wave guide: a North Atlantic case study. *Quarterly Journal of the Royal Meteorological Society*, in press.

Harr, P. A., 2010: The extratropical transition of tropical cyclones: Structural characteristics, downstream impacts, and forecast challenges. *Global Perspectives on Tropical Cyclones - From Science to Mitigation*, Chan, J. C. L. and J. D. Kepert, Eds., World Scientific Publishing Co. Pte. Ltd., 149–174.

Harr, P. A., D. Anwender, and S. C. Jones, 2008: Predictability Associated with the Downstream Impacts of the Extratropical Transition of Tropical Cyclones: Methodology and a Case Study of Typhoon Nabi (2005). *Monthly Weather Review*, **136**, 3205–+, doi:10.1175/2008MWR2248.1.

Harr, P. A. and R. L. Elsberry, 2000: Extratropical Transition of Tropical Cyclones over the Western North Pacific. Part I: Evolution of Structural Characteristics during the Transition Process. *Monthly Weather Review*, **128**, 2613–+, doi:10.1175/1520-0493(2000)128<2613:ETOTCO>2.0.CO;2.

Harr, P. A., R. L. Elsberry, and T. F. Hogan, 2000: Extratropical Transition of Tropical Cyclones over the Western North Pacific. Part II: The Impact of Midlatitude Circulation Characteristics. *Monthly Weather Review*, **128**, 2634–+, doi:10.1175/1520-0493(2000)128<2634:ETOTCO>2.0.CO;2.

Hart, R. E., 2003: A Cyclone Phase Space Derived from Thermal Wind and Thermal Asymmetry. *Monthly Weather Review*, **131**, 585–+, doi:10.1175/1520-0493(2003)131<0585:ACPSDF>2.0.CO;2.

Hart, R. E. and J. L. Evans, 2001: A Climatology of the Extratropical Transition of Atlantic Tropical Cyclones. *Journal of Climate*, **14**, 546–564, doi:10.1175/1520-0442(2001)014<0546:ACOTET>2.0.CO;2.

Holton, J. R., 2004: *An Introduction to Dynamic Meteorology Fourth Edition*, International Geophysics Series, Vol. 88. Elsevier Academic Press, Burlington, San Diego, London.

Hoskins, B. J. and P. Berrisford, 1988: A potential vorticity perspective of the storm of 15-16 October 1987. *Weather*, **43**, 122–129.

Hoskins, B. J., I. Draghici, and H. C. Davies, 1978: A new look at the $\omega$-equation. *Quarterly Journal of the Royal Meteorological Society*, **104**, 31–38, doi:10.1002/qj.49710443903.

Hoskins, B. J., M. E. McIntyre, and A. W. Robertson, 1985: On the use and significance of isentropic potential vorticity maps. *Quarterly Journal of the Royal Meteorological Society*, **111**, 877–946, doi:10.1002/qj.49711147002.

Jones, S. C., 2002: The dynamics of tropical cyclones in vertical shear and of their extratropical transition. Ph.D. thesis, Universität München, Germany., habilitationsschrift.

Jones, S. C., P. A. Harr, J. Abraham, L. F. Bosart, P. J. Bowyer, J. L. Evans, D. E. Hanley, B. N. Hanstrum, R. E. Hart, F. Lalaurette, M. R. Sinclair, R. K. Smith, and C. Thorncroft, 2003: The Extratropical Transition of Tropical Cyclones: Forecast Challenges, Current Understanding, and Future Directions. *Weather and Forecasting*, **18**, 1052–+, doi:10.1175/1520-0434(2003)018<1052:TETOTC>2.0.CO;2.

Jones, S. C. and A. J. Thorpe, 1992: The Three-Dimensional Nature of 'Symmetric' Instability. *Quarterly Journal of the Royal Meteorological Society*, **118**, 227–258, doi:10.1002/qj.49711850404.

Kessler, E., 1969: On the distribution and continuity of water substance in atmospheric circulation models. *Meteorological Monographs*, **10**.

Klein, P. M., 1997: Extratropical transition of western North Pacific tropical cyclones. Ph.D. thesis, Naval Postgraduate School Monterey, CA, USA, m.S. thesis [NTIS ADA-341-420.].

Klein, P. M., P. A. Harr, and R. L. Elsberry, 2000: Extratropical Transition of Western North Pacific Tropical Cyclones: An Overview and Conceptual Model of the Transformation Stage. *Weather and Forecasting*, **15**, 373–396, doi:10.1175/1520-0434(2000)015<0373:ETOWNP>2.0.CO;2.

———, 2002: Extratropical Transition of Western North Pacific Tropical Cyclones: Midlatitude and Tropical Cyclone Contributions to Reintensification. *Monthly Weather Review*, **130**, 2240–+, doi:10.1175/1520-0493(2002) 130<2240:ETOWNP>2.0.CO;2.

Kleinschmidt, E., 1950a: Über Aufbau und Entstehung von Zyklonen (1.Teil). *Meteorologische Rundschau*, **3**, 1–6.

———, 1950b: Über Aufbau und Entstehung von Zyklonen (2.Teil). *Meteorologische Rundschau*, **3**, 54–61.

———, 1951: Über Aufbau und Entstehung von Zyklonen (3.Teil). *Meteorologische Rundschau*, **4**, 89–96.

———, 1955: Die Entstehung einer Höhenzyklone über Nordamerika. *Tellus*, **7**, 96–+.

Klemp, J. B. and R. B. Wilhelmson, 1978: The Simulation of Three-Dimensional Convective Storm Dynamics. *Journal of Atmospheric Sciences*, **35**, 1070–1096.

Knabb, R. D., J. R. Rhome, and D. P. Brown, 2005: Tropical Cyclone Report: Hurricane Katrina 23-30 August 2005. Tech. rep., National Hurricane Center.

Lang, S. D., 2011: Extratropical transition and Singular vectors (preliminary). Ph.D. thesis, Karlsruhe Institute of Technology (Karlsruhe, Germany).

Martius, O., C. Schwierz, and H. C. Davies, 2008: Far-upstream precursors of heavy precipitation events on the Alpine south-side. *Quarterly Journal of the Royal Meteorological Society*, **134**, 417–428, doi:10.1002/qj.229.

Matano, H. and M. Sekioka, 1971: Some aspects of the extratropical transformation of a tropical cyclone. *Journal of the Meteorlogical Society of Japan*, **49**, 736–743.

McTaggart-Cowan, R., J. R. Gyakum, and M. K. Yau, 2001: Sensitivity Testing of Extratropical Transitions Using Potential Vorticity Inversions to Modify Initial Conditions: Hurricane Earl Case Study. *Monthly Weather Review*, **129**, 1617–+, doi:10.1175/1520-0493(2001)129<1617:STOETU>2.0.CO;2.

———, 2003: The Influence of the Downstream State on Extratropical Transition: Hurricane Earl (1998) Case Study. *Monthly Weather Review*, **131**, 1910–+, doi:10.1175//2589.1.

———, 2004: The Impact of Tropical Remnants on Extratropical Cyclogenesis: Case Study of Hurricanes Danielle and Earl (1998). *Monthly Weather Review*, **132**, 1933–+, doi:10.1175/1520-0493(2004)132<1933:TIOTRO>2.0.CO;2.

Mellor, G. L. and T. Yamada, 1982: Development of a turbulence closure model for geophysical fluid problems. *Reviews of Geophysics and Space Physics*, **20**, 851–875, doi:10.1029/RG020i004p00851.

Moore, R. W. and M. T. Montgomery, 2004: Reexamining the Dynamics of Short-Scale, Diabatic Rossby Waves and Their Role in Midlatitude Moist Cyclogenesis. *Journal of Atmospheric Sciences*, **61**, 754–768, doi:10.1175/1520-0469(2004)061<0754:RTDOSD>2.0.CO;2.

———, 2005: Analysis of an Idealized, Three-Dimensional Diabatic Rossby Vortex: A Coherent Structure of the Moist Baroclinic Atmosphere. *Journal of Atmospheric Sciences*, **62**, 2703–2725.

Parker, D. J. and A. J. Thorpe, 1995: Conditional Convective Heating in a Baroclinic Atmosphere: A Model of Convective Frontogenesis. *Journal of Atmospheric Sciences*, **52**, 1699–1711, doi:10.1175/1520-0469(1995)052<1699:CCHIAB>2.0.CO;2.

Persson, A. and F. Grazzini, 2007: User Guide to ECMWF forecast products. *Meteorological Bulletin*, **M3.2**, `http://www.ecmwf.int/products/forecasts/guide/user_guide.pdf`.

Petterssen, S. and Smebye, 1971: On the development of extratropical cyclones. *Quarterly Journal of the Royal Meteorological Society*, **97**, 457–482, doi:10.1002/qj.49709741407.

Pinto, J. G., M. Klawa, U. Ulbrich, R. Rudari, and P. Speth, 2001: Extreme precipitation events over northwest Italy and their relationship with tropical-extratropical interactions over the Atlantic. *Proceedings of the 3rd, ECS Plinius Conference on Mediterranean Storms - Baja Sardinia, Italy, October 2001*.

Pomroy, H. R. and A. J. Thorpe, 2000: The Evolution and Dynamical Role of Reduced Upper-Tropospheric Potential Vorticity in Intensive Observing Period One of FASTEX. *Monthly Weather Review*, **128**, 1817–+, doi:10.1175/1520-0493(2000)128<1817:TEADRO>2.0.CO;2.

Raschendorfer, M., 2001: The new turbulence parametrization of LM. *COSMO Newsletter*, **1**, 89–97.

Raschendorfer, M., C. Simmer, and P. Gross, 2003: Parameterisation of turbulent transport in the atmosphere. *Dynamics of multiscale earth systems.*, Neugebauer, H. J. and C. Simmer, Eds., Springer Berlin Heidelberg, 167–185.

Riemer, M., 2007: Außertropische Umwandlung tropischer Wirbelstürme: Einfluss auf das Strömungsmuster in den mittleren Breiten. Ph.D. thesis, Universität Karlsruhe, Germany.

Riemer, M. and S. C. Jones, 2010: The downstream impact of tropical cyclones on a developing baroclinic wave in idealized scenarios of extratropical transition. *Quarterly Journal of the Royal Meteorological Society*, **136**, 617–637, doi:10.1002/qj.605.

Riemer, M., S. C. Jones, and C. A. Davis, 2008: The impact of extratropical transition on the downstream flow: An idealized modelling study with a straight jet. *Quarterly Journal of the Royal Meteorological Society*, **134**, 69–91, doi:10.1002/qj.189.

Ritchie, E. A. and R. L. Elsberry, 2001: Simulations of the Transformation Stage of the Extratropical Transition of Tropical Cyclones. *Monthly Weather Review*, **129**, 1462–+, doi:10.1175/1520-0493(2001)129<1462:SOTTSO>2.0.CO;2.

———, 2003: Simulations of the Extratropical Transition of Tropical Cyclones: Contributions by the Midlatitude Upper-Level Trough to Reintensification. *Monthly Weather Review*, **131**, 2112–+, doi:10.1175/1520-0493(2003)131<2112:SOTETO>2.0.CO;2.

———, 2007: Simulations of the Extratropical Transition of Tropical Cyclones: Phasing between the Upper-Level Trough and Tropical Cyclones. *Monthly Weather Review*, **135**, 862–+, doi:10.1175/MWR3303.1.

Ritter, B. and J. F. Geleyn, 1992: A comprehensive radiation scheme for numerical weather prediction models with potential applications in climate simulations. *Monthly Weather Review*, **120**, 303–325.

Schäfler, A., A. Dörnbrack, C. Kiemle, S. Rahm, and M. Wirth, 2010: Tropospheric Water Vapor Transport as Determined from Airborne Lidar Measurements. *Journal of Atmospheric and Oceanic Technology*, **27**, 2017–2030, doi:10.1175/2010JTECHA1418.1.

Scheck, L., S. C. Jones, and M. Juckes, 2011a: The resonant interaction of a tropical cyclone and a tropopause front in a barotropic model. part i: Zonally oriented front. *Journal of the Atmospheric Sciences*, **68 (3)**, 405–419, doi:10.1175/2010JAS3482.1.

———, 2011b: The resonant interaction of a tropical cyclone and a tropopause front in a barotropic model. part ii: Frontal waves. *Journal of the Atmospheric Sciences*, **68 (3)**, 420–429, doi:10.1175/2010JAS3483.1.

Schäfler, A., A. Dörnbrack, H. Wernli, C. Kiemle, and S. Pfahl, 2011: Airborne lidar observations in the inflow region of a warm conveyor belt. *Quarterly Journal of the Royal Meteorological Society*, **137 (658)**, 1257–1272, doi:10.1002/qj.827.

Stoelinga, M. T., 1996: A Potential Vorticity-Based Study of the Role of Diabatic Heating and Friction in a Numerically Simulated Baroclinic Cyclone. *Monthly Weather Review*, **124**, 849–+, doi:10.1175/1520-0493(1996)124<0849:APVBSO>2.0.CO;2.

Thorncroft, C. and S. C. Jones, 2000: The Extratropical Transitions of Hurricanes Felix and Iris in 1995. *Monthly Weather Review*, **128**, 947–+, doi:10.1175/1520-0493(2000)128<0947:TETOHF>2.0.CO;2.

Thorpe, A. J., 1985: Diagnosis of Balanced Vortex Structure Using Potential Vorticity. *Journal of Atmospheric Sciences*, **42**, 397–406, doi:10.1175/1520-0469(1985)042<0397:DOBVSU>2.0.CO;2.

Thorpe, A. J., B. J. Hoskins, and V. Innocentini, 1989: The Parcel Method in a Baroclinic Atmosphere. *Journal of Atmospheric Sciences*, **46**, 1274–1284, doi:10.1175/1520-0469(1989)046<1274:TPMIAB>2.0.CO;2.

Tiedtke, M., 1989: A Comprehensive Mass Flux Scheme for Cumulus Parameterization in Large-Scale Models. *Monthly Weather Review*, **117**, 1779–+, doi:10.1175/1520-0493(1989)117<1779:ACMFSF>2.0.CO;2.

Velden, C., J. Daniels, D. Stettner, D. Santek, J. Key, J. Dunion, K. Holmlund, G. Dengel, W. Bresky, and P. Menzel, 2005: Recent innovations in deriving tropospheric winds from meteorological satellites. *Bulletin of the American Meteorological Society*, **86 (2)**, 205–223, doi:10.1175/BAMS-86-2-205.

Velden, C. S., 1996: Winds derived from geostationary satellite moisture channel observations: Applications and impact on numerical weather prediction. *Meteorology and Atmospheric Physics*, **60**, 37–46, doi:10.1007/BF01029784.

Velden, C. S., C. M. Hayden, S. J. Nieman, W. P. Menzel, S. Wanzong, and J. S. Goerss, 1997: Upper-tropospheric winds derived from geostationary satellite water vapor observations. *Bulletin of the American Meteorological Society*, **78 (2)**, 173–195, doi:10.1175/1520-0477(1997)078<0173:UTWDFG>2.0.CO;2.

Velden, C. S. and L. M. Leslie, 1991: The Basic Relationship between Tropical Cyclone Intensity and the Depth of the Environmental Steering Layer in the Australian Region. *Weather and Forecasting*, **6**, 244–253, doi:10.1175/1520-0434(1991)006<0244:TBRBTC>2.0.CO;2.

Weissmann, M., R. Busen, A. Dörnbrack, S. Rahm, and O. Reitebuch, 2005: Targeted Observations with an Airborne Wind Lidar. *Journal of Atmospheric and Oceanic Technology*, **22**, 1706–+, doi:10.1175/JTECH1801.1.

Weissmann, M., F. Harnisch, C.-C. Wu, P.-H. Lin, Y. Ohta, K. Yamashita, Y.-H. Kim, T. Nakazawa, and S. Aberson, 2010: The Influence of Assimilating Dropsonde Data on Typhoon Track and Midlatitude Forecasts. *Monthly Weather Review*, **139**, 908–920, doi:10.1175/2010MWR3377.1.

Wernli, H., 1997: A Lagrangian-based analysis of extratropical cyclones. II: A detailed case-study. *Quarterly Journal of the Royal Meteorological Society*, **123**, 1677–1706, doi:10.1256/smsqj.54210.

Wernli, H. and H. C. Davies, 1997: A Lagrangian-based analysis of extratropical cyclones. I: The method and some applications. *Quarterly Journal of the Royal Meteorological Society*, **123**, 467–489, doi:10.1256/smsqj.53810.

Wilks, D. S., 1995: *Statistical Methods in the Atmospheric Sciences*, International Geophysics Series, Vol. 59. Academic Press, San Diego, London.

Wirth, M., A. Fix, P. Mahnke, H. Schwarzer, F. Schrandt, and G. Ehret, 2009: The airborne multi-wavelength water vapor differential absorption lidar WALES: system design and performance. *Applied Physics B: Lasers and Optics*, **96**, 201–213, doi:10.1007/s00340-009-3365-7.

Wu, C.-C. and Y. Kurihara, 1996: A Numerical Study of the Feedback Mechanisms of Hurricane-Environment Interaction on Hurricane Movement from the Potential Vorticity Perspective. *Journal of Atmospheric Sciences*, **53**, 2264–2282, doi:10.1175/1520-0469(1996)053<2264:ANSOTF>2.0.CO;2.

# Wissenschaftliche Berichte des Instituts für Meteorologie und Klimaforschung des Karlsruher Instituts für Technologie (0179-5619)

Bisher erschienen:

**Nr. 1:** *Fiedler, F. / Prenosil, T.*
Das MESOKLIP-Experiment. (Mesoskaliges Klimaprogramm im Oberrheintal).
August 1980

**Nr. 2:** *Tangermann-Dlugi, G.*
Numerische Simulationen atmosphärischer Grenzschichtströmungen über langgestreckten mesoskaligen Hügelketten bei neutraler thermischer Schichtung.
August 1982

**Nr. 3:** *Witte, N.*
Ein numerisches Modell des Wärmehaushalts fließender Gewässer unter Berücksichtigung thermischer Eingriffe.
Dezember 1982

**Nr. 4:** *Fiedler, F. / Höschele, K. (Hrsg.)*
Prof. Dr. Max Diem zum 70. Geburtstag.
Februar 1983 (vergriffen)

**Nr. 5:** *Adrian, G.*
Ein Initialisierungsverfahren für numerische mesoskalige Strömungsmodelle.
Juli 1985

**Nr. 6:** *Dorwarth, G.*
Numerische Berechnung des Druckwiderstandes typischer Geländeformen.
Januar 1986

**Nr. 7:** *Vogel, B.; Adrian, G. / Fiedler, F.*
MESOKLIP-Analysen der meteorologischen Beobachtungen von mesoskaligen Phänomenen im Oberrheingraben.
November 1987

**Nr. 8:** *Hugelmann, C.-P.*
Differenzenverfahren zur Behandlung der Advektion.
Februar 1988

**Nr. 9:** *Hafner, T.*
Experimentelle Untersuchung zum Druckwiderstand der Alpen.
April 1988

**Nr. 10:** *Corsmeier, U.*
Analyse turbulenter Bewegungsvorgänge in der maritimen atmosphärischen Grenzschicht.
Mai 1988

**Nr. 11:** *Walk, O. / Wieringa, J.(eds)*
Tsumeb Studies of the Tropical Boundary-Layer Climate.
Juli 1988

**Nr. 12:** *Degrazia, G. A.*
Anwendung von Ähnlichkeitsverfahren auf die turbulente Diffusion in der konvektiven und stabilen Grenzschicht.
Januar 1989

**Nr. 13:** *Schädler, G.*
Numerische Simulationen zur Wechselwirkung zwischen Landoberflächen und atmophärischer Grenzschicht.
November 1990

**Nr. 14:** *Heldt, K.*
Untersuchungen zur Überströmung eines mikroskaligen Hindernisses in der Atmosphäre.
Juli 1991

**Nr. 15:** *Vogel, H.*
Verteilungen reaktiver Luftbeimengungen im Lee einer Stadt – Numerische Untersuchungen der relevanten Prozesse.
Juli 1991

**Nr. 16:** *Höschele, K.(ed.)*
Planning Applications of Urban and Building Climatology – Proceedings of the IFHP / CIB-Symposium Berlin, October 14-15, 1991.
März 1992

**Nr. 17:** *Frank, H. P.*
Grenzschichtstruktur in Fronten.
März 1992

**Nr. 18:** *Müller, A.*
Parallelisierung numerischer Verfahren zur Beschreibung von Ausbreitungs- und chemischen Umwandlungsprozessen in der atmosphärischen Grenzschicht.
Februar 1996

**Nr. 19:** *Lenz, C.-J.*
Energieumsetzungen an der Erdoberfläche in gegliedertem Gelände.
Juni 1996

**Nr. 20:** *Schwartz, A.*
Numerische Simulationen zur Massenbilanz chemisch reaktiver Substanzen im mesoskaligen Bereich.
November 1996

**Nr. 21:** *Beheng, K. D.*
Professor Dr. Franz Fiedler zum 60. Geburtstag.
Januar 1998

**Nr. 22:** *Niemann, V.*
Numerische Simulation turbulenter Scherströmungen mit einem Kaskadenmodell.
April 1998

**Nr. 23:** *Koßmann, M.*
Einfluß orographisch induzierter Transportprozesse auf die Struktur der atmosphärischen Grenzschicht und die Verteilung von Spurengasen.
April 1998

**Nr. 24:** *Baldauf, M.*
Die effektive Rauhigkeit über komplexem Gelände – Ein Störungstheoretischer Ansatz.
Juni 1998

**Nr. 25:** *Noppel, H.*
Untersuchung des vertikalen Wärmetransports durch die Hangwindzirkulation auf regionaler Skala.
Dezember 1999

**Nr. 26:** *Kuntze, K.*
Vertikaler Austausch und chemische Umwandlung von Spurenstoffen über topographisch gegliedertem Gelände.
Oktober 2001

**Nr. 27:** *Wilms-Grabe, W.*
Vierdimensionale Datenassimilation als Methode zur Kopplung zweier verschiedenskaliger meteorologischer Modellsysteme.
Oktober 2001

**Nr. 28:** *Grabe, F.*
Simulation der Wechselwirkung zwischen Atmosphäre, Vegetation und Erdoberfläche bei Verwendung unterschiedlicher Parametrisierungsansätze.
Januar 2002

Nr. 29: *Riemer, N.*
Numerische Simulationen zur Wirkung des Aerosols auf die troposphärische Chemie und die Sichtweite.
Mai 2002

**Nr. 30:** *Braun, F. J.*
Mesoskalige Modellierung der Bodenhydrologie.
Dezember 2002

**Nr. 31:** *Kunz, M.*
Simulation von Starkniederschlägen mit langer Andauer über Mittelgebirgen.
März 2003

**Nr. 32:** *Bäumer, D.*
Transport und chemische Umwandlung von Luftschadstoffen im Nahbereich von Autobahnen – numerische Simulationen.
Juni 2003

**Nr. 33:** *Barthlott, C.*
Kohärente Wirbelstrukturen in der atmosphärischen Grenzschicht.
Juni 2003

**Nr. 34:** *Wieser, A.*
Messung turbulenter Spurengasflüsse vom Flugzeug aus.
Januar 2005

**Nr. 35:** *Blahak, U.*
Analyse des Extinktionseffektes bei Niederschlagsmessungen mit einem C-Band Radar anhand von Simulation und Messung.
Februar 2005

**Nr. 36:** *Bertram, I.*
Bestimmung der Wasser- und Eismasse hochreichender konvektiver Wolken anhand von Radardaten, Modellergebnissen und konzeptioneller Betrachtungen.
Mai 2005

**Nr. 37:** *Schmoeckel, J.*
Orographischer Einfluss auf die Strömung abgeleitet aus Sturmschäden im Schwarzwald während des Orkans „Lothar".
Mai 2006

**Nr. 38:** *Schmitt, C.*
Interannual Variability in Antarctic Sea Ice Motion: Interannuelle Variabilität antarktischer Meereis-Drift.
Mai 2006

**Nr. 39:** *Hasel, M.*
Strukturmerkmale und Modelldarstellung der Konvektion über Mittelgebirgen.
Juli 2006

Ab Band 40 erscheinen die Wissenschaftlichen Berichte des Instituts für Meteorologie und Klimaforschung bei KIT Scientific Publishing (ISSN 0179-5619).

Die Bände sind unter www.ksp.kit.edu als PDF frei verfügbar oder als Druckausgabe bestellbar.

**Nr. 40:** *Lux, R.*
Modellsimulationen zur Strömungsverstärkung von orographischen Grundstrukturen bei Sturmsituationen. (2007)
ISBN 978-3-86644-140-8

**Nr. 41:** *Straub, W.*
Der Einfluss von Gebirgswellen auf die Initiierung und Entwicklung konvektiver Wolken. (2008)
ISBN 978-3-86644-226-9

**Nr. 42:** *Meißner, C.*
High-resolution sensitivity studies with the regional climate model COSMO-CLM. (2008)
ISBN 978-3-86644-228-3

**Nr. 43:** *Höpfner, M.*
Charakterisierung polarer stratosphärischer Wolken mittels hochauflösender Infrarotspektroskopie. (2008)
ISBN 978-3-86644-294-8

**Nr. 44:** *Rings, J.*
Monitoring the water content evolution of dikes. (2009)
ISBN 978-3-86644-321-1

**Nr. 45:** *Riemer, M.*
Außertropische Umwandlung tropischer Wirbelstürme: Einfluss auf das Strömungsmuster in den mittleren Breiten. (2012)
ISBN 978-3-86644-766-0

**Nr. 46:** *Anwender, D.*
Extratropical Transition in the Ensemble Prediction System of the ECMWF: Case Studies and Experiments. (2012)
ISBN 978-3-86644-767-7

**Nr. 47:** *Rinke, R.*
Parametrisierung des Auswaschens von Aerosolpartikeln durch Niederschlag. (2012)
ISBN 978-3-86644-768-4

**Nr. 48:** *Stanelle, T.*
Wechselwirkungen von Mineralstaubpartikeln mit thermodynamischen und dynamischen Prozessen in der Atmosphäre über Westafrika. (2012)
ISBN 978-3-86644-769-1

**Nr. 49:** *Peters, T.*
Ableitung einer Beziehung zwischen der Radarreflektivität, der Niederschlagsrate und weiteren aus Radardaten abgeleiteten Parametern unter Verwendung von Methoden der multivariaten Statistik. (2012)
ISBN 978-3-86644-323-5

**Nr. 50:** *Khodayar Pardo, S.*
High-resolution analysis of the initiation of deep convection forced by boundary-layer processes. (2012)
ISBN 978-3-86644-770-7

**Nr. 51:** *Träumner, K.*
Einmischprozesse am Oberrand der konvektiven atmosphärischen Grenzschicht. (2012)
ISBN 978-3-86644-771-4

**Nr. 52:** *Schwendike, J.*
Convection in an African Easterly Wave over West Africa and the Eastern Atlantic: A Model Case Study of Hurricane Helene (2006) and its Interaction with the Saharan Air Layer. (2012)
ISBN 978-3-86644-772-1

**Nr. 53:** *Lundgren, K.*
Direct Radiative Effects of Sea Salt on the Regional Scale. (2012)
ISBN 978-3-86644-773-8

**Nr. 54:** *Sasse, R.*
Analyse des regionalen atmosphärischen Wasserhaushalts unter Verwendung von COSMO-Simulationen und GPS-Beobachtungen. (2012)
ISBN 978-3-86644-774-5

**Nr. 55:** *Grenzhäuser, J.*
Entwicklung neuartiger Mess- und Auswertungsstrategien für ein scannendes Wolkenradar und deren Anwendungsbereiche. (2012)
ISBN 978-3-86644-775-2

**Nr. 56:** *Grams, C.*
Quantification of the downstream impact of extratropical transition for Typhoon Jangmi and other case studies. (2013)
ISBN 978-3-86644-776-9